Preparation, Characterization, Properties, Simulation, and Applications of Nanostructured Materials

Preparation, Characterization, Properties, Simulation, and Applications of Nanostructured Materials

Guest Editor

Thomas Dippong

MDPI

Basel • Beijing • Wuhan • Barcelona • Belgrade • Novi Sad • Cluj • Manchester

Guest Editor
Thomas Dippong
Chemistry and Biology
Department
Technical University of
Cluj-Napoca
Baia Mare
Romania

Editorial Office
MDPI AG
Grosspeteranlage 5
4052 Basel, Switzerland

This is a reprint of the Special Issue, published open access by the journal *Nanomaterials* (ISSN 2079-4991), freely accessible at: https://www.mdpi.com/journal/nanomaterials/special_issues/4QT40EVRNU.

For citation purposes, cite each article independently as indicated on the article page online and as indicated below:

Lastname, A.A.; Lastname, B.B. Article Title. *Journal Name* **Year**, *Volume Number*, Page Range.

ISBN 978-3-7258-7414-9 (Hbk)
ISBN 978-3-7258-7415-6 (PDF)
https://doi.org/10.3390/books978-3-7258-7415-6

Contents

About the Editor

Thomas Dippong

Thomas Dippong (Professor at the Technical University of Cluj-Napoca) is a chemical engineer with a Ph.D. and habilitation in chemistry. His current research activities are related to the characterization of nanoparticles for various applications as part of an ongoing research project in partnership with the Technical University of Cluj-Napoca within the field of ferrite embedded in silica matrix. He is an expert in analytical chemistry, organic/inorganic chemistry, thermal treatment, instrumental analysis, and the synthesis of magnetic nanomaterials and nanocomposites and characterization techniques, food thermal behavior, volatile profile, and water quality. Prof. Dippong has published 166 peer-reviewed publications (110 papers in high-ranked scientific ISI-Thomson journals (65 Q1, 20 Q2, and 25 Q3) and 56 in other national and international journals), with 3715 citations and an h-index of 49 (WoS). He has given 37 lectures at international conferences. He has also published 2 books with international publishing houses and 15 books with national publishing houses. In the last 5 consecutive years (2020–2024), Prof. Thomas Dippong was included in the prestigious list of the World's Top 2% researchers. He has been a contract manager for many projects and is currently an active member of four projects. Prof. Dippong has reviewed 800 scientific articles for 150 ISI-Thomson journals. He has been a Guest Editor for 10 Special Issues published by three prestigious Q1 ISI-Thomson journals.

Preface

The Special Issue presented the development of composite materials, which is a frontier research field with applications in tissue engineering scaffolds, drug delivery, supercapacitors, food packaging, water remediation and decontamination, energy conversion, sensing, storage filters, lithium-ion batteries, nanoelectronics, sensors, energy-storage devices, supercapacitors, and food packaging, etc. Functional nanocomposites with specific functionalities such as drag reduction, anti-wear, self-healing, anti-fouling, and stimulus response have emerged as highly attractive materials. In this regard, inorganic quantum dots/nanoparticles, nanorods/nanotubes, and 2D materials (i.e., graphene-based nanosheets, carbon nanotubes, carbon nanospheres, bentonites, nano-sized silica particles, and zeolites) can be decorated in the inorganic/organic matrix via chemical synthesis or physical blending for improved characteristics.

The Special Issue focused on enhanced thermal stability of nanomaterials or nanocomposites, resistance to water and chemicals, transparency, thermal conductivity and tensile strength. It presented the latest applications in the hybridization of SiO2 with suitable materials, as ferrite/SiO2 composites can be tailored to meet the needs of new, next-generation technologies. These nanomaterial systems have remarkable properties for removing harmful contaminants such as heavy metals, organic pollutants, and pharmaceuticals from various water sources.

The Special Issue aimed to present original research focused on the development, preparation, chemical synthesis, structural design, material selection, characterization and application of advanced nanocomposites. New syntheses and design principles for nanocomposites (clays, nanoparticles, ferrite incorporated into the SiO2/PVA matrix, catalysts, dyes, drugs, etc.), formation mechanisms, their physicochemical and morphological characterization and applications of nanocomposites incorporated into SiO2 were addressed. Also, progress in the characterization of the structure and properties of nanocomposites, the processing and simulation of nanocomposite deterioration and failures were followed.

In the Special Issue, new applications of functional nanocomposites in biomedicine, healthcare, green energy, coatings, magnetic materials, optoelectronics and sensors, as well as industrial implementations and technological innovations using nanocomposites, were presented.

Thomas Dippong
Guest Editor

nanomaterials

MDPI

Article

Thermal, Structural, and Morphological Analysis of $ZnFe_2O_4$ Embedded and Non-Embedded in a SiO_2 Matrix for Magnetic and Photocatalytic Applications

Thomas Dippong [1,*], Anamaria-Magdalena Savolszki-Madaras [1], Raul Marius Reiz [1], Ioan Petean [2] and Oana Cadar [3]

1 Faculty of Science, Technical University of Cluj-Napoca, 76 Victoriei Street, 430122 Baia Mare, Romania
2 Faculty of Chemistry and Chemical Engineering, Babes-Bolyai University, 11 Arany Janos Street, 400084 Cluj-Napoca, Romania; ioan.petean@ubbcluj.ro
3 National Institute for Research and Development of Optoelectronics INOE 2000, Research Institute for Analytical Instrumentation, 67 Donath Street, 400293 Cluj-Napoca, Romania; oana.cadar@icia.ro
* Correspondence: dippong.thomas@yahoo.ro

Abstract: This study compares the structural, morphological, magnetic, and photocatalytic properties of a pure SiO_2 matrix, a $ZnFe_2O_4$-doped SiO_2 nanocomposite (both synthesized via the sol-gel method), and bulk $ZnFe_2O_4$ produced by thermal decomposition. Thermogravimetric analysis (TGA) reveals that metal oxalates form below 200 °C and decompose into metal oxides, which subsequently form ferrite. Fourier-transform infrared (FTIR) spectroscopy confirms the embedding of both undoped and $ZnFe_2O_4$-doped nanoparticles into the SiO_2 matrix at all investigated annealing temperatures. X-ray diffraction (XRD) consistently reveals the formation of crystalline $ZnFe_2O_4$, with the crystallite size increasing from 48 to 93 nm upon annealing. Atomic force microscopy (AFM) shows spherical ferrite nanoparticles surrounded by an amorphous layer, with particle growth observed at higher temperatures. Structural parameters derived from XRD (e.g., crystallite size, density, porosity, lattice constant, unit cell volume) and AFM (e.g., particle size, coating thickness) as well as magnetic parameters (saturation magnetization, remanence, anisotropy, coercivity) demonstrate clear dependence on both dopant presence and annealing temperature. Magnetic measurements reveal enhanced properties with increasing ferrite content and heat treatment, with a transition from superparamagnetic behavior at 700 °C to ferrimagnetic behavior above 1000 °C. Scavenger experiments confirmed the involvement of holes, hydroxyl radicals, and superoxide radicals in the photocatalytic process. The photocatalytic efficiency, as evaluated by the Rhodamine B degradation under visible light, highlights the promising potential of the obtained nanocomposite for advanced environmental and technological applications.

Keywords: nanoparticles; sol-gel; crystallinity; morphology; VSM; photocatalysis

1. Introduction

Nanostructured ferrites exhibit outstanding magnetic properties, including high saturation magnetization (*Ms*), low coercivity (*Hc*), high permeability, large specific surface area, high electrical resistivity, low eddy current and dielectric losses, as well as mechanical hardness, chemical stability, physical flexibility, and biocompatibility [1]. These attributes make them suitable for a broad range of applications, including environmental remediation [2], biomedical uses (e.g., anticancer, antibacterial, bioimaging, and biosensors) [3,4], energy storage, spintronics, ferrofluids, imaging and therapy, electronics, high-frequency devices, and military technologies [3–6]. Their chemical and thermal stability, mechanical robustness [7], non-toxicity [8], corrosion resistance, and low cost further enhance their applicability. However, the structural and magnetic properties of ferrites are highly sensitive to factors such as chemical composition, precursor type, synthesis route, thermal treatment conditions, and the presence of dopants, impurities, or secondary phases [4–12].

Cubic ferrites, with the general formula MFe_2O_4 (where M = Co, Cu, Mg, Mn, Ni, Zn, or their combinations), are soft magnetic materials that crystallize in a spinel structure in which M^{2+} ions typically occupy tetrahedral (A) sites and Fe^{3+} reside in octahedral (B) sites [4,7,9]. The cation distribution plays a crucial role in determining the structural, magnetic, and electronic properties of these materials [7]. Zinc ferrite ($ZnFe_2O_4$) adopts a normal spinel structure, with Zn^{2+} ions occupying A- sites and Fe^{3+} ions residing in B- sites [3]. In ceramics with well-developed porosity, such as nanostructured $MgAl_2O_4$ spinel, positron annihilation occurs through the positron trapping channel and ortho-positronium decay. This approach can be applied to a wide range of porous functional nanomaterials [13].

The synthesis of uniform nanostructured ferrite particles is critical, as particle size significantly influences their optical, electrical, and magnetic properties [10–12]. These properties are strongly affected by factors such as the synthesis method, cation stoichiometry, type of divalent cations, and heat treatment conditions [7,14]. Various techniques are employed to prepare nanoferrites, including sol-gel, heat treatment temperature, hydrothermal, coprecipitation, sonochemical processing, dry and wet grinding, and microemulsion methods [15]. Among these, heat treatment temperature plays a pivotal role in microstructural evolution, namely lower temperatures typically result in higher porosity and limited densification, whereas higher temperatures promote grain growth and improved densification [11,16–18]. The sol-gel method offers several advantages, including rapid reaction kinetics, high chemical homogeneity, cost-effectiveness, energy efficiency, compositional flexibility, low processing temperatures, monodisperse particle distribution, and well-controlled surface morphologies [7,19,20].

Embedding ferrites in a SiO_2 matrix further improves material performance by reducing nanoparticle agglomeration, as SiO_2 acts as a chemically stable, non-toxic, inert spacer and protective coating owing to its network-forming ability [11,17,18]. Although SiO_2 is non-magnetic, it can indirectly affect the magnetic and photocatalytic behavior of ferrites by modifying their structure and morphology [16–18,21]. SiO_2 attends as a versatile component in photocatalytic materials, enhancing surface properties, charge separation, and stability in order to boost efficiency [22,23]. For example, in TiO_2/SiO_2 composites, SiO_2 improves TiO_2 dispersion, stability, and surface area, facilitating contaminant adsorption [22]. In $ZnFe_2O_4/SiO_2$ nanocomposites, SiO_2 prevents nanoparticle agglomeration, increases the specific surface area (SSA), and enhances structural and thermal stability, thereby improving photocatalytic performance [22,23]. In g-C_3N_4/SiO_2 composites, SiO_2 increases SSA, enabling contaminant adsorption and providing more reactive sites. It also promotes electron–hole pair separation and migration, reducing charge recombination and improving photocatalytic efficiency [22].

$ZnFe_2O_4$ nanoparticles have demonstrated effective catalytic activity in the degradation of organic dyes and the removal of toxic metals, dyes, and organic contaminants, including methylene blue, Direct Red 81, Evans blue, and Malachite green [2]. Their magnetic properties enable efficient recovery and reuse, supporting their applicability in large-scale catalytic and depollution processes [10]. The effectiveness of $ZnFe_2O_4$ as a photocatalyst and nanosorbent is attributed to its regenerability, versatility, and facile separation, maintaining high photocatalytic performance over at least five consecutive cycles without significant loss of activity [2,24]. Photocatalytic efficiency is strongly influenced by surface area, particle size, and dopant concentration [1]. Due to their narrow band gaps, $ZnFe_2O_4$ nanoparticles are particularly advantageous for visible-light-driven applications, such as pollutant degradation in wastewater treatment, where performance is closely linked to morphological features such as surface area and particle size [25]. The magnetic properties of nanostructured ferrites depend on several factors, including particle size, synthesis route, surface energy, and thermal treatment conditions [7,26]. In ferrites, ferromagnetism arises from Fe^{3+} ions occupying both A- and B- sites, enabling strong superexchange interactions, which tend to weaken as particle size increases. $ZnFe_2O_4$ typically exhibits paramagnetic behavior and relatively high coercivity at room temperature [26].

The objectives of this study are (i) to compare the thermal decomposition and sol-gel methods for synthesizing $ZnFe_2O_4$, both embedded and non-embedded in a SiO_2 matrix; (ii) to investigate thermal transformation processes and the formation of succinate precursors via thermal analysis; (iii) to perform structural and morphological characterization using Fourier-transform infrared spectroscopy (FT-IR), X-ray diffraction (XRD), and atomic force microscopy (AFM), with particular emphasis on matrix formation and its influence on the ferrite structure; (iv) to synthesize $ZnFe_2O_4$ nanoparticles via a modified sol-gel method and evaluate their photocatalytic performance in degrading Rhodamine B; and (v) to analyze the magnetic properties (saturation magnetization (M_S), remanence magnetization (M_R), coercivity (H_C), and anisotropy (K)) as a function of the SiO_2 matrix and thermal temperature.

2. Materials and Methods

2.1. Synthesis

SiO_2 and 50%$ZnFe_2O_4$/50%SiO_2 samples were synthesized via a modified sol-gel method using zinc nitrate hexahydrate ($Zn(NO_3)_2 \cdot 6H_2O$), iron nitrate nonahydrate ($Fe(NO_3)_3 \cdot 9H_2O$), tetraethyl orthosilicate (TEOS), 1,4-butanediol, and ethanol. The synthesis involved the dropwise addition of an ethanolic TEOS solution to a nitric acid-acidified diol–ethanol mixture under vigorous magnetic stirring at room temperature. The resulting clear solution was stirred for an additional 30 min and then allowed to gel under ambient conditions. After gelation, the xerogels were crushed and dried at 40 °C for 6 h. Thermal treatments were conducted at 200, 400, 700, and 1000 °C for 6 h, with a heating rate of 20 °C/min, using an LT9 muffle furnace (Nabertherm, Lilienthal, Germany) (Figure 1a,b).

The $ZnFe_2O_4$ sample was synthesized via thermal decomposition of Zn(II) and Fe(III) succinate precursors, which were obtained through a redox reaction among the corresponding metal nitrates and 1,4-butanediol. The resulting material was dried at 40 °C for 6 h, followed by thermal decomposition at 200 °C for 6 h. Subsequent annealing was performed at 400, 700, and 1000 °C for 6 h, with a heating rate of 20 °C/min, in an LT9 muffle furnace (Nabertherm) (Figure 1c).

2.2. Characterization

The formation and decomposition of carboxylation precursors were monitored by thermogravimetric (TG) and differential thermal (DTA) analyses in air up to 1000 °C at a heating rate of 5 °C·min^{-1} using alumina standards and an SDT Q600 (TA Instruments, New Castle, DE, USA).

FT-IR spectra were recorded from 1% KBr pellets using a Prestige-21 (Shimadzu, Tokyo, Japan).

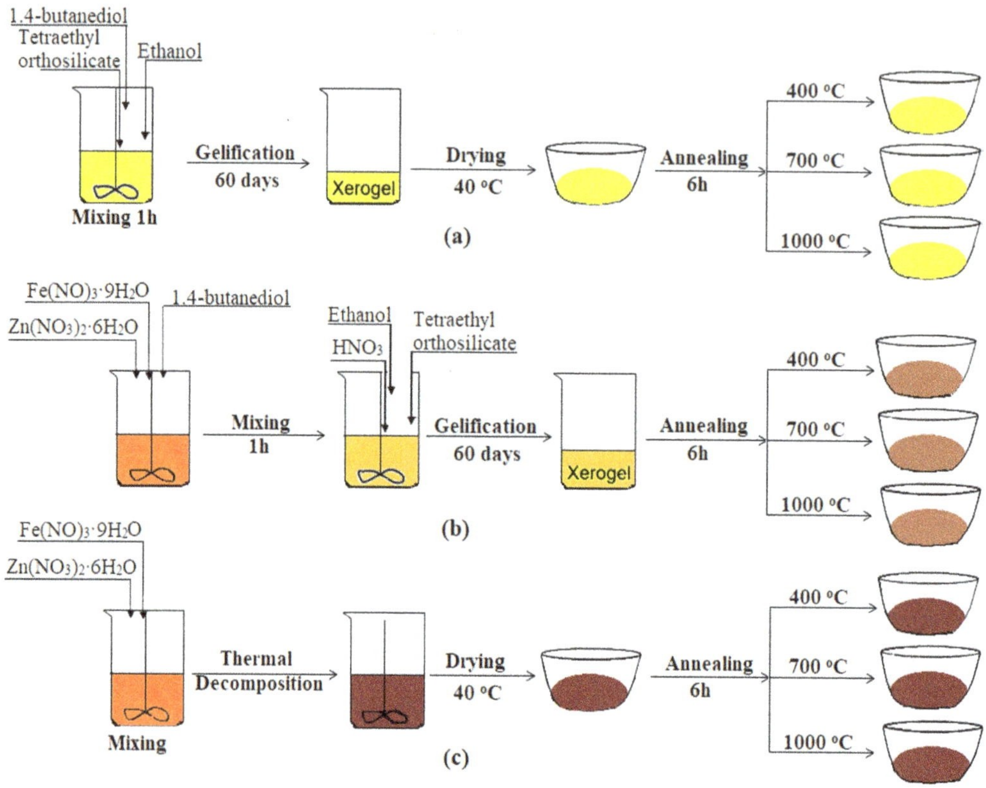

Figure 1. Flowchart of the synthesis for SiO_2 (**a**), $(ZnFe_2O_4)_{0.5}/(SiO_2)_{0.5}$ (**b**), and $ZnFe_2O_4$ (**c**).

The XRD measurements were conducted at room temperature using a D8 Advance diffractometer (Bruker, Karlsruhe, Germany) employing a $CuK_{\alpha1}$, radiation source (λ = 1.5418 Å) operated at 40 kV and 40 mA.

The specific surface area (SSA) was determined at liquid nitrogen temperature (−196 °C) using a Sorptomatic 1990 (Thermo Scientific, Waltham, MA, USA) analyzer. Prior to the measurements, the samples were degassed at 200 °C under a vacuum pressure of 2 Pa for 4 h.

For AFM, thin films were prepared by dispersing 1 g of powder in 10 mL deionized water and stirring magnetically for 15 min. Clean glass slides were vertically immersed in the suspension for 30 s, then slowly withdrawn, and dried horizontally at room temperature to allow for the self-assembly of nanoparticles. The films were characterized using a JSPM-4210 scanning probe microscope (JEOL, Tokyo, Japan) in AC mode, employing NSC15-Hard probes (MikroMasch, Sofia, Bulgaria) at a resonant frequency of 325 kHz and a spring constant of 40 nN/nm over a 1 μm^2 scan area. Topographic images were analyzed using WinSPM 2.0 software (JEOL) to assess surface roughness and particle size. Topographic images were analyzed using WinSPM 2.0 software (JEOL) to assess surface roughness and particle size. At least three different macroscopic sites were investigated, and the mean values were provided with the standard deviation (±).

SEM microscopy was effectuated with a Hitachi SU8230 (Hitachi Company, Tokyo, Japan) operated in high vacuum mode at an acceleration voltage of 30 kV. TEM microscopy was effectuated with an HD-2700 (Hitachi, Tokyo, Japan) on the dispersed ferrite powder samples. Electron microscopy images were analyzed with the professional software Image J

1.53t (National Institute of Health, Bethesda, Rockville, MD, USA), assessing the pores and particle distribution.

Magnetic measurements were conducted using a CFM-12 T cryogen-free VSM magnetometer (Cryogenic Ltd., London, UK). M_S was determined under magnetic fields up to 10 T, and hysteresis loops were recorded between −2 T and 2 T at 300 K. To prevent particle movement during measurements, powder samples were embedded in epoxy resin.

The photocatalytic performance was assessed via Rhodamine B (RhB) degradation under visible light in a laboratory-scale reactor equipped with a 400 W halogen lamp (Osram, Premstätten, Austria) and an ultrasonic bath. To investigate the active species generated in the photocatalytic system (PS) consisting of a photocatalyst, RhB, and H_2O_2, scavenger experiments were performed. Ethylenediaminetetraacetic acid (EDTA-2Na, 10 mM), benzoquinone (p-BQ, 1 mM), $AgNO_3$ (100 mM), and tert-butyl alcohol (t-BuOH) (1:20 vol.) were used as scavengers introduced into the PS in the amount of 1 mL to capture holes (h•), superoxide radicals ($•O_2^-$), electrons (e^-), and hydroxyl radicals (OH•), respectively. Tests of cyclic photocatalysis were carried out using TiO_2@0.2%Fe_2O_3 and TiO_2@2%Fe_2O_3 heterostructures. After dark adsorption, a 90 min photocatalytic decomposition process of RhB was carried out and repeated four times. After each decomposition process, the photocatalyst was separated from the solution of RhB by centrifugation and washed with ethanol three times. After that, the powder was dried for 4 h at 70 °C and used again.

3. Results and Discussions

3.1. Thermal Analysis

Figure 2 shows the TG–DTA curves of (a) SiO_2, (b) 50%$ZnFe_2O_4$/50%SiO_2, and (c) $ZnFe_2O_4$ non-embedded in SiO_2. The SiO_2 sample exhibits two endothermic effects at 53 °C and 164 °C, corresponding to a ~20% mass loss, attributed to the removal of adsorbed water, condensation reactions, and the elimination of unbound 1,4-butanediol [18,27,28]. A pronounced exothermic peak at 294 °C indicates the combustion of organic chains chemically bound into the SiO_2 matrix [18,27]. The bulk $ZnFe_2O_4$ sample decomposes in four distinct steps: (*i*) an endothermic peak at 85 °C (~9% mass loss), associated with the release of crystallization water and residual moisture; (*ii*) an endothermic peak at 202 °C (~28% mass loss), associated with the formation of Zn^{2+} and Fe^{3+} succinates via a redox reaction between the corresponding nitrates and 1,4-butanediol; and (*iii*) two successive exothermic peaks at 255 °C and 289 °C (~30% total mass loss), corresponding to the decomposition of Zn^{2+} and Fe^{3+} succinates into amorphous ZnO and Fe_2O_3. The intermediate oxides subsequently react in situ to form the $ZnFe_2O_4$ spinel phase [16–18,27,28].

The DTA curve of the 50%$ZnFe_2O_4$/50%SiO_2 gels dried at 40 °C (Figure 2) reveals the following thermal effects: (*i*) an endothermic peak at 75 °C with ~15% mass loss, attributed to the removal of residual moisture and physically adsorbed water; (*ii*) two additional endothermic peaks at 135 °C and 200 °C, corresponding to the formation of Zn^{2+} and Fe^{3+} succinates, with associated mass losses of 18% and 10%. This process is driven by the higher acidity of $[Fe(H_2O)_6]^{3+}$ compared to $[Zn(H_2O)_6]^{2+}$, which facilitates redox reactions between metal nitrates and 1,4-butanediol, accompanied by the evolution of volatile products (H_2O, NO_2) [16–18,27,28]; and (*iii*) two exothermic peaks at 252 °C and 288 °C (~20% mass loss), indicating the oxidative decomposition of metal succinates into ZnO and Fe_2O_3, followed by in situ formation of $ZnFe_2O_4$ spinel.

The total mass loss was highest for bulk $ZnFe_2O_4$ (67%), slightly lower for the $ZnFe_2O_4$/SiO_2 composite (63%), and lowest for pure SiO_2 (48%). The thermal behavior of the composite reflects overlapping processes within the SiO_2 matrix, which makes it difficult to unambiguously attribute the observed thermal effects solely to succinate decomposition [27].

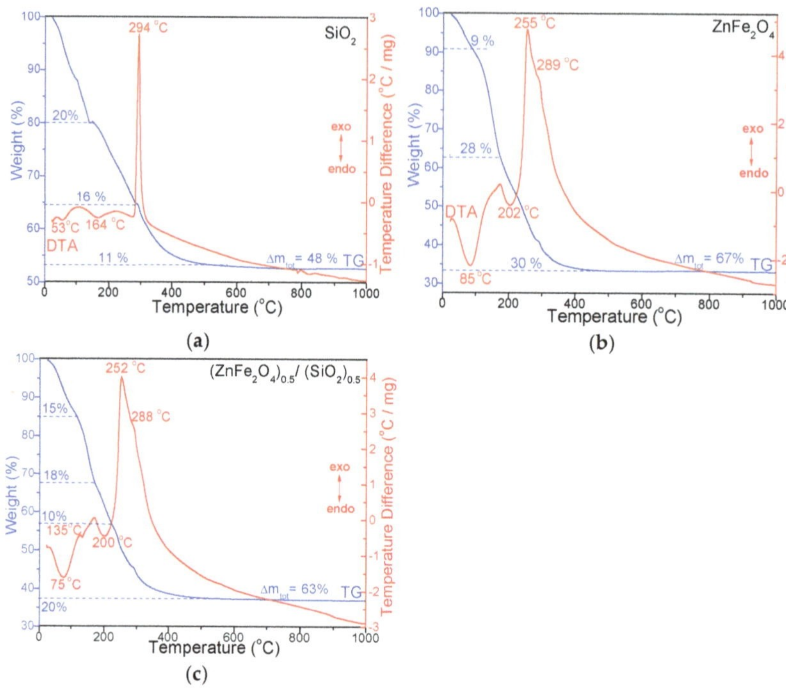

Figure 2. TG and DTA curves of SiO_2 (**a**), 50%$ZnFe_2O_4$/50%SiO_2 (**b**), and $ZnFe_2O_4$ (**c**) dried at 40 °C.

3.2. FT-IR Analysis

Figure 3 displays the FT-IR spectra of the gels dried at 40 °C and 200 °C. All samples dried at 40 °C, except pure SiO_2 (which lacks nitrate precursors), exhibit a characteristic nitrate band at 1384–1387 cm^{-1} [16–18,27,28]. This band disappears after heating to 200 °C, indicating complete nitrate decomposition of nitrates at this temperature. Absorption bands at 1673–1679 cm^{-1} correspond to the symmetric and asymmetric –OH stretching vibrations, originating from 1,4-butanediol and crystallization water in the metal nitrates. The formation of the SiO_2 matrix in both the pure SiO_2 and the 50% $ZnFe_2O_4$/50% SiO_2 samples is confirmed by the presence of characteristic Si–O stretching bands at 1192–1216 cm^{-1}, Si–O–Si stretching at 1049–1069 cm^{-1}, Si–OH deformation at 943–955 cm^{-1}, and Si–O chain vibrations within SiO_4 tetrahedra at 799–826 cm^{-1} [16–18,27,28]. Bands observed at 569–582 cm^{-1} are assigned to tetrahedral M–O (metal–oxygen) stretching and cyclic Si–O–Si structures, while bands at 433–448 cm^{-1} correspond to octahedral M–O stretching and Si–O vibrations [16–18,27]. In bulk $ZnFe_2O_4$, bands at 572–575 cm^{-1} and 405–442 cm^{-1} are associated with Zn–O bending and Fe–O vibrations, respectively [16–18,27,28].

The formation of the SiO_2 matrix in both the pure SiO_2 and the 50% $ZnFe_2O_4$/50% SiO_2 samples is confirmed at 400, 700, and 1000 °C by characteristic absorption bands. These bands correspond to symmetric stretching and bending vibrations of Si–O–Si chains (791–806 cm^{-1}), in-plane and out-of-plane Si–O–Si stretching (1064–1112 cm^{-1}), the Si–OH deformation (971–977 cm^{-1}), the cyclic Si–O–Si structures (560–595 cm^{-1}), and the Si–O bond vibrations (444–470 cm^{-1}). A bending vibration of adsorbed H–O–H groups, attributable to atmospheric moisture, is observed at 1634–1691 cm^{-1}. M–O vibrations are indicated by two intrinsic bands: Fe–O stretching vibrations in A-sites at 410–458 cm^{-1} and Zn–O vibrations in B-sites at 560–595 cm^{-1} [16–18,27,28].

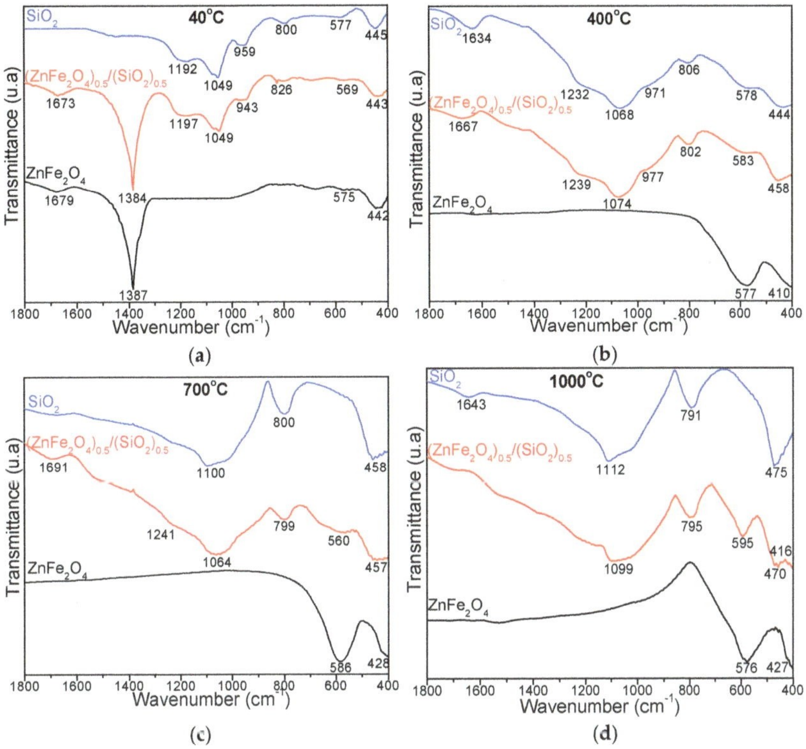

Figure 3. FT-IR spectra of SiO_2, $(ZnFe_2O_4)_{0.5}/(SiO_2)_{0.5}$ and $ZnFe_2O_4$ heat treated at 40 °C (**a**) and annealed at 400 °C (**b**), 700 °C (**c**), and 1000 °C (**d**).

3.3. XRD Analysis

At all temperatures, the XRD pattern of SiO_2 (Figure 4) displays a broad amorphous halo between 15 and 25° 2θ, characteristic of its amorphous structure, with no crystalline phases of the SiO_2 matrix identified. At 400 °C (Figure 4a), the $(ZnFe_2O_4)_{0.5}/(SiO_2)_{0.5}$ displays broad, low intensity reflections corresponding to $ZnFe_2O_4$, indicating incomplete crystallization, reduced crystallite size, and the dilution effect caused by the amorphous SiO_2 matrix. In contrast, pure $ZnFe_2O_4$ exhibits distinct diffraction peaks indexed to the (111), (220), (311), (400), and (511) planes, consistent with the spinel structure (JCPDS No. 52-0278) [28]. The sharper and more intense peaks observed in the pure phase suggest larger crystallite sizes compared to the $ZnFe_2O_4$ embedded in the SiO_2 matrix.

At 700 °C (Figure 4b), the XRD pattern of $(ZnFe_2O_4)_{0.5}/(SiO_2)_{0.5}$ shows $ZnFe_2O_4$ diffraction peaks that are clearer than those at 400 °C. However, the peaks remain broadened and of lower intensity, reflecting the influence of the amorphous SiO_2 matrix and small crystallite size. In contrast, the pure $ZnFe_2O_4$ sample displays sharper and more intense peaks, indicating increased crystallinity and crystallite growth at this temperature.

At 1000 °C (Figure 4c), the XRD pattern of $(ZnFe_2O_4)_{0.5}/(SiO_2)_{0.5}$ shows the characteristic diffraction peaks of pure $ZnFe_2O_4$, while the intensity remains low due to the dilution effect of the amorphous SiO_2 matrix [28]. Continued peak broadening indicates limited crystallite growth and suggests that the inert SiO_2 surface layer constrains grain coarsening.

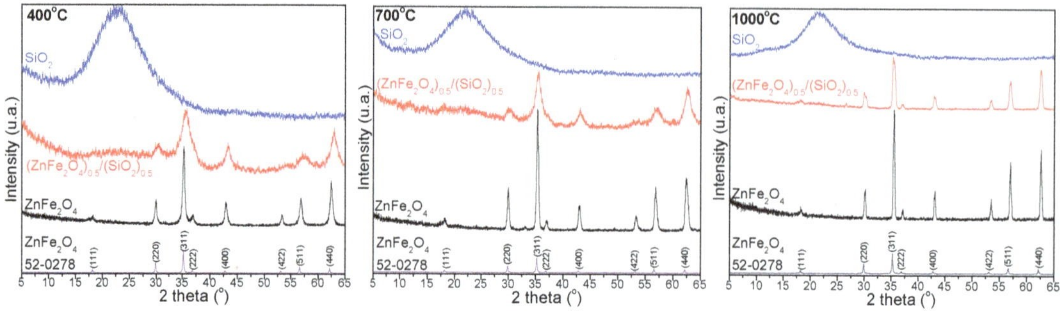

(a) (b) (c)

Figure 4. XRD patterns for $ZnFe_2O_4$, $(ZnFe_2O_4)_{0.5}/(SiO_2)_{0.5}$, and SiO_2 annealed at 400 (**a**), 700 (**b**), and 1000 ° C (**c**).

The average crystallite size (D_{CS}) was estimated by applying the Scherrer equation (Equation (1)) [16–18,27–30]:

$$D_{CS} = \frac{0.9 \cdot \lambda}{\beta \cdot \cos\theta} \tag{1}$$

where λ is the wavelength of CuK_α radiation (1.5418 Å), β is the broadening of full width at half the maximum intensity (FWHM), and θ is the Bragg diffraction angle (°).

The D_Cs increases with increasing annealing temperature, ranging from 35 to 79 nm in the $(ZnFe_2O_4)_{0.5}/(SiO_2)_{0.5}$ nanocomposite and from 48 to 95 nm in bulk $ZnFe_2O_4$. This behavior indicates that crystal growth is closely linked to the degree of cation disorder within the ferrite lattice, as incomplete cation occupancy can hinder grain coalescence. Furthermore, variations in cation valence states and ionic radii contribute to increased lattice strain and generate additional nucleation sites, promoting the formation of smaller crystallites [16–18,27]. Beyond thermal and instrumental contributions, the variation in D_{Cs} may also arise from peak broadening effects due to lattice strain and the physical constraint on grain growth imposed by the SiO_2 matrix. Notably, D_{CS} plays a pivotal role in governing the magnetic properties of the nanocomposites, particularly when the grain size approaches the crystallite size [16–18,27].

The lattice constant, a (Å), was calculated from Bragg's law with the Nelson–Riley function (Equation (2)) [16–18,27]:

$$a = \frac{\lambda\sqrt{h^2 + k^2 + l^2}}{2 \cdot sin\theta} \tag{2}$$

where λ is the wavelength of $CuK\alpha$ radiation (1.5406 Å), *hkl* are the Miller indices, and θ is the Bragg angle (°).

The lattice parameter (a) increases linearly from 8.345 to 8.411 Å with increasing Zn content, reflecting lattice expansion attributed to the larger ionic radius of Zn^{2+} (0.74 Å) compared to Fe^{3+} ions (0.64 Å (Table 1)). Deviations between experimental and theoretical lattice parameters arise from the simplifying assumption that ions behave as rigid spheres fixed within an ideal lattice. Zn^{2+} ions predominantly occupy the A- sites, whereas Fe^{3+} ions are distributed between both tetrahedral (A) and B- sites, depending on cation disorder and synthesis conditions [16–18,27–30].

Table 1. The structural parameters of $(ZnFe_2O_4)_{0.5}/(SiO_2)_{0.5}$ and $ZnFe_2O_4$ samples estimated by XRD.

Sample	Temp [°C]	D_{CS}, [nm]	a, [Å]	V, [Å^3]	d_p [g/cm^3]	d_{XRD} [g/cm^3]	P [%]	d_A [nm]	d_B [nm]
$(ZnFe_2O_4)_{0.5}/(SiO_2)_{0.5}$	400	35	8.345	581	4.554	5.512	17.4	3.613	2.950
	700	54	8.352	583	4.586	5.493	16.5	3.617	2.953
	1000	79	8.367	586	4.625	5.465	15.4	3.623	2.958
$ZnFe_2O_4$	400	32	8.375	587	4.388	5.456	19.6	3.626	2.961
	700	48	8.389	590	4.375	5.428	19.4	3.632	2.966
	1000	69	8.411	595	4.362	5.383	19.0	3.642	2.974
Errors	-	±2.0	±0.015	±1.0	±0.011	±0.016	±1.0	±0.010	±0.010

The unit cell volume (V) and the X-ray density (d_{XRD}), which reflect atomic packing within the crystal lattice, were derived from the XRD data using Avogadro's number ($6.022 \cdot 10^{23}$ mol/L) and the ferrite molar mass according to Equations (3) and (4) [29,30]. Both a and V increase with annealing temperature, consistent with the trends observed in related ferrite systems [16–18,27–30]. The physical density (d_p) was calculated according to the Archimedes' principle by measuring the dry and suspended weights in xylene (Equation (5)), with lower d_p values indicative of increased porosity [18,27–30]:

$$V = a^3 \tag{3}$$

$$d_{XRD} = \frac{8 \cdot M_w}{N_A \cdot a^3} \tag{4}$$

$$d_p - \frac{w \cdot \rho}{w - w'} \tag{5}$$

where w is the sample weight in air (g), w′ is the sample weight in xylene (g), ρ is the xylene density (g/cm^3), 8 is the quantity of molecules within the unit cell, Mw is the molecular weight, and N_A is Avogadro's number ($6.022 \cdot 10^{23}$ mol^{-1}).

Embedding $ZnFe_2O_4$ within the SiO_2 matrix along with increasing the annealing temperature results in an increased V, ranging from 581 to 586 Å^3 for the $(ZnFe_2O_4)_{0.5}/(SiO_2)_{0.5}$ nanocomposite and 587–595 Å^3 for bulk $ZnFe_2O_4$. This expansion is attributed to the generation of oxygen vacancies, a reduction in crystallization temperature, and an increase in molecular weight, which collectively outweigh the expected volumetric contraction during densification [16–18,27–30].

The X-ray density (d_{XRD}) of the samples is calculated using Equation (4) [29,30]. Our results indicate that the d_{XRD} is highest in bulk $ZnFe_2O_4$ and decreases with increasing annealing temperature (5.512–5.465 g/cm^3 for $(ZnFe_2O_4)_{0.5}/(SiO_2)_{0.5}$, 5.456–5.383 g/cm^3 for $ZnFe_2O_4$). This decrease is likely due to the formation of microstructural pores at higher temperatures [16–18,27].

The porosity (*P*) was evaluated using *dp* and d_{XRD} using Equation (6) [15–17,24–27]:

$$P = \left(1 - \frac{d_p}{d_{XRD}}\right)100 \tag{6}$$

The P values were higher in bulk $ZnFe_2O_4$ compared to $ZnFe_2O_4$ embedded in the SiO_2 matrix, attributable to increased vacancy and pore formation during synthesis. Conversely, increased annealing temperatures promote densification, thereby reducing porosity (17.4–15.4% for $(ZnFe_2O_4)_{0.5}/(SiO_2)_{0.5}$ and 19.6–19.0% for bulk $ZnFe_2O_4$). Low porosity is critical for achieving high-performance magnetic materials, as excess porosity degrades magnetic properties. The observed variations in P values are also influenced by the presence of Zn^{2+} ions, which generate additional oxygen vacancies and lower the effective cation concentration. These effects contribute to lattice expansion while preserving overall structural symmetry [16–18,27–30].

The hopping length (d) for A (d_A) and B (d_B) sites was calculated according to Equations (7) and (8) [26,27]:

$$d_A = 0.25{\cdot}a\sqrt{3} \quad (7)$$

$$d_B = 0.25{\cdot}a\sqrt{2} \quad (8)$$

The d_A and d_B are closely related to the a, primarily due to the larger ionic radius of the Zn^{2+} ions compared to the Fe^{3+} ions, which induce lattice expansion. Higher d_A and d_B values imply increased separation between cations, which raises the energy barrier for charge carrier transfer between sites, thereby impacting electronic and magnetic behavior [16–18,27–30].

3.4. BET Analysis

The BET measurements did not allow the determination of the SSA of the samples annealed at 1000 °C. One possible explanation is that the stable form of the SiO_2 matrix, formed after annealing at 1000 °C, has been reported to exhibit an SSA value below 0.5 m^2/g, which is below the detection limit of the method used in this study. The SSA values vary in the following order: 375 m^2/g (SiO_2) > 245 cm^2/g ($(ZnFe_2O_4)_{0.5}/(SiO_2)_{0.5}$) > 37.5 m^2/g ($ZnFe_2O_4$) at 400 °C and 215 m^2/g (SiO_2) > 115 m^2/g ($(ZnFe_2O_4)_{0.5}/(SiO_2)_{0.5}$) ($ZnFe_2O_4$ SSA value < 0.5 m^2/g) at 700 °C. The observed SSA values can be explained by the porous SiO_2 matrix, which not only contributes to the high SSA but also disperses and stabilizes the ferrite nanoparticles, thereby preventing agglomeration [20]. This behavior is typical of oxide materials and aligns with the XRD and AFM results, where higher crystallinity and larger particle sizes result in fewer interparticle pores and alterations in the crystalline phases [20].

3.5. AFM Analysis

Figure 5a shows the morphology of $ZnFe_2O_4$ nanoparticles annealed at 400 °C, revealing predominantly spherical particles with an average diameter of 35 nm uniformly dispersed onto the glass substrate. This uniform adsorption promotes a high degree of particle individualization, minimizing their interactions at the nanoscale. However, small nanoparticles still tend to coalesce, forming local clusters of around 100–150 nm, primarily located at the edges of the image. Increasing the annealing temperature to 700 °C activates fusion between smaller nanoparticles, resulting in a particle size of around 52 nm. The growth of the ferrite crystallite induces slightly squared shapes of the particles in good agreement with the observation in the literature [31]. Their size is very close to that resulting from XRD calculations, proving that these nanoparticles are identical to the observed crystallites. Although the adsorption of the nanoparticles is slightly irregular due to a tendency towards local self-assembly, the particles appear more individualized, and the clustering tendency is significantly reduced. At an annealing temperature of 1000 °C, a well-developed ferrite is observed, imparting a square morphology to the nanoparticles. They resemble small rectangles with rounded edges (Figure 5c). The particle size increases further to 72 nm (in good agreement with XRD results, proving their identification with the observed crystallites), and the nanoparticles are arranged in well-defined positions within the self-assembled thin film.

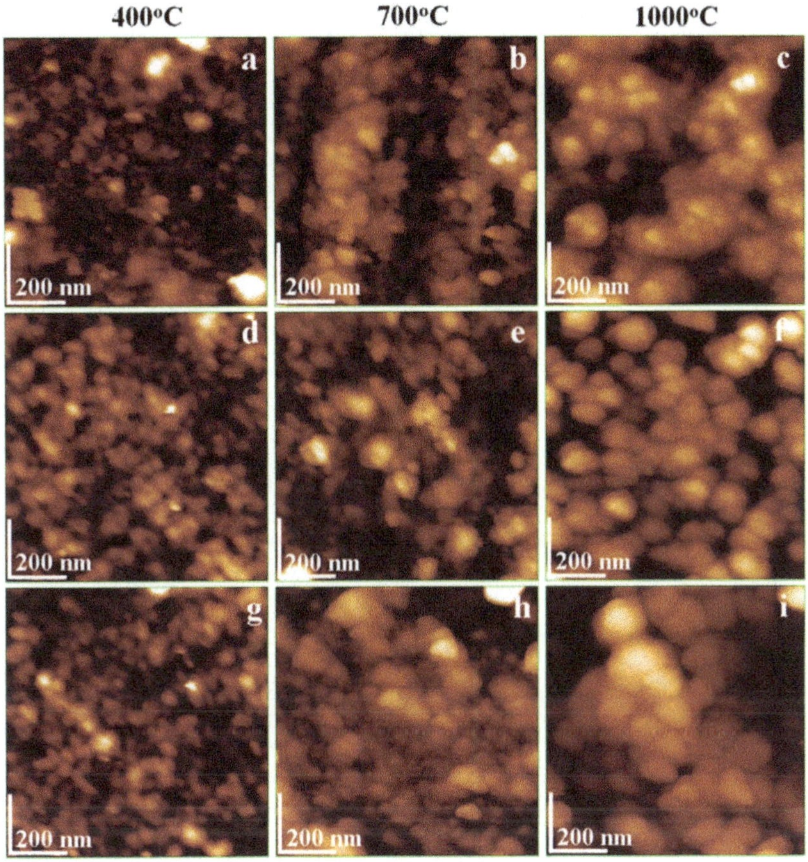

Figure 5. AFM topographic images of the samples: $ZnFe_2O_4$ annealed at (**a**) 400 °C, (**b**) 700 °C, and (**c**) 1000 °C; SiO_2 annealed at (**d**) 400 °C, (**e**) 700 °C, and (**f**) 1000 °C; and $(ZnFe_2O_4)_{0.5}/(SiO_2)_{0.5}$ annealed at (**g**) 400 °C, (**h**) 700 °C, and (**i**) 1000 °C.

After annealing at 400 °C, SiO_2 facilitates the formation of well-individualized, rounded nanoparticles with diameters of about 40 nm. These particles adsorb uniformly onto the glass slide without exhibiting any noticeable clustering tendency (Figure 5d). Upon increasing the annealing temperature to 700 °C, lateral diffusion among smaller nanoparticles leads to the formation of larger aggregates, as shown in Figure 5e. The nanoparticles' diameter increases to 58 nm, with several clusters measuring approximately 80 nm clearly visible near the center of the image. These nanoparticles generally retain a rounded morphology because of the amorphous silica clustering. Further annealing at 1000 °C results in a significant increase in particle size to approximately 84 nm. The rounded forms become boulder like (partly elongated with deeply rounded sides), as is typical for the natural form of the silica sand found in various technical applications [32–34]. They are highly individualized and uniformly adsorbed across the glass substrate (Figure 5f). The behavior of the $(ZnFe_2O_4)_{0.5}/(SiO_2)_{0.5}$ nanocomposite at a low annealing temperature closely resembles that of pure SiO_2, as shown in Figure 5g. At this stage, the nanoparticles are rounded, well-individualized, and uniformly adsorbed, forming a continuous thin film with an average diameter of approximately 55 nm. Increasing the annealing temperature to 700 °C promotes significant development of the ferrite core, in good agreement with XRD data, generating well-developed nanoparticles featuring a distinct square-shaped core with rounded margins and an average size of 65 nm. XRD shows that the ferrite crystallite has a diameter of about 54 nm; the difference to the particle diameter of 65 nm is given by the amorphous SiO_2 glaze forming a thin crust over the ferrite crystallite. Further annealing at 1000 °C causes the nanoparticles to grow, reaching an average size of 98 nm. Their growth is related to the prominent development of the ferrite core having a diameter of about 79 nm, as observed by XRD, and the outer amorphous SiO_2 layer leads to a final particle diameter of about 98 nm. They remain well-individualized and exhibit a relatively square morphology with rounded edges, indicative of a strongly developed ferrite phase. The SiO_2 crust acts as a barrier to the ferrite core's fusion into larger clusters, allowing a better distribution of magnetic centers within the composite powder.

AFM images (Figure 5) confirm that the nanoparticle size increases with annealing temperature and composition, progressing from $ZnFe_2O_4$ to SiO_2. This trend is consistent with previous reports on ferrite nanoparticle behavior [32,33]. Notably, the amorphous SiO_2 coating acts as a physical barrier, effectively isolating the ferrite crystallites and inhibiting the formation of larger clusters. As a result, it promotes the development of well-dispersed and uniformly distributed nanoparticles [34,35].

The topographical aspects are more clearly observed in the three-dimensional profiles presented in Figure 6.

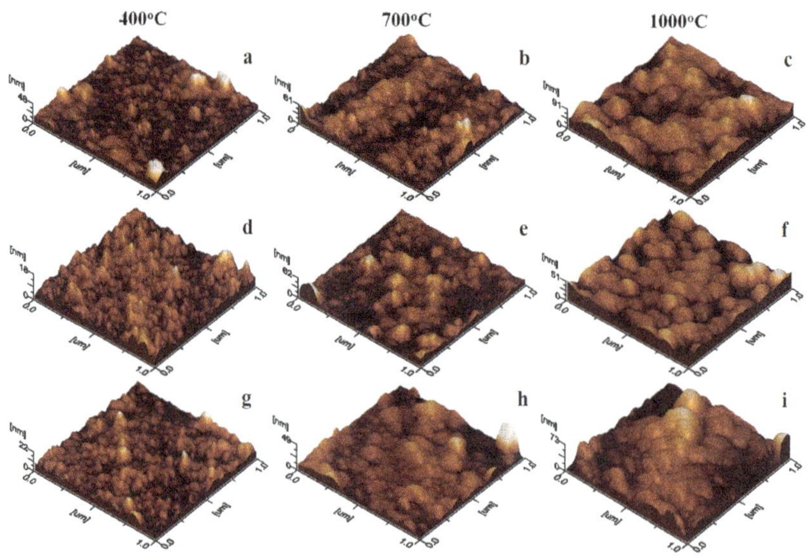

Figure 6. Three-dimensional profiles of the samples: $ZnFe_2O_4$ annealed at (**a**) 400 °C, (**b**) 700 °C, and (**c**) 1000 °C; SiO_2 annealed at (**d**) 400 °C, (**e**) 700 °C, and (**f**) 1000 °C; and $(ZnFe_2O_4)_{0.5}/(SiO_2)_{0.5}$ annealed at (**g**) 400 °C, (**h**) 700 °C, and (**i**) 1000 °C.

The uniformly adsorbed small nanoparticles, such as SiO_2 and $(ZnFe_2O_4)_{0.5}/(SiO_2)_{0.5}$ annealed at 400 °C (Figure 6d,g), form smooth thin films, with surface roughness ranging from 3.15 to 5.10 nm, as reported in Table 2. In contrast, the nanoclusters present in the $ZnFe_2O_4$ sample annealed at 400 °C (Figure 6a) lead to an increased surface roughness of 4.01 nm due to ferrite crystallite coalescence, which induces protrusions causing localized height variations. Thus, amorphous SiO_2 acts as a refining agent of nanostructure. Surface roughness further increases progressively with the development of larger, well-individualized nanoparticles formed after annealing at 700 and 1000 °C. Thus, the $(ZnFe_2O_4)_{0.5}/(SiO_2)_{0.5}$ thin film exhibits the highest roughness, as observed in Figure 6i.

This variation in surface roughness plays a critical role in the performance of ferrite thin films in technological applications, including sustainable energy and biomechanical energy conversion [7,33].

Table 2. $ZnFe_2O_4$, SiO_2 and $(ZnFe_2O_4)_{0.5}/(SiO_2)_{0.5}$ sample parameters measured with AFM.

Sample	Temperature [°C]	Height [nm]	Rq Roughness [nm]	Average Particle Diameter [nm]
$ZnFe_2O_4$	400	48 ± 1.44	4.01 ± 0.24	35 ± 2.1
	700	61 ± 1.83	6.83 ± 0.40	52 ± 3.12
	1000	91 ± 2.73	12.6 ± 0.75	72 ± 4.32
SiO_2	400	18 ± 0.54	2.04 ± 0.12	40 ± 2.4
	700	62 ± 1.86	5.10 ± 0.30	58 ± 3.48
	1000	51 ± 1.53	7.37 ± 0.44	84 ± 5.04
$(ZnFe_2O_4)_{0.5}/(SiO_2)_{0.5}$	400	22 ± 0.66	1.89 ± 0.11	55 ± 3.3
	700	49 ± 1.47	3.15 ± 0.18	65 ± 3.9
	1000	73 ± 2.19	11.2 ± 0.67	98 ± 5.88

3.6. SEM Analysis

Powders annealed at 1000 °C feature the best development of ferrite crystallites and, thus, are most important for magnetic applications. $ZnFe_2O_4$ particles are predominantly small, in good agreement with AFM observations. The absence of insulation material predisposes the ferrite particles to agglomeration due to interparticle attraction forces (Figure 7a). Thus, white clusters are distributed over the surface surrounded by finer ferrite particles. Their square-based crystallites and particle shapes give angular interfaces, resulting in randomly distribute polyhedral pores. The result is that the total porosity of ferrite powder is about 29%, which is in good agreement with the XRD. The microstructural porosity is slightly higher than the one calculated from XRD because of the random disposal of the particles regarding the spreading on the sample holder during the SEM investigation. The mean pore size class is situated in the range of 50–100 nm, as observed in the distribution histogram. Larger pores are completely absent.

The SiO_2 sample is characterized predominantly by rounded particles, with the shape resembling river gravel. Thus, their connectivity is lower than in pure ferrite allowing for more uniform dispersion on the sample holder (Figure 7b). The mutual tangency points between amorphous SiO_2 particles ensure a curvy interface generating dendritic pores detected by Image J software. The clusters are less aggregated than in ferrite but larger in size (about 200 nm), and the pores' dominant size class ranges between 150 and 200 nm, depending on the uniform distribution of the clusters. Blue and green areas dominate the microstructure, and thus, resulting in a total porosity of 17%.

The amorphous SiO_2 coating formed upon the ferrite core within the $(ZnFe_2O_4)_{0.5}/(SiO_2)_{0.5}$ composite prevents agglomeration, which are relatively sparse but with larger sizes of about 250–300 μm (Figure 7b). The particles are well-dispersed on the specimen holder, ensuring a proper interface between them, but the squared shape of the ferrite core influences the disposed angle, increasing the pore sizes. The mean representative pore class is 300–350 μm, which is partially dependent on both cluster size and the spatial arrangement of the individual particles. The squared corner of individualized particles stacks over the rounded side of the clusters, enhancing microstructural pore development. Thus, the total porosity is 21%, which is slightly larger than the theoretical resulting from the XRD calculations. This can be explained by the interaction between the different geometries of the well individualized particles and microstructural clusters.

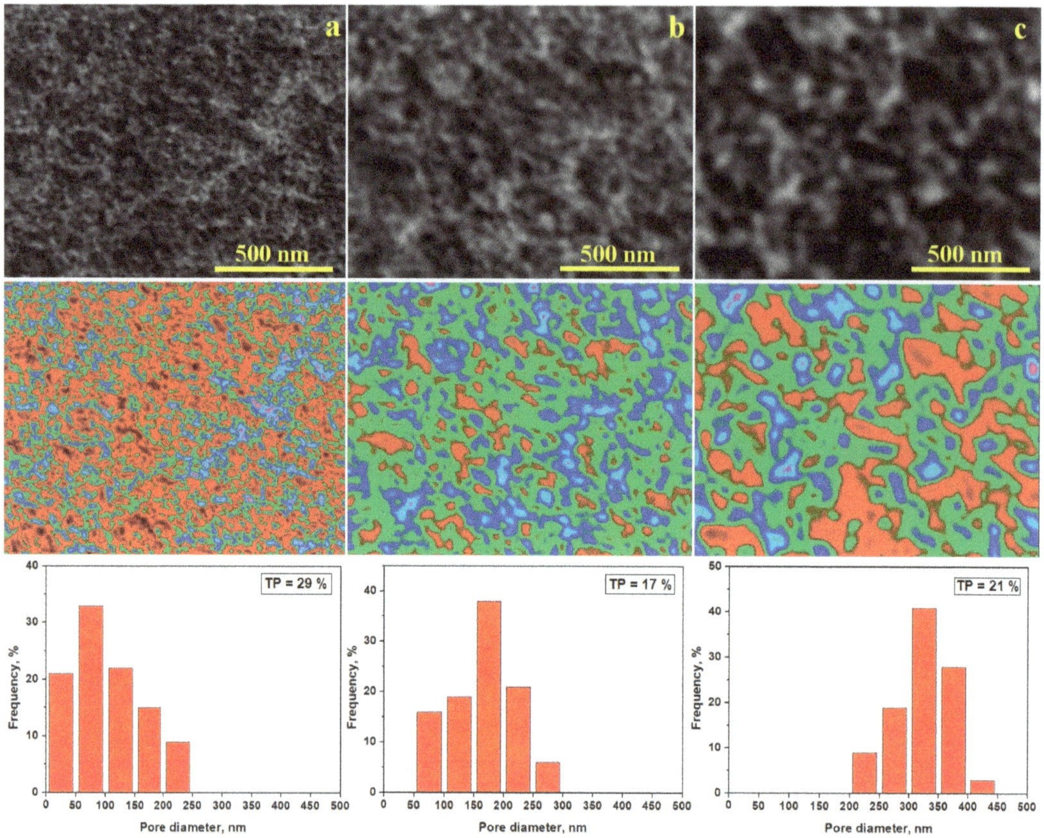

Figure 7. SEM images with pore distribution analysis for the samples annealed at 1000 °C: (**a**) $ZnFe_2O_4$, (**b**); SiO_2, and (**c**) $(ZnFe_2O_4)_{0.5}/(SiO_2)_{0.5}$. Red areas correspond to pores, green areas correspond to well dispersed particles, and blue areas belong to clusters.

The samples were randomly dispersed on the TEM grid, to ensure a representative imaging. Ferrite particles and the ferrite core within the composite appear in regions of intense dark contrast, while amorphous SiO_2 particles and coating surrounding of the composite particles appear in intense grey contrast. Pure ferrite particles (Figure 8a) are well individualized with a high coalescence tendency. The particle size distribution analysis reveals a mean size of 70 nm. Both clusters and well individualized particles of amorphous SiO_2 are observed in Figure 8b. The particle size distribution reveals a mean value of 85 nm, which is comparable to that of the composite particles being situated at 100 nm, consistent with the AFM measurements.

3.7. VSM Analysis

The hysteresis loops of $(ZnFe_2O_4)_{0.5}/(SiO_2)_{0.5}$ and pure $ZnFe_2O_4$ annealed at 700 °C and 1000 °C are shown in Figure 9. The key magnetic parameters, including saturation magnetization (M_S), remanent magnetization (M_R), coercivity (H_c), and crystalline anisotropy constant (K), were extracted from these curves. The results reveal notable differences between the pure and embedded $ZnFe_2O_4$, highlighting the influence of the non-magnetic matrix on the overall magnetic behavior. These differences further emphasize that the magnetic properties of the nanoparticles are strongly dependent on their crystallinity and morphology [1,11]. Smaller nanoparticles typically exhibit superparamagnetic behavior, as thermal energy is sufficient to overcome their magnetic anisotropy energy barriers. In contrast, larger particles, particularly those with a higher Zn^{2+} content in the spinel lattice, tend to display ferromagnetic behavior, whereas bulk $ZnFe_2O_4$ remains paramagnetic at room temperature [1,11]. The $ZnFe_2O_4$ nanoparticles embedded in the SiO_2 matrix demonstrate reduced magnetic properties compared to uncoated $ZnFe_2O_4$ reported in previous studies. This attenuation is primarily attributed to an increased surface defect density and magnetic dilution effects, both of which are influenced by the synthesis conditions and the presence of the non-magnetic SiO_2 layer [1,11].

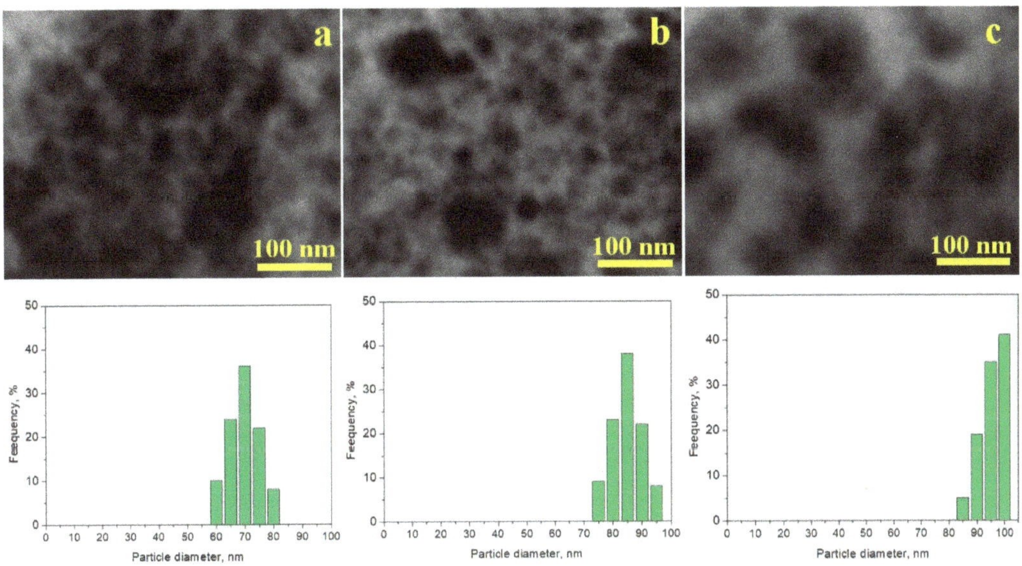

Figure 8. TEM images with particle size distribution analysis for the samples annealed at 1000 °C: (**a**) $ZnFe_2O_4$, (**b**); SiO_2, and (**c**) $(ZnFe_2O_4)_{0.5}/(SiO_2)_{0.5}$.

The coercivity (H_C) of $(ZnFe_2O_4)_{0.5}/(SiO_2)_{0.5}$ increases from 160 Oe at 700 °C to 200 Oe at 1000 °C, while in uncoated $ZnFe_2O_4$, H_C rises from 260 Oe to 300 Oe over the same temperature range [1,11]. The H_c of ferrite nanoparticles is governed by a combination of factors, including particle size and shape distribution, magnetocrystalline anisotropy (K), internal strain, and interparticle magnetic interactions [1,11]. The lowest H_C values observed for the $ZnFe_2O_4$ embedded in the SiO_2 matrix at 700 °C are attributed to smaller crystallite sizes and enhanced magnetic disorder. These conditions promote superparamagnetic behavior and result in partial loss of M_S in the absence of an external field [1,11]. The highest H_C values observed in bulk $ZnFe_2O_4$ at elevated annealing temperatures suggest a reduction in both structural and magnetic disorders. In this case, thermal energy appears insufficient to activate the magnetic domain wall motion, thereby limiting magnetization reversal under an applied magnetic field and promoting improved alignment of magnetic moments. Furthermore, low Hc values may indicate crystallite coalescence into larger nanostructures, which enhances magnetic coupling and contributes to the observed increase in M_S. Additionally, surface strain induced by the SiO_2 matrix can reduce particle size and hinder the rotation of both magnetic particles and their surface magnetic moments, further influencing H_C [1,11].

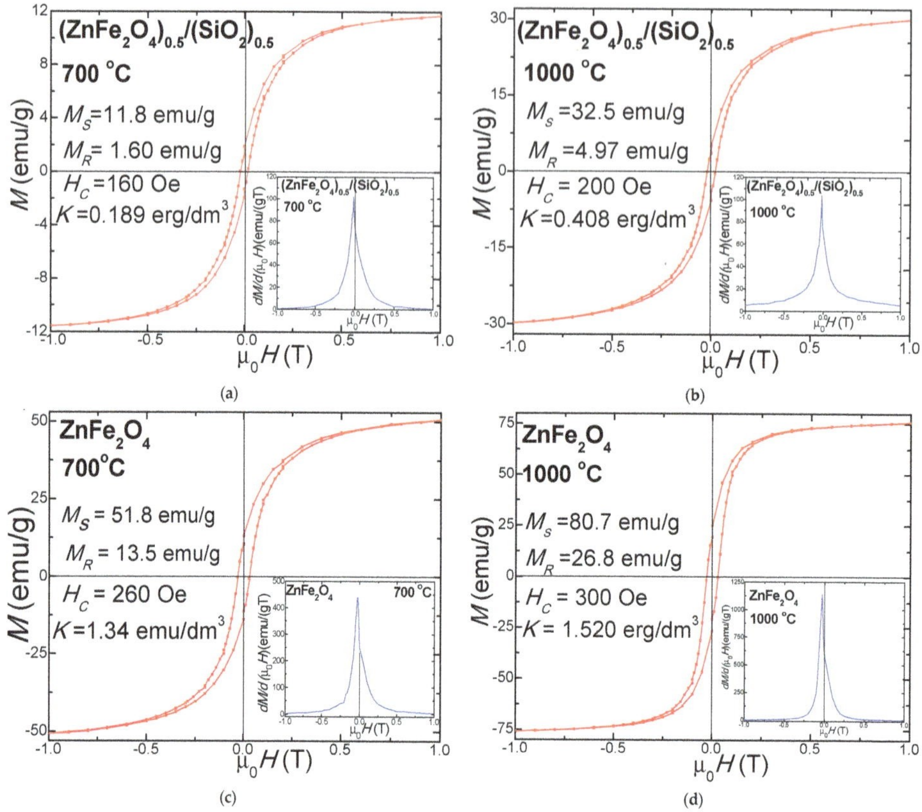

Figure 9. Magnetic hysteresis loops (red) and magnetization derivative (blue) of $ZnFe_2O_4$ and $ZnFe_2O_4)_{0.5}/(SiO_2)_{0.5}$ SiO_2 annealed at 700 ° C (**a**,**c**) and 1000 ° C (**b**,**d**).

The saturation magnetization (M_S) of $(ZnFe_2O_4)_{0.5}/(SiO_2)_{0.5}$ increases from 11.8 emu/g at 700 °C to 32.5 emu/g at 1000 °C, while bulk $ZnFe_2O_4$ shows a corresponding increase from 51.8 emu/g to 80.7 emu/g. This enhancement with increasing annealing temperature is primarily due to the growth in crystallite size and reduction in lattice defects, which facilitate a more effective magnetic moment alignment [1,11,14]. Additionally, changes in the M_S are affected by the formation of antisite defects, which affect the magnetic response independently of the applied magnetic field direction [1,11,36,37]. The significantly lower M_S values observed in the embedded $ZnFe_2O_4$ are due to a higher surface defect density and the magnetic dilution caused by the presence of the non-magnetic matrix [1,11,37]. According to the Néel model of ferrimagnetism, in bulk $ZnFe_2O_4$ spinel, half of the Fe^{3+} ions occupy the A-sites, while the remaining Fe^{3+} ions and the non-magnetic Zn^{2+} ions reside in the B-sites. The magnetic moments of the Fe^{3+} ions at the A-sites are aligned antiparallel to those at the B sites, resulting in partial cancellation and a reduced net magnetic moment [1,11,37,38]. However, at the nanoscale, cation redistribution and thermal effects during annealing can induce a net magnetic moment by altering site occupancy [37]. Additionally, magnetic moment suppression and reduced M_S can occur due to cation disorder and weakened superexchange interactions between A- and B-sites in the spinel structure. In addition to magnetic dilution, the presence of the SiO_2 matrix introduces surface disorder and increases the density of broken chemical bonds on the particle surfaces. These defects promote spin canting, which disrupts magnetic alignment and further influences the particle size, ultimately affecting the overall magnitude of the M_S [1,11,37,38].

The remanent magnetization (M_R) of $(ZnFe_2O_4)_{0.5}/(SiO_2)_{0.5}$ increases from 1.60 emu/g at 700 °C to 4.97 emu/g at 1000 °C, while for bulk $ZnFe_2O_4$, M_R rises from 13.5 emu/g to 26.8 emu/g over the same temperature range [1,11]. The low M_R values observed in the embedded $ZnFe_2O_4$ are indicative of superparamagnetic behavior, where thermal fluctuations prevent stable magnetic alignment in the absence of an external magnetic field [1,11]. The observed increase in M_R with higher annealing temperature is attributed to crystal growth and a reduction in magnetic disorder, leading to enhanced magnetic moment retention [1,11,14].

The crystalline anisotropy constant (K) of $(ZnFe_2O_4)_{0.5}/(SiO_2)_{0.5}$ rises from 0.189 erg/dm^3 at 700 °C to 0.408 erg/dm^3 at 1000 °C, while for non-embedded $ZnFe_2O_4$, K rises from 1.34 erg/dm^3 to 1.520 erg/dm^3 over the same temperature range [1,11]. Higher K values are typically associated with larger particle sizes and are influenced by factors such as particle shape, crystalline symmetry, and the surface distribution of magnetic ions. These factors collectively determine the preferred magnetization direction along specific crystallographic axes [1,11]. The lower K value observed in the embedded sample is attributed to the dilution effect imposed by the non-magnetic SiO_2 matrix. The increase in K with annealing temperature is linked to enhanced crystallinity and a reduction in structural defects [1,11]. Magnetic anisotropy serves as an energy barrier that resists changes in the direction of magnetization, making it a critical parameter in the superparamagnetic behavior of nanoparticles [1,11,37,38]. When thermal energy becomes comparable to or exceeds the anisotropy energy, it can induce magnetization reversal, thereby promoting superparamagnetic behavior [1,11,37,38].

3.8. Photocatalytic Analysis

The optical properties of the samples annealed at 1000 °C were characterized using UV–Vis spectroscopy (Figure 10a). The optical response of nanoparticles is governed by factors such as quantum confinement, lattice strain, structural defects, impurities, surface effects, the light source, the separation efficiency of electron–hole pairs, the surface structure of the catalyst, microstructure, component characterization, phase structure, and the reaction conditions [21,28,39–55]. A pronounced absorption peak in the visible region indicates promising visible-light photocatalytic activity. The active species verification experiments showed that the hole, hydroxyl radical, and superoxide radical were the main active species in the degradation of the photocatalyst [28,31].

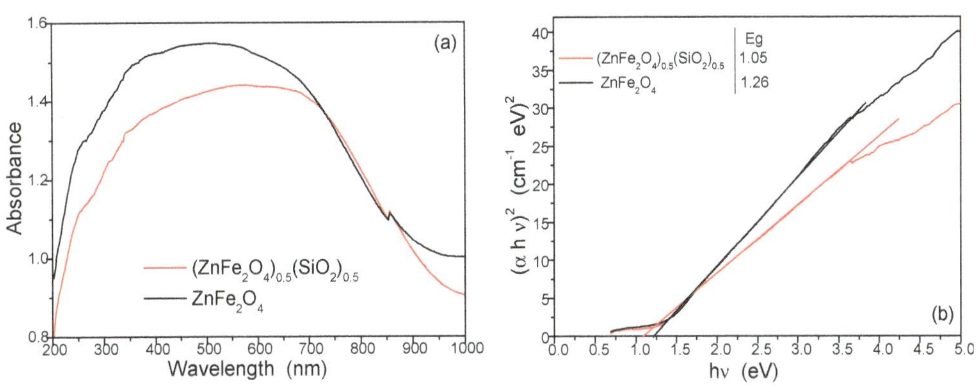

Figure 10. (**a**) UV–Vis absorbance spectra and (**b**) Tauc's plot of $ZnFe_2O_4$, $(ZnFe_2O_4)_{0.5}/(SiO_2)_{0.5}$, and SiO_2 annealed at 1000 ° C.

Photocatalytic activity is strongly influenced by the band gap energy, estimated here via Tauc's method considering the UV–Vis absorption spectra. Figure 10b shows the Tauc plots for samples annealed at 1000 °C. The sample embedded in the SiO_2 matrix exhibits a reduced band gap of 1.05 eV compared to 1.26 eV for the unembedded samples, attributable to their higher ferrite content. The prominent absorption arises from electron excitation across the band gap, primarily involving transitions from the valence band to the conduction band. Variations in the absorption edge are linked to interface and point defects as well as interactions involving photogenerated electrons [39–53].

The photocatalytic process of RhB was analyzed using the first-order kinetic model based on its absorbance (Equation (9)) [21,28,39–53]:

$$-ln\frac{A_t}{A_0^* x} = k_i * t, \quad (9)$$

where A_t is the absorbance of RhB at time t, A_0^* is the absorbance of RhB after dark adsorption, t is the irradiation time, and k is the apparent kinetic constant.

The removal rate of the RhB solution in the case of all the samples is illustrated in Figure 9a. The removal rate involves both adsorption and photocatalysis processes. The existence of $ZnFe_2O_4$ in the SiO_2 matrix enhances the adsorption capacity of the samples. After 240 min of irradiation, Rhodamine B degradation was 98% ($(ZnFe_2O_4)_{0.5}(SiO_2)_{0.5}$) and 64% ($ZnFe_2O_4$), respectively. The photodegradation process is well described by a pseudo-first-order kinetic model and varies linearly with time t (Figure 11b). The linear dependence of photocatalytic degradation on irradiation time is illustrated in Figure 11b.

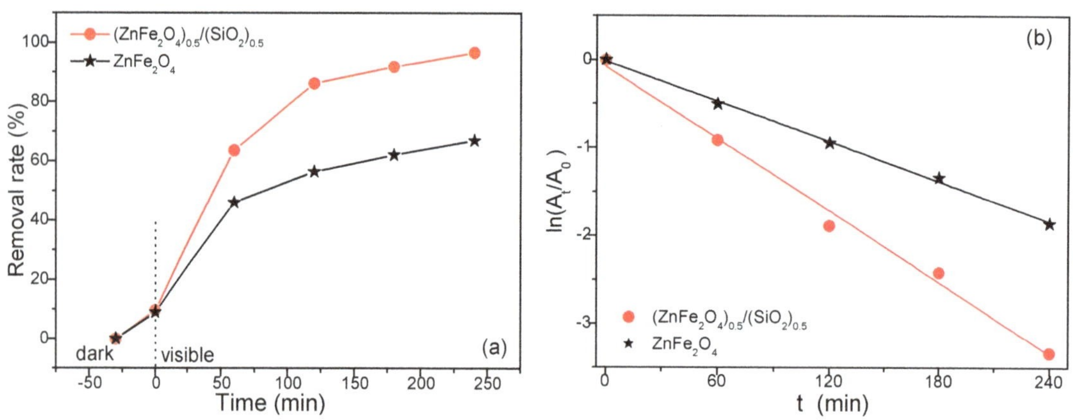

Figure 11. Removal rate of RhB under visible-light irradiation (**a**) and photodegradation kinetic of $ZnFe_2O_4$, $(ZnFe_2O_4)_{0.5}/(SiO_2)_{0.5}$, and SiO_2 annealed at 1000 °C (**b**).

Upon light irradiation, electron–hole pairs are generated, initiating the photocatalytic process. Suppressing charge recombination allows the excited electrons in the conduction band to reduce surface-adsorbed O_2 molecules, forming superoxide radicals. Meanwhile, the holes in the valence band oxidize water or hydroxyl groups, producing hydroxyl radicals. The apparent first-order rate constants (*k*) are summarized in Table 3. Based on the *k* values, the photocatalytic activity follows the order $(ZnFe_2O_4)_{0.5}(SiO_2)_{0.5}$ (k = 14.1 × 10^{-3} min^{-1}, R^2 = 0.9991) > $ZnFe_2O_4$ (k = 4.12 × 10^{-3} min^{-1}, R^2 = 0.9033), as compared to previously reported data (Table 2). The enhanced photocatalytic performance observed for both $(ZnFe_2O_4)_{0.5}(SiO_2)_{0.5}$ and $ZnFe_2O_4$ is primarily attributed to their mesoporous structure and the effective separation of photogenerated electron–hole pairs.

Table 3. Comparison of first-order rate constants for various $ZnFe_2O_4$ photocatalysts with Rhodamine B under visible light.

Sample	Lights	Dyes	$k \times 10^{-3}$	Reference
$(ZnFe_2O_4)_{0.5}(SiO_2)_{0.5}$	Visible	Rhodamine B	14.1	This work
$ZnFe_2O_4$/ ZnO	Visible	Rhodamine B	3.5	[36]
$Co_{0.6}Zn_{0.4}Fe_2O_4$	Visible	Rhodamine B	15.1	[37]
$Co_3O_4/ZnFe_2O_4$	Visible	Rhodamine B	9.32	[38]
$ZnFe_2O_4$/Cd-Se_5GO	Visible	Rhodamine B	16.40	[54]
$ZnS/K_2S_2O_8$	Visible	Rhodamine B	11.98	[40]
20%$ZnFe_2O_4$/80%SnO_2	Visible	Rhodamine B	38	[41]
33.3%$ZnFe_2O_4$/66.7%SnO_2	Visible	Rhodamine B	33	[41]
THCRT/ $ZnFe_2O_4$/ZnO	Visible	Rhodamine B	32.9	[42]
THCRT/ $ZnFe_2O_4$/ZnO	Visible	Rhodamine B	22.6	[42]
$ZnFe_2O_4$/ZnO-0.1	Visible	Rhodamine B	18.5	[42]
$ZnFe_2O_4/NaNbO_3$	Visible	Rhodamine B	21.5	[43]
ZnO–$ZnFe_2O_4$	Visible	Rhodamine B	18.5	[44]
ZnO/$ZnFe_2O_4$/zeolite/ PMS	Visible	Rhodamine B	10	[45]
$ZnFe_2O_4$/BiOBr	Visible	Rhodamine B	7.8	[46]
$ZnFe_2O_4$@TiO_2@Ag_2O	Visible	Rhodamine B	10.2	[47]

Analysis of the obtained values shows that $(ZnFe_2O_4)_{0.5}(SiO_2)_{0.5}$ exhibits the highest photocatalytic activity, with a constant approximatively five times greater than that of $ZnFe_2O_4$. The photocatalytic performance of $(ZnFe_2O_4)_{0.5}(SiO_2)_{0.5}$ annealed at 1000 °C is comparable to that of $ZnFe_2O_4$ and other Zn-based ferrite systems reported in previous literature (Table 3).

Scavenger experiments were conducted to identify the reactive species involved in RhB degradation. To selectively reduce holes and hydroxyl and superoxide radicals, 5 mM solutions of ethylenediaminetetraacetic acid (EDTA), isopropyl alcohol, phenol, and vitamin C were added to the initial RhB solution. As shown in Figure 12a, the presence of the scavengers significantly reduced the degradation rate compared to the control group, indicating that all three species are important in the photocatalytic process. The tests were performed using the $(ZnFe_2O_4)_{0.5}(SiO_2)_{0.5}$ nanocomposite, which contains $ZnFe_2O_4$ embedded within the SiO_2 matrix that facilitates charge separation by mediating electron–hole transfer. Additionally, this composite effectively generates reactive oxygen species (ROS) capable of degrading the RhB molecule [28]. To assess photocatalyst stability, the $(ZnFe_2O_4)_{0.5}(SiO_2)_{0.5}$ nanocomposite was subjected to four consecutive degradation cycles. After each cycle, the catalyst was magnetically recovered, washed with ethanol and water, and reused. The degradation efficiency remained nearly unchanged over the four cycles, confirming the excellent photostability and reusability of the photocatalyst (Figure 12b).

Based on the results, the photocatalytic mechanism can be attributed to the generation of electron–hole pairs under light irradiation. When recombination is suppressed, the photogenerated electrons in the conduction band reduce the adsorbed oxygen to form superoxide radicals, while holes in the valence band oxidize water to produce hydroxyl radicals [28,50–56]. These reactive oxygen species degrade organic pollutants such as RhB. The enhanced photocatalytic activity observed for the $(ZnFe_2O_4)_{0.5}(SiO_2)_{0.5}$ nanocomposite may result from its improved adsorption capacity and the formation of intermediate $ZnFe_2O_4$ energy levels within the SiO_2 matrix, which promote charge separation. $ZnFe_2O_4$ domains act as mediators of charge transfer, facilitating efficient generation of ROS and contributing to the high degradation efficiency [52–56]. These results confirm both the efficacy of the synthesized material and the suitability of the sol-gel method for fabricating high-performance photocatalysts.

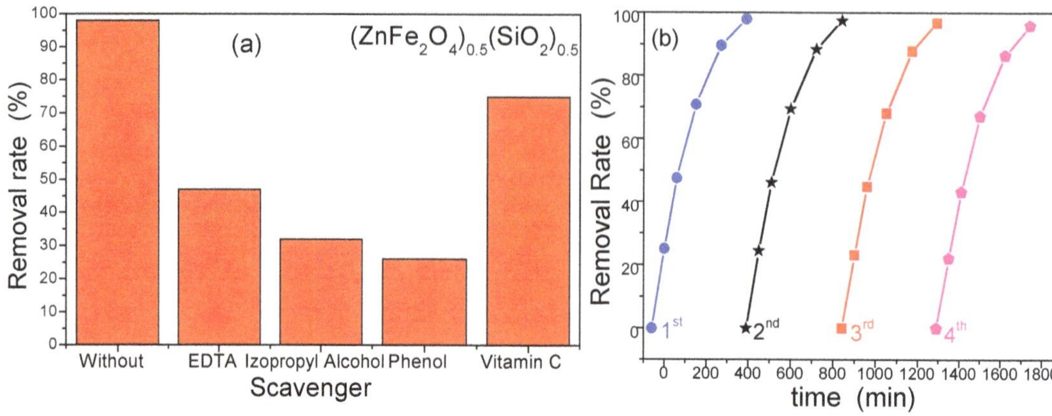

Figure 12. The effect of scavengers on the RhB degradation (**a**) and the reusability of the $(ZnFe_2O_4)_{0.5}(SiO_2)_{0.5}$ nanocomposite used in the removal experiments (**b**).

4. Conclusions

This study investigated the synthesis of $ZnFe_2O_4$ via thermal decomposition as well as the preparation of a $(ZnFe_2O_4)_{0.5}(SiO_2)_{0.5}$ nanocomposite and pure SiO_2 via the sol-gel method. Thermal analysis revealed distinct decomposition of the Zn^{2+} and Fe^{3+} succinate precursors, leading to the formation of $ZnFe_2O_4$ both non-embedded and embedded within the SiO_2 matrix. The highest total mass loss was observed for bulk $ZnFe_2O_4$ (67%), followed by the $ZnFe_2O_4/SiO_2$ nanocomposite (63%) and pure SiO_2 (48%). FT-IR spectroscopy confirmed the formation of the SiO_2 matrix and reaction progress. FT-IR and XRD indicated the formation of a single-phase spinel structure and the successful embedding of $ZnFe_2O_4$ into the SiO_2 matrix. Crystallite sizes increased with the annealing temperature, ranging from 35 to 79 nm for the nanocomposites and 48 to 95 nm for bulk $ZnFe_2O_4$. Magnetic properties improved with annealing temperature and $ZnFe_2O_4$ content: saturation magnetization (Ms) from 11.8 to 80.7 emu/g, remanent magnetization (Mr) from 1.6 to 26.8 emu/g, coercivity (Hc) from 160 to 300 Oe, and anisotropy constant (K) from 0.189×10^{-3} to 1.520×10^{-3} erg/cm^3. These enhancements were attributed to increased particle size, crystallinity, and reduced surface spin disorder. The best photocatalytic performance was recorded for the $(ZnFe_2O_4)_{0.5}(SiO_2)_{0.5}$ nanocomposite, which exhibited the best photocatalytic performance, presumably due to the synergistic effect of $ZnFe_2O_4$ energy levels within the silica matrix and an optimized Zn–Fe balance in the ferrite structure. This is likely due to additional energy levels in the conduction band that promote the recombination of photogenerated carriers. Scavenger experiments show that holes and hydroxyl and superoxide radicals are implicated in photocatalytic processes.

Author Contributions: Conceptualization, T.D.; methodology, T.D., A.-M.S.-M., R.M.R., I.P. and O.C.; software, T.D.; validation, T.D., I.P. and O.C.; formal analysis, T.D., A.-M.S.-M., R.M.R., I.P. and O.C.; investigation, T.D., A.-M.S.-M., I.P. and O.C.; resources, T.D., I.P. and O.C.; data curation, T.D.; writing—original draft preparation, T.D., I.P. and O.C.; writing—review and editing, T.D. and O.C.; visualization, T.D., A.-M.S.-M., R.M.R., I.P. and O.C.; supervision, T.D. All authors have read and agreed to the published version of the manuscript.

Funding: This research received no external funding.

Data Availability Statement: Data are available on request from the corresponding author.

Conflicts of Interest: The authors declare no conflicts of interest.

References

1. Jadhav, S.A.; Somvanshi, S.B.; Khedkar, M.V.; Patade, S.R.; Jadhav, K.M. Magneto-structural and photocatalytic behavior of mixed Ni–Zn nano-spinel ferrites: Visible light-enabled active photodegradation of rhodamine B. *J. Mater. Sci. Mater. Electron.* **2020**, *31*, 11352–11365. [CrossRef]
2. Divakara, S.G.; Mahesh, B. A comprehensive review on current trends in greener and sustainable synthesis of ferrite nanoparticles and their promising applications. *Results Eng.* **2024**, *21*, 101702. [CrossRef]
3. Ibrahim, A.A.; Sa'ed, M.; Abdullah, S.A.; Dellinger, K.; Obare, S.O.; Pathiraja, G. Magnetic zinc ferrite nanostructures: Recent advancements for environmental and biomedical applications. *J. Alloys Compd.* **2024**, *4*, 100038. [CrossRef]
4. Ali, S.S.L.; Selvaraj, S.; Batoo, K.M.; Chauhan, A.; Rana, G.; Kalaichelvan, S.; Radhakrishnan, A. Green synthesis of cubic spinel ferrites and their potential biomedical applications. *Ceram. Int.* **2024**, *50*, 52159–52189. [CrossRef]
5. Askarzadeh, N.; Shokrollahi, H. A review on synthesis, characterization and properties of lithium ferrites. *Results Chem.* **2024**, *10*, 101679. [CrossRef]
6. Riaz, H.; Anjum, S.; Ejaz, S.R.; Khan, M.I.; Shanableh, A.; Luque, R. Multifunctional Co^{2+} and Cr^{3+} Substituted Ni-Cu Spinel Ferrites for Wastewater Remediation and Supercapacitor Applications. *Results Eng.* **2025**, *25*, 104209. [CrossRef]
7. Abidin, M.Z.U.; Ikram, M.; Moeen, S.; Nazir, G.; Kanoun, M.B.; Goumri-Said, S. A comprehensive review on the synthesis of ferrite nanomaterials via bottom-up and top-down approaches advantages, disadvantages, characterizations and computational insights. *Coord. Chem. Rev.* **2024**, *520*, 216158. [CrossRef]
8. Hamed, A.; Fitzgerald, A.G.; Wang, L.; Gueorguieva, M.; Malik, R.; Melzer, A. Synthesis, characterization and surface modification of $ZnCrFeO_4$ nanoparticles. *Mater. Sci. Eng. C* **2013**, *33*, 1623–1628. [CrossRef]
9. Ardrey, K.D.; McAuliffe, R.D.; Armstrong, B.L. The role of ferrites in wireless charging systems for electric vehicles. *J. Magn. Magn. Mater.* **2024**, *610*, 172576. [CrossRef]
10. Akbar, A.; Lakshmi, M.B.; Das, T.K.; Ghosh, M. Spinel ferrites in the photocatalytic and adsorptive remediation of dyes and heavy metals: A review. *J. Water Proc. Eng.* **2025**, *71*, 107259. [CrossRef]
11. Iqbal, B.; Laybourn, A.; ul-Hamid, A.; Zaheer, M. Size-controlled synthesis of spinel nickel ferrite nanorods by thermal decomposition of a bimetallic Fe/Ni-MOF. *Ceram. Int.* **2021**, *47*, 12433–12441. [CrossRef]
12. Monfared, A.H.; Zamanian, A.; Beygzadeh, M.; Sharifi, I.; Mozafari, M. A rapid and efficient thermal decomposition approach for the synthesis of manganese-zinc/oleylamine core/shell ferrite nanoparticles. *J. Alloys Compd.* **2017**, *693*, 1090–1095. [CrossRef]
13. Klym, K.; Karbovnyk, I.; Piskunov, S.; Popov, A.I. Positron annihilation lifetime spectroscopy insight on free volume conversion of nanostructured $MgAl_2O_4$ ceramics. *Nanomaterials* **2021**, *11*, 3373. [CrossRef]
14. Munir, S.; Warsi, M.F.; Zulfiqar, S.; Ayman, I.; Haider, S.; Alsafari, I.A.; Agboola, P.O.; Shakir, I. Nickel ferrite/zinc oxide nanocomposite: Investigating the photocatalytic and antibacterial properties. *J. Saudi Chem. Soc.* **2021**, *25*, 101388. [CrossRef]
15. Hossain, M.F.; Paul, T.C.; Khan, M.N.I.; Islam, S.; Bala, P. Magnetic and dielectric properties of $ZnFe_2O_4$/nanoclay composites synthesized via sol-gel autocombustion. *Mater. Chem. Phys.* **2021**, *271*, 124914. [CrossRef]
16. Dippong, T.; Goga, F.; Levei, E.-A.; Cadar, O. Influence of zinc substitution with cobalt on thermal behaviour, structureand morphology of zinc ferrite embedded in silica matrix. *J. Solid State Chem.* **2019**, *275*, 159–166. [CrossRef]
17. Dippong, T.; Mereu, R.A. Effect of La^{3+} on thermal, structural and morphological properties of Zn–Co ferrite spinel based pigments. *Ceram. Int.* **2024**, *50*, 10314–10324. [CrossRef]
18. Stefanescu, M.; Stoia, M.; Dippong, T.; Stefanescu, O.; Barvinschi, P. Preparation of $Co_XFe_{3-X}O_4$ Oxydic System Starting from Metal Nitrates and Propanediol. *Acta Chim. Slov.* **2009**, *56*, 379–385.
19. Karadi, I.; Hiremath, V.J.; Timmanagoudar, S.; Mathad, S.N.; Mehulkumar, G. Nano Ferrites: Synthesis, Proprietes and Emerging Applications—A Comperhensive Review. *J. Advis. Mater. Eng.* **2025**, *10*, 1–21. [CrossRef]
20. Shirsath, S.E.; Wang, D.; Jadhav, S.; Mane, M.L.; Li, S. Ferrites Obtained by Sol-Gel Method. In *Handbook of Sol-Gel Science and Technology*; Springer: Cham, Switzerland, 2017; pp. 1–41. [CrossRef]
21. Dippong, T.; Toloman, D.; Lazar, M.D.; Petean, I. Effects of Lanthanum Substitution and Annealing on Structural, Morphologic, and Photocatalytic Properties of Nickel Ferrite. *Nanomaterials* **2023**, *13*, 3096. [CrossRef]
22. Kibombo, H.; Ntakarutimana, F.; Nsanzumuhire, A.; Niyigena, J. Synthesis, characterization and photocatalytic behavior of SiO_2@nitrized-TiO_2 nanocomposites obtained by a straightforward novel approach. *Chem. Open* **2021**, *10*, 1001–1012. [CrossRef]
23. Ge, Y.; Li, C.; Waterhouse, G.I.N.; Zhang, Z.; Yu, L. $ZnFe_2O_4$@SiO_2@Polypyrrole nanocomposites with efficient electromagnetic wave absorption properties in the K and K_a band regions. *Ceram. Int.* **2021**, *47*, 1728–1739. [CrossRef]
24. Suppuraj, P.; Thirunarayanan, G.; Swaminathan, M.; Muthuvel, I. Synthesis of Spinel Nanocrystalline $ZnFe_2O_4$: Enhanced Photocatalytic and Microbial Applications. *Mater. Sci. Appl. Chem.* **2017**, *34*, 5–11. [CrossRef]
25. Padmapriya, G.; Manikandan, A.; Krishnasamy, V.; Jaganathan, S.K.; Antony, S.A. Enhanced catalytic activity and magnetic properties of spinel $Mn_xZn_{1-x}Fe_2O_4$ ($0.0 \leq x \geq 1.0$) nano-photocatalysts by microwave irradiation route. *J. Supercond. Nov. Magn.* **2016**, *29*, 2141–2149. [CrossRef]

26. Yadav, R.S.; Kuřitka, I.; Vilcakova, J.; Urbánek, P.; Machovsky, M.; Masař, M.; Holek, M. Structural, magnetic, optical, dielectric, electrical and modulus spectroscopic characteristics of $ZnFe_2O_4$ spinel ferrite nanoparticles synthesized via honey-mediated sol-gel combustion method. *J. Phys. Chem. Solids* **2017**, *110*, 87–99. [CrossRef]
27. Dippong, T.; Cadar, O.; Deac, I.G.; Barbu Tudoran, L.; Levei, E.A. Influence of Ni^{2+} substitution by Co^{2+} on the morphology and magnetic properties of single domain $Co_{\alpha}Ni_{0.9-\alpha}Zn_{0.1}Fe_2O_4$ nanoparticles. *J. Alloys Compd.* **2023**, *952*, 170074. [CrossRef]
28. Dippong, T.; Cadar, O.; Goga, F.; Toloman, D.; Levei, E.A. Impact of Ni Content on the Structure and Sonophotocatalytic Activity of Ni-Zn-Co Ferrite Nanoparticles. *Int. J. Mol. Sci.* **2022**, *23*, 14167. [CrossRef] [PubMed]
29. Powar, R.R.; Phadtare, V.D.; Parale, V.G.; Park, H.H.; Pathak, S.; Kamble, P.R.; Piste, P.B.; Zambare, D.N. Structural, morphological, and magnetic properties of $Zn_xCo_{1-x}Fe_2O_4$ ($0 < x < 1$) prepared using a chemical co-precipitation method. *Ceram. Int.* **2018**, *44*, 20782–20789.
30. Renuka, L.; Anantharaju, K.S.; Sharma, S.C.; Vidya, Y.S.; Nagaswarupa, H.P.; Prashantha, S.C.; Nagabhushana, H. Synthesis of $ZnFe_2O_4$ Nanoparticle by Combustion and Sol Gel Methods and their Structural, Photoluminescence and Photocatalytic Performance. *Mater. Today Proc.* **2018**, *5*, 20819–20826. [CrossRef]
31. Wang, S.; Zhang, Y.; Han, Y.; Yu, X.; Deng, L.; Hu, L.; Gao, H.; Angadi, V.J.; Shaik, S.F.; Ubdaidullah, M. Double internal magnetic fields significantly improve photocatalytic activity over $Ba_{0.5}Sr_{0.5}TiO_3/BaFe_{12}O_{19}/SrFe_{12}O_{19}$ photocatalysts. *Colloids Surf. A Physicochem. Eng. Asp.* **2025**, *705*, 135608. [CrossRef]
32. Avram, S.E.; Birle, B.V.; Tudoran, L.B.; Borodi, G.; Petean, I. Investigation of Used Water Sediments from Ceramic Tile Fabrication. *Water* **2024**, *16*, 1027. [CrossRef]
33. Badji, M.; Belalem, K.A.; Bakha, Y.; Zerdali, M. The Effect of Deposition Cycles on the Morphological Properties of Bismuth Ferrite Nanostructured Thin Films. *Proceedings* **2024**, *105*, 66. [CrossRef]
34. Avram, S.E.; Tudoran, L.B.; Borodi, G.; Petean, I. Microstructural Characterization of the Mn Lepidolite Distribution in Dark Red Clay Soils. *Appl. Sci.* **2025**, *15*, 6445. [CrossRef]
35. Andriyanti, W.; Nur, M.A.C.H.; Puspitarum, D.L.; Sujitno, T.; Suprihatin, H.; Purwanto, S.; Suharyadi, E. Microstructures, magnetic properties and microwave absorption of ion-implanted bismuth ferrite thin films. *Phys. B Condens. Matter.* **2024**, *676*, 415690. [CrossRef]
36. Thulasi, V.; Lakshmi, P.; Jebaseelan Samuel, E.J.; Mohana Roopan, S. Fabrication of ferrite nanoparticle based improved composite piezoelectric energy harvester for biomechanical energy conversion. *Mater. Sci. Eng. B* **2025**, *313*, 117915. [CrossRef]
37. Islam, S.; Rahman, M.L.; Moni, M.R.; Biswas, B.; Ahmed, M.F.; Sharmin, N. Impacts of annealing temperature on microstructure, optical and electromagnetic properties of zinc ferrites nanoparticles synthesized by polymer assisted sol-gel method. *Arab. J. Chem.* **2023**, *16*, 105186. [CrossRef]
38. Lassoued, A.; Li, J.F. Structure and optical, magnetic and photocatalytic properties of Cr^{3+} substituted zinc nano-ferrites. *J. Mol. Struct.* **2022**, *1262*, 133021. [CrossRef]
39. Hu, P.; Chen, C.; Song, J.; Tang, Z. Efficient visible-light photocatalysis of ZIF-derived mesoporous $ZnFe_2O_4$/ ZnO nanocomposite prepared by a two-step calcination method. *Mater. Sci. Semicond. Process.* **2018**, *77*, 40–49. [CrossRef]
40. Sundararajan, M.; Sailaja, V.; Kennedy, L.J.; Vijaya, J.J. Photocatalytic degradation of rhodamine B under visible light using nanostructured zinc doped cobalt ferrite: Kinetics and mechanism. *Ceram. Int.* **2017**, *43*, 540–548. [CrossRef]
41. Nagesh, T.; Ramesh, K.; Ashok, B.; Jyothi, B.; Kumar, B.V.; Upender, G. Insights into charge transfer via Z-scheme for Rhodamine B degradation over novel $Co_3O_4/ZnFe_2O_4$ nanocomposites. *Opt. Mater.* **2023**, *143*, 114140. [CrossRef]
42. Miao, Z.; Tian, Y.; Li, S.; Ding, Z.; Chen, X.; Ma, W.; Chang, Y. Photocatalytic degradation of Rhodamine B over popcorn-like $ZnFe_2O_4$/CdSeGO ternary composite. *J. Mater. Res. Technol.* **2022**, *21*, 1863–1877. [CrossRef]
43. Zhu, B.; Cheng, H.; Ma, J.; Kong, Y.; Komarneni, S. Efficient degradation of rhodamine B by magnetically separable $ZnSeZnFe_2O_4$ composite with the synergistic effect from persulfate. *Chemosphere* **2019**, *237*, 124547. [CrossRef]
44. Sonia; Kumari, H.; Sharma, S.; Sheetal; Chahal, S.; Kumar, P.; Kumar, A. Magnetic $ZnFe_2O_4/SnO_2$ nanocomposites for efficient wastewater Treatment: Structural, optical, and photocatalytic investigations. *Ceram. Int.* **2025**, *51*, 31478–31486. [CrossRef]
45. Du, B.; Ye, H.; Wang, Y.; Shen, M.; Zhou, H. Tetrasulfonatomethyl-n-hexyl calix[4]resorcinarene tetrasodium assisted preparation of $ZnFe_2O_4$/ZnO composites with enhanced photodegradation performance for organic contaminants. *Colloids Surf. A Physicochem. Eng. Asp.* **2024**, *699*, 134733. [CrossRef]
46. Zhou, D.; Zhou, X.; Hu, Z.; Zheng, L.; Tian, Y.; Tu, Y.; Hua, C.; Xue, L.; Xiong, J. Pyroelectric field drived photocatalysis by $ZnFe_2O_4/NaNbO_3$ heterojunction for dye degradation through integration of solar and thermal energy. *Arab. J. Chem.* **2024**, *17*, 105996. [CrossRef]
47. AlSalhi, M.S.; Devanesan, S.; Asemi, N.; Ahamed, A. Concurrent fabrication of ZnO–$ZnFe_2O_4$ hybrid nanocomposite for enhancing photocatalytic degradation of organic pollutants and its bacterial inactivation. *Chemosphere* **2023**, *318*, 137928. [CrossRef]
48. Hu, T.; Li, J.; Wang, L.; Wang, H.; Zhang, Z.; Jiang, W.; Xue, C. ZnO/$ZnFe_2O_4$/zeolite composite catalyst for peroxymonosulfate oxidation and photocatalysis. *Mater. Lett.* **2023**, *330*, 133310. [CrossRef]

49. Meng, X.; Yang, Y.; Zhang, L.; Liu, D.; Zheng, H.; Huo, S.; You, Y. Preparation and visible light catalytic degradation of magnetically recyclable $ZnFe_2O_4$/BiOBr flower-like microspheres. *J. Alloys Compd.* **2023**, *954*, 169981. [CrossRef]
50. Ren, X.; Chen, R.; Ding, S.; Fu, N. Preparation and photocatalytic performance of a magnetically recyclable $ZnFe_2O_4$@TiO_2@Ag_2O p-n/Z-type tandem heterojunction photocatalyst: Degradation pathway and mechanism. *Colloids Surf. A Physicochem. Eng. Asp.* **2023**, *658*, 130604. [CrossRef]
51. Kumar, R.; Sudhaik, A.; Sonu; Raizada, P.; Nguyen, V.-H.; Le, Q.V.; Ahamad, T.; Thakur, S.; Hussain, C.M.; Singh, P. Integrating K and P co-doped g-C_3N_4 with $ZnFe_2O_4$ and graphene oxide for S-scheme-based enhanced adsorption coupled photocatalytic real wastewater treatment. *Chemosphere* **2023**, *337*, 139267. [CrossRef] [PubMed]
52. Syed, J.A.S.; Huang, Y.; Han, N.; Fang, J.-B.; Zi, T.-Q.; Li, A.-D. Magnetically separable Z-scheme $ZnFe_2O_4$/MoS_2 photocatalyst for boosting photo-Fenton degradation activity by a hole-mediated mechanism. *J. Alloys Compd.* **2024**, *990*, 174508. [CrossRef]
53. Zhao, W.; Wei, Z.; Li, C.; Ding, M. Composite $ZnFe_2O_4$/$SrWO_4$ hollow microspheres as catalyst for high-performance photo-Fenton degradation. *Ceram. Int.* **2024**, *50*, 24063–24069. [CrossRef]
54. Yu, X.; Wang, S.; Zhang, Y.; Gao, H.; Zhou, X.; Li, D.; Yang, H.; Fang, L.; Zhang, H.; Syed, A. Novel high entropy alloy/$NiAl_2O_4$ photocatalysts for the degradation of tetracycline hydrochloride: Heterojunction construction, performance evaluation and mechanistic insights. *Ceram. Int.* **2024**, *50*, 29528–29546. [CrossRef]
55. Yu, X.; Wang, S.; Zhang, Y.; Yu, X.; Gao, H.; Yang, H.; Fang, L.; Zhang, H.; Syed, A. Utilization of stable and efficient high-entropy $(Ni_{0.2}Zn_{0.2}Mg_{0.2}Cu_{0.2}Co_{0.2})Al_2O_4$ catalyst with polyvalent transition metals to boost peroxymonosulfate Activation toward pollutant degradation. *Small* **2025**, *21*, 2410819. [CrossRef]
56. Pathania, A.; Thakur, P.; Trukhanov, A.V.; Trukhanov, S.V.; Panina, L.V.; Luders, U.; Thakur, A. Development of tungsten doped Ni-Zn nano-ferrites with fast response and recovery time for hydrogen gas sensing application. *Results Phys.* **2019**, *15*, 10253. [CrossRef]

nanomaterials

MDPI

Article

Systematic Evaluation of Zn^{2+}, Ca^{2+}, and Co^{2+} Doping for Tailoring the Thermal, Structural, Morphological and Magnetic Performance of $CdBi_{0.1}Fe_{1.9}O_4$@SiO_2 Nanocomposites

Thomas Dippong [1,*], Ioan Petean [2] and Oana Cadar [3]

1 Faculty of Science, Technical University of Cluj-Napoca, 76 Victoriei Street, 430122 Baia Mare, Romania
2 Faculty of Chemistry and Chemical Engineering, Babes-Bolyai University, 11 Arany Janos Street, 400084 Cluj-Napoca, Romania; ioan.petean@ubbcluj.ro
3 INCDO-INOE 2000, Research Institute for Analytical Instrumentation, 67 Donath Street, 400293 Cluj-Napoca, Romania; oana.cadar@icia.ro
* Correspondence: dippong.thomas@yahoo.ro

Abstract

The influence of Zn^{2+}, Ca^{2+} and Co^{2+} doping on the thermal, structural, morphological, and magnetic characteristics of $CdBi_{0.1}Fe_{1.9}O_4$ nanoparticles synthetized via the sol–gel technique and calcined at 300, 600, 900 and 1200 °C was investigated. Thermal analysis revealed the initial formation of metallic glyoxylates up to 300 °C, followed by their decomposition into metal oxides and subsequent ferrite formation. X-ray diffraction revealed that the ferrites were poorly crystallized at lower temperatures, whereas at higher calcination temperatures all nanocomposites exhibited well-crystalized ferrites coexisting with the SiO_2 matrix, except for the $Co_{0.1}Cd_{0.9}Bi_{0.1}Fe_{1.9}O_4$@$SiO_2$ nanocomposite, which formed a single, well-defined crystalline phase. Atomic force microscopy images revealed spherical ferrite particles encapsulated within an amorphous layer, with particle size, surface area, and coating thickness influenced by both the type of dopant ion and the calcination temperature. The structural parameters estimated by X-ray diffraction, as well as the magnetic characteristics, were strongly influenced by the dopant type and thermal treatment. These results demonstrate that the structural and magnetic characteristics of $CdBi_{0.1}Fe_{1.9}O_4$ ferrites can be effectively tuned through controlled doping and calcination, providing insights for the design of tailored functional applications.

Keywords: cadmium-bismuth doped ferrite; divalent ions; SiO_2 matrix; structural-property relationship; magnetic characteristics

1. Introduction

Nanostructured ferrites exhibit advantageous magnetic characteristics, such as reduced coercivity (H_c), high saturation magnetization (M_s), elevated magnetic permeability, large surface area, high electrical resistivity, and low dielectric and eddy current losses. In addition, their mechanical rigidity, chemical and thermal stability, non-toxicity, physical flexibility, corrosion resistance, and cost effectiveness further enhance their technological significance [1–9]. Nevertheless, the structural and magnetic characteristics of ferrites are strongly dependent on their chemical composition, precursor selection, synthesis strategy, thermal processing conditions, and the incorporation of dopants, impurities, or secondary phases [6–12]. Owing to their multifunctional characteristics, ferrites find applications across several technological sectors, including (i) biomedical applications such as anti-cancer and antibacterial therapies, bioimaging, biosensors), (ii) energy and environmental applications, including energy storage, microwave absorption, magnetic refrigeration, corrosion protection, and water decontamination, (iii) electronic and magnetic technologies, such as spintronics, humidity sensors, ferrofluids, high-capacity information storage media, and (iv) industrial applications, including ceramic pigments, ferroceramics, and magnetic inks [13–18].

$CdFe_2O_4$ crystallizes in a normal spinel structure, in which Cd^{2+} ions preferentially occupy the tetrahedral (A) sites and Fe^{3+} ions on the octahedral (B) sites within a face-centered cubic lattice [1–4]. Moreover, $CdFe_2O_4$ is a typical example of geometrically frustrated antiferromagnet, exhibiting unique magnetic behavior that arise from superexchange interactions among its magnetic ions [1–5]. Although $CdFe_2O_4$ is non-magnetic, the relatively large radius of the Cd^{2+} ion induces a lattice distortion in the spinel structure, which can lead to spin canting and perturbation of the magnetic moment [1,3,5,8]. Compared to other magnetic materials, $CdFe_2O_4$ exhibits high electrical resistivity, a characteristic that is particularly advantageous for the design of soft ferrites, as it minimizes eddy-current losses and provides a clear advantage over magnetic materials such as metal alloys and metal oxides [1–7]. Owing to the synergy of its chemical stability, structural, mechanical integrity, and multifunctional electrical, magnetic, and optical characteristics, $CdFe_2O_4$ is suitable for a wide range of applications including transformer core materials, microwave components, and biomedical technologies [1–6].

The incorporation of non-magnetic Cd^{2+} ions into the spinel ferrite lattice is known to modify the cation distribution and enhance magnetic performance, notably by increasing the net magnetic moment and M_s, enabling applications in loading coils, magnetic recording heads, antenna cores and microwave devices [3,5,6]. Furthermore, co-doping with divalent ions Zn^{2+} or Co^{2+} can further improve magnetic characteristics, electrical conductivity, permeability, density, photocatalytic activity, and biological activity [7–9]. Additionally, substituting Bi^{3+} or Fe^{3+} ions with other transition metal ions can induce lattice distortions, modify the cycloidal spin structure and enhance ferrimagnetism [19–22].

Numerous synthetic strategies have been developed for ferrite preparation, including wet chemical techniques (co-precipitation, hydrothermal/solvothermal methods), sonochemical, sol-gel, solid-state methods (solid-state reaction, ball milling/mechanochemical processing), mechanical treatments (milling, grinding), thermal and combustion-based methods (decomposition, flash combustion, auto-combustion), as well as vapor phase and aerosol-assisted techniques (vapor phase deposition and surfactant-mediated aerosol processing) [23–29]. Each method provides distinct advantages in tuning particle size, morphology, and functional characteristics, with wet-chemical routes favoring nanosized particles and solid-state routes typically used for bulk ferrite systems [23,27,28]. Among these approaches, the sol-gel method provides a versatile and cost-effective strategy for producing ferrites with well-controlled composition, structural uniformity, and tailored functional characteristics [23,27–31]. Nanoparticle agglomeration, which can hinder the formation of single-phase ferrites, can be mitigated by applying homogenous ultrathin coatings or dispersing particles within non-magnetic matrices. In this context, the sol-gel method using tetraethylorthosilicate (TEOS) enables the formation of robust inorganic networks with moderate reactivity, facilitating the incorporation of a wide range of organic and inorganic species. Among potential coating materials, mesoporous SiO_2 is particularly attractive due to its hydrophilicity, chemical stability, non-toxicity and biocompatibility, and its ability to control particle growth and size distribution [23,27,28]. Embedding ferrite nanoparticles within a SiO_2 matrix suppresses agglomeration, promotes controlled crystallite development, and can enhance magnetic performance. Moreover, the bio-inert SiO_2

https://doi.org/10.3390/nano16040259

matrix improves the biocompatibility of nanoferrites and mitigates inflammatory risks in biological environments [23,27,28].

The magnetic behavior of nanoferrites is primarily determined by the combined effects of cation distribution at the A and B sites and surface-related spin disorder, both of which are influenced by particle size, synthesis method, and post-synthesis thermal processing conditions [32–34]. Additionally, the magnetic characteristics of $CoFe_2O_4$, $ZnFe_2O_4$ and $CdFe_2O_4$ are influenced by size, shape, purity, and magnetic stability of the constituent particles [27]. Multiferroic systems have attracted significant interest due to the coexistence of switchable ferroelectric, ferrimagnetic, and ferroelastic order parameters within a single phase [19,35]. However, when ferrimagnetic ferrite particles decrease below a critical size, thermal fluctuations can disrupt single-domain magnetic ordering, leading to a substantially reduction in M_s and coercivity (H_c) [27].

This study investigates the influence of Zn^{2+} ($Zn_{0.1}Cd_{0.9}Bi_{0.1}Fe_{1.9}O_4$), Ca^{2+} ($Ca_{0.1}Cod_{0.9}Bi_{0.1}Fe_{1.9}O_4$), and Co^{2+} ($Co_{0.1}Cd_{0.9}Bi_{0.1}Fe_{1.9}O_4$) doping on the structural, morphological, and magnetic characteristics of $CdBi_{0.1}Fe_{1.9}O_4$ nanoparticles embedded within a SiO_2 matrix, synthetized via the sol-gel route and calcined at various temperatures. The thermal evolution and decomposition of glyoxylate precursors were analyzed using thermal analysis and Fourier transform infrared (FT-IR) spectroscopy. The crystalline phase development and crystallinity were assessed using X-ray diffraction (XRD) which was used to study, while particle shape, morphology, size and agglomeration were investigated using atomic force microscopy (AFM). The magnetic characteristics, including the variation of M_s vs. H_c were evaluated for both undoped and Zn-, Ca- and Co-doped $CdBi_{0.1}Fe_{1.9}O_4$ samples.

The novelty of the present study arises from (i) a comprehensive assessment of the combined effect of the SiO_2 matrix and divalent dopant ions on the formation of metal-glyoxylate precursors and their subsequent thermal transformation into ferrite phases embedded in the SiO_2 matrix; (ii) a systematic comparative investigation of Zn^{2+}, Ca^{2+}, and Co^{2+} doping in $CdBi_{0.1}Fe_{1.9}O_4@SiO_2$ nanocomposites (NCs) to elucidate their influence on morphological, structural, and magnetic characteristics, identifying pathways to enhance functional performance and broaden potential applications; (iii) the development of new trimetallic ferrites embedded in a SiO_2 matrix; (iv) achieving single-phase crystalline ferrites at relatively low calcination temperatures (300 °C) for both undoped and doped systems, demonstrating the efficiency of the proposed synthesis; and (v) showing that dopant incorporation, particularly Co^{2+} ion, suppresses the formation of secondary phases at high calcination temperatures (1200 °C), highlighting the stabilizing role of selective doping.

2. Materials and Methods

2.1. Reagents

Zinc nitrate hexahydrate ($Zn(NO_3)_2{\cdot}6H_2O$, 98%), calcium nitrate tetrahydrate ($Ca(NO_3)_2{\cdot}4H_2O$, 99%), cobalt nitrate hexahydrate ($Co(NO_3)_2{\cdot}6H_2O$, 98%), and cadmium nitrate tetrahydrate ($Cd(NO_3)_2{\cdot}4H_2O$, 99%), were purchased from Carlo Erba (Milan, Italy). Ferric nitrate nonahydrate ($Fe(NO_3)_3{\cdot}9H_2O$, 98%), bismuth nitrate pentahydrate ($Bi(NO_3)_2{\cdot}5H_2O$ (98%)), ethylene glycol (EG, 99%), TEOS (99%) and ethanol (96%) were obtained from Merck, Darmstadt, Germany.

2.2. Synthesis

$CdBi_{0.1}Fe_{1.9}O_4$ and $M_{0.1}Cd_{0.9}Bi_{0.1}Fe_{1.9}O_4$ (M = Zn, Ca, Co) embedded within a SiO_2 matrix were prepared using a sol-gel technique. The resulting NCs consisted of 45 wt.% ferrite and 55 wt.% SiO_2. Appropriate molar ratios of metal nitrates were selected to achieve M/Cd/Bi/Fe of 0/10/1/19 for the undoped composition and 1/9/1//19 for the Zn-, Ca-, and Co-doped samples. The precursor sols were prepared by dissolving

https://doi.org/10.3390/nano16040259

the corresponding metal nitrates in a mixed solvent system containing EG, TEOS, and ethanol using a NO_3^-/EG/TEOS molar ratio of 1/1/1. The solutions were magnetically stirred for 30 min to ensure homogeneity and subsequently aged under ambient conditions until complete gelation, which occurred after approximately 30 days. The resulting gels entrapped uniformly distributed metal ions within the organic–inorganic network. To enhance the phase purity and promote the crystallite growth, the dried gels were finely ground and subjected to a thermal pretreatment at 40 °C for 4 h. Calcination was then carried out subsequentially at 300, 600, 900 and 1200 °C for 4 h at each temperature using a Nabertherm LT9 muffle furnace (Lilienthal, Germany).

2.3. Characterization

The thermal evolution of the samples was investigated by thermogravimetric (TG) and differential thermal analysis (DTA) in air, using a heating rate of at 10 $°C \cdot min^{-1}$ up to 1200 °C; measurements were performed SDT Q600 analyzer (TA Instruments, New Castle, DE, USA), employing alumina as the reference material. FT-IR spectra were recorded on KBr pellets containing 1 wt.% of the sample using a Prestige-21 (Shimadzu, Tokyo, Japan). Phase identification was performed by XRD using a D8 Advance diffractometer (Bruker, Karlsruhe, Germany) with CuKα radiation (λ = 1.5418 Å) operated at 40 kV and 35 mA. Atomic force microscopy (AFM) measurements were performed using a high-resolution scanning probe microscope JSPM 4210 produced by JEOL Company (Tokyo, Japan) operating in tapping mode with a NSC 15 Hard cantilever (resonant frequency of 325 kHz and a force constant of 40 N/m). The acquired topographic images were analyzed with WinSPM software (version 2.0, JEOL, Tokyo, Japan) to determine surface height, roughness, and particle diameter. The magnetic characterization was carried out using a 7400 vibrating-sample magnetometer (VSM, Lake Shore, Carson, CA, USA). Hysteresis loops were measured at room temperature under applied fields up to 2 T, while the magnetization measurements were performed in fields up to 5 T.

3. Results and Discussion

3.1. Formation and Decomposition of Glyoxylate Precursor and SiO_2 Development

Figure 1 presents the formation and decomposition of metal-glyoxylate precursors within the ferrite system, as well as the development of the SiO_2 matrix, as evidenced by the TG DTA profiles of the gels heated at 40 °C and the comparative FT-IR spectra of the gels heated at 40 and 200 °C.

The TG and DTA curves of all gels reveal five distinct thermal processes: (i) removal of residual water, including physically adsorbed moisture, manifested by an endothermic peak at approximatively 36 °C, corresponding to a mass loss of 2.5%; (ii) formation of Cd, Zn-, Ca-, Co- glyoxylates, evidenced by an endothermic effect near 153 °C with a mass loss of ~6.5%, accompanied by the release of H_2O and NO (these redox reactions likely yield a homogenous mixture of homonuclear Fe^{3+}, Bi^{3+} and Cd^{2+}, Zn^{2+}, Ca^{2+}, Co^{2+} glyoxylates); (iii) formation of Fe- and Bi- glyoxylates occurring around 191 °C, marked by an endothermic effect and a mass loss of ~13%, due to the release of crystallization water from the nitrates and volatile species (H_2O, NO_2), driving redox reactions [13,15]; (iv) oxidative decomposition of Cd-, Zn-, Ca-, Co- glyoxylates into their corresponding metal oxides (CdO, ZnO, CaO, CoO) along with the concomitant formation of ferrites, appearing as an exothermic effect near 273 °C with the release of CO, CO_2 and H_2O, corresponding to a mass loss ~12%; and (v) oxidative decomposition of Fe- and Bi- glyoxylates into Fe_2O_3 and Bi_2O_3 occurring around 318 °C, producing an exothermic effect with a mass loss ~22% due to the elimination of CO, CO_2 and H_2O (Figure 1a,b) [13–15,28].

https://doi.org/10.3390/nano16040259

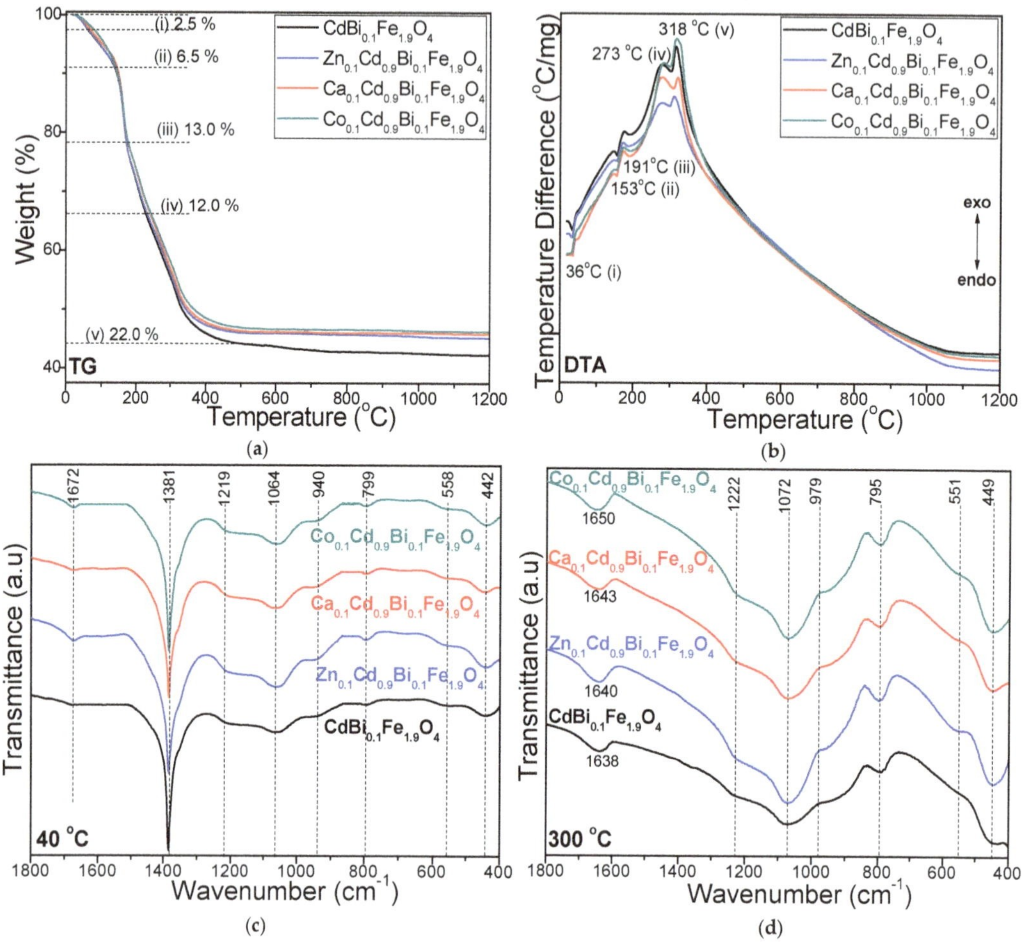

Figure 1. (**a**) TG and (**b**) DTA curves of $CdBi_{0.1}Fe_{1.9}O_4$, $Zn_{0.1}Cd_{0.9}Bi_{0.1}Fe_{1.9}O_4$, $Ca_{0.1}Cod_{0.9}Bi_{0.1}Fe_{1.9}O_4$, $Co_{0.1}Cd_{0.9}Bi_{0.1}Fe_{1.9}O_4$ gels heated at 40 °C; (**c**) FT-IR spectra of gels heated at 40 °C and (**d**) FT-IR spectra of gels calcinated at 300 °C.

The two-stage progress (at 153 °C and 191 °C) of the redox reaction between metal nitrates and EG is attributed to the higher acidity of the $[Fe(H_2O)_6]^{3+}$ and $[Bi(H_2O)_6]^{3+}$ cations compared to $[Cd(H_2O)_6]^{2+}$, $[Zn(H_2O)_6]^{2+}$, $[Ca(H_2O)_6]^{2+}$ and $[Co(H_2O)_6]^{2+}$ cations [13,15]. Consequently, the redox reaction leading to the formation of the Fe carboxylates occurs at lower temperatures, producing a mixture of homonuclear Fe^{3+} and Co^{2+} carboxylates [13,15]. In contrast, the Fe^{3+} and Bi^{3+} glyoxylates form at higher temperatures [13,15]. All other synthesized gels display similar TG and DTA curves, with the position and intensity of the exothermic effects depending on the specific $Fe(NO_3)_3/Bi(NO_3)_3/Co(NO_3)_2$, $Fe(NO_3)_3/Bi(NO_3)_3/Co(NO_3)_2/M^{2+}(NO_3)$ (M = Zn, Ca or Co) ratio. The highest total mass loss is observed for $CdBi_{0.1}Fe_{1.9}O_4@SiO_2$ (58.0%), while slightly lower mass losses are recorded for $Zn_{0.1}Cd_{0.9}Bi_{0.1}Fe_{1.9}O_4@SiO_2$ (54.7%), $Ca_{0.1}Cd_{0.9}Bi_{0.1}Fe_{1.9}O_4@SiO_2$ (54.3%), $Co_{0.1}Cd_{0.9}Bi_{0.1}Fe_{1.9}O_4@SiO_2$ (53.7%). During thermal treatment, the SiO_2 matrix undergoes multiple transformations, which makes it difficult to make the distinction between processes associated with the formation and decomposition of glyoxylate precursors [13–15,28].

https://doi.org/10.3390/nano16040259

The FT-IR spectra indicate the formation of metal glyoxylate precursors from metal nitrates, as evidenced by the disappearance of the nitrate-related band at 1381 cm^{-1} upon calcination at 300 °C (Figure 1c,d) [13–15,28]. The SiO_2 matrix already begins to develop at 40 °C and becomes more clearly defined after calcination at 300 °C, as confirmed by the presence of Si–O–Si stretching vibrations observed around 1064 cm^{-1} (40 °C) and 1072 cm^{-1} (300 °C), together with Si–O vibrational vibrations appearing as shoulders near 1219 cm^{-1} (40 °C) and 1222 cm^{-1} (300 °C) [13–16,23–28]. Deformation vibrations of the Si–OH groups, arising from the hydrolysis of $-Si(OCH_2CH_3)_4$ (TEOS), are observed at 940 cm^{-1} (40 °C) and 972 cm^{-1} (300 °C), while Si–O stretching within SiO_4 tetrahedra appear at 799 cm^{-1} (40 °C) and 795 cm^{-1} (300 °C) [13–16,23–28]. O–H vibrations from EG and adsorbed water appear at 1672 cm^{-1} at 40 °C; after calcination at 300 °C, these bands broaden and shift to 1638–1650 cm^{-1}, increasing in intensity by Zn^{2+}, Ca^{2+}, or Co^{2+} doping. This behavior is attributed to $-COO^-$ vibrations, indicating the formation of chelated complexes through coordination of glyoxylate ligands with the metal ions [13–16,23–28].

The FT-IR spectra exhibit absorption bands at 558 cm^{-1} (40 °C) and 551 cm^{-1} (300 °C), which are assigned to the stretching vibrations of tetrahedral Cd–O, Zn–O, Ca–O, and Co–O bonds, as well as to cyclic Si–O–Si structures. The bands observed at 442 and 449 cm^{-1} are associated with the octahedral Fe–O, Bi–O, and Si–O vibrational modes [13–16,23–28]. The shifts in the A- and B- site band positions with different dopants and calcination temperatures indicate lattice distortions arising from changes in M–O bond lengths [35,36]. These variations are in full agreement with the inverse relationship between bond length, reduced atomic mass, and stretching vibration frequency [37–39].

3.2. Formation of the Crystalline Structure of Ferrites

At low calcination temperatures, oxidic phases typically exhibit poor crystallinity or may remain partially amorphous; thus, the development of the desired surface characteristics and crystallinity strongly depends on optimizing the calcination conditions [13,14,16,19]. In addition, the pronounced chemical reactivity of the amorphous SiO_2 matrix facilitates its participation in various chemical transformations during the calcination process. Figure 2 shows the XRD patterns of the NCs calcined at 300, 600, 900, and 1200 °C.

At 300 °C, the diffraction peaks corresponding to the (220), (311), (222), (400), (422), (511), and (440) planes confirm the formation of single, low-crystallized $CdBi_{0.1}Fe_{1.9}O_4$ phase, consistent with $CdFe_2O_4$ (JCPDS card no. 65-3115) and $Bi_2Fe_4O_9$ (JCPDS card no. 74-1098), exhibiting a cubic spinel structure with space group *Fd3m* [26–28]. No secondary or impurity-phases are remarked at this temperature, indicating that the dopant ions are successfully incorporated into the spinel ferrite lattice without compromising its structural integrity.

At 600 °C, the XRD pattern of the undoped $CdBi_{0.1}Fe_{1.9}O_4@SiO_2$ NC is dominated by the CdO phase (JCPDS card no. 05-0640), while the characteristic diffraction peaks of the $CdBi_{0.1}Fe_{1.9}O_4$ phase appear only weakly, indicating incomplete ferrite formation and poor crystallinity. In contrast, doping $CdBi_{0.1}Fe_{1.9}O_4$ with Zn^{2+}, Ca^{2+}, and Co^{2+} ions promotes the formation of low crystallized single-phase ferrites. Among the doped samples, $Zn_{0.1}Cd_{0.9}Bi_{0.1}Fe_{1.9}O_4@SiO_2$ NC exhibits the most intense diffraction peaks, indicating the highest degree of crystallinity. These results suggest that incorporation of Zn^{2+}, Ca^{2+}, and Co^{2+} ions at 600 °C promotes the formation of single-phase ferrite structures and improves structural ordering compared with the undoped sample. The presence of secondary phases can be attributed to the higher cation mobility and lattice strain variations induced during calcination, which also slightly shift the 2θ positions, broadens the peaks, and increases crystallite sizes [37,40]. At 600 °C, doping $CdBi_{0.1}Fe_{1.9}O_4$

https://doi.org/10.3390/nano16040259

with Zn^{2+}, Ca^{2+}, or Co^{2+} ions, restricts Cd^{2+} diffusion and enhances thermodynamic stability of $Zn_{0.1}Cd_{0.9}Bi_{0.1}Fe_{1.9}O_4$, $Ca_{0.1}Cd_{0.9}Bi_{0.1}Fe_{1.9}O_4$, $Co_{0.1}Cd_{0.9}Bi_{0.1}Fe_{1.9}O_4$ phases embedded in SiO_2 matrix. The homogeneity of the metal oxide particles may also contribute to an increased defect density and higher pore volume in the final composites [40,41].

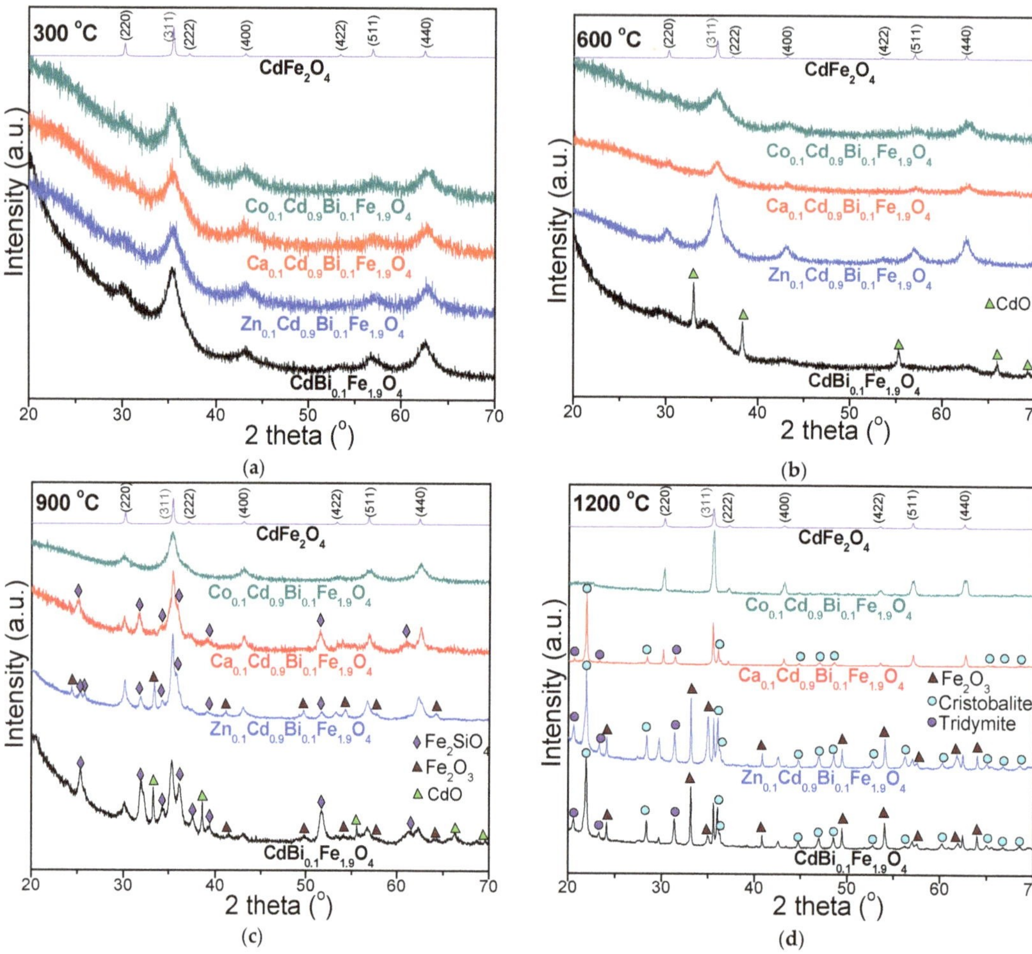

Figure 2. XRD patterns of $CdBi_{0.1}Fe_{1.9}O_4@SiO_2$, $Zn_{0.1}Cd_{0.9}Bi_{0.1}Fe_{1.9}O_4@SiO_2$, $Ca_{0.1}Cd_{0.9}Bi_{0.1}Fe_{1.9}O_4@SiO_2$, $Co_{0.1}Cd_{0.9}Bi_{0.1}Fe_{1.9}O_4@SiO_2$ NCs calcined at (**a**) 300 °C, (**b**) 600 °C, (**c**) 900 °C and (**d**) 1200 °C.

At 900 °C, the XRD pattern of the undoped $CdBi_{0.1}Fe_{1.9}O_4@SiO_2$ NC reveals the coexistence of the crystalline ferrite phase with secondary crystalline phases of CdO, α-Fe_2O_3 (JCPDS card no. 33-0664), and Fe_2SiO_4 (JCPDS card no. 70-1861). For the Zn-doped $CdBi_{0.1}Fe_{1.9}O_4@SiO_2$ NC, CdO is no longer detected, while the diffraction pattern consists of the crystalline ferrite phase accompanied by α-Fe_2O_3 and Fe_2SiO_4. In the case of Ca-doped $CdBi_{0.1}Fe_{1.9}O_4@SiO_2$, α-Fe_2O_3 also disappears, resulting in a diffraction pattern containing the ferrite phase together with Fe_2SiO_4 as single secondary phase. The decomposition of carboxylate precursors generates a reducing environment that partially converts Fe^{3+} ions into Fe^{2+} ions, which subsequently interact with the SiO_2 matrix, resulting in the formation of Fe_2SiO_4 at 900 °C in all NCs [27,28,39]. The formation of α-Fe_2O_3 instead of pure crystalline ferrite may result from the incomplete embedding of the ferrite within the SiO_2 matrix and from suboptimal calcinations conditions [19,20]. The presence of α-Fe_2O_3 indicates the decomposition of $Fe(NO_3)_3$ into α-Fe_2O_3, which promotes spinel ferrite formation, while the excess of metal oxides in insoluble secondary phases may enhance densification by increasing pore volume and generating demagnetizing fields [37,40]. The formation of secondary phases is further influenced by the cation mobility and the lattice strain induced during the calcination process. These effects also display as peak broadening, increased crystallite size, and slight shifts in 2θ positions [14,27,28].

Following calcination at 600 and 900 °C, the ferrite phases remain poorly crystalline, while higher calcination temperatures promote the formation of larger crystallites and accelerate the crystallite growth due to improved nucleation and growth kinetics [27,28]. The presence of sharp diffraction peaks at elevated temperature confirms the development of well-crystallized spinel ferrite structures. At 1200 °C, both undoped $CdBi_{0.1}Fe_{1.9}O_4$@SiO_2 and $Zn_{0.1}Cd_{0.9}Bi_{0.1}Fe_{1.9}O_4$@$SiO_2$ display four crystalline phases: $CdBi_{0.1}Fe_{1.9}O_4$, α-Fe_2O_3, cristobalite (JCPDS card no. 74-9378), and tridymite (JCPDS card no. 42-1401), indicating partial crystallization of the SiO_2 matrix in the absence of ferrite. Ca^{2+} doping of $CdBi_{0.1}Fe_{1.9}O_4$@SiO_2 leads to the disappearance of the α-Fe_2O_3 phase, resulting in the ferrite phase coexisting only with the crystalline SiO_2 phases (cristobalite and tridymite). In contrast, Co^{2+} doping results in the formation of a single-phase ferrite, $Co_{0.1}Cd_{0.9}Bi_{0.1}Fe_{1.9}O_4$, corresponding to $CoFe_2O_4$ (JCPDS card no. 22-1086), $CdFe_2O_4$ (JCPDS card no. 65-3115), and $Bi_2Fe_4O_9$ (JCPDS card no. 74-1098). This behavior is attributed to the atomic level cation mixing, which promotes the formation of a dense, fine-grained, and homogeneous ferrite structure [40–43].

The diffraction peaks of the ferrite phases become more intense at 1200 oC, indicating high crystallinity, uniform nucleation, and efficient crystal growth, with minimal influence of from surface coatings [14,27,28]. Variations in peak intensity and signal-to-noise ratio corresponds to the variations in crystallite size and degree of crystallinity [32].

Table 1 summarizes the structural parameters of the NCs calcined at 300, 600, 900 and 1200 °C, including average crystallite size (D_{CS}), lattice parameter (a), unit cell volume (V), physical (d_p) and X-ray (d_{XRD}) densities, hopping length at the A (d_A) and B (d_B) sites, and porosity (P), as determined by XRD [44,45].

Table 1. Structural parameters and quantitative analysis of $CdBi_{0.1}Fe_{1.9}O_4$@SiO_2, $Zn_{0.1}Cd_{0.9}Bi_{0.1}Fe_{1.9}O_4$@$SiO_2$, $Ca_{0.1}Cd_{0.9}Bi_{0.1}Fe_{1.9}O_4$@$SiO_2$, $Co_{0.1}Cd_{0.9}Bi_{0.1}Fe_{1.9}O_4$@$SiO_2$ NCs estimated by XRD.

Sample	Temp [°C]	D_{CS} [nm]	a (Å)	V (Å³)	d_A (Å)	d_B (Å)	d_p (g/cm³)	d_{XRD} (g/cm³)	P (%)	Quantitative Analysis
$CdBi_{0.1}Fe_{1.9}O_4$@SiO_2	300	7.3	8.379	588	3.628	2.962	5.342	6.855	22.1	100% $CdBi_{0.1}Fe_{1.9}O_4$
	600	14.9	8.415	596	3.644	2.975	5.379	6.763	20.5	74% CdO/26% $CdBi_{0.1}Fe_{1.9}O_4$
	900	31.6	8.457	605	3.662	2.990	5.411	6.662	18.8	34%$CdBi_{0.1}Fe_{1.9}O_4$/23% Fe_2SiO_4/6%Fe_2O_3/37%CdO
	1200	48.2	8.498	614	3.680	3.004	5.467	6.564	16.7	32% $CdBi_{0.1}Fe_{1.9}O_4$/35% Fe_2O_3/33% SiO_2
$Zn_{0.1}Cd_{0.9}Bi_{0.1}Fe_{1.9}O_4$@$SiO_2$	300	8.3	8.344	581	3.613	2.950	5.349	6.830	21.7	100% $Zn_{0.1}Cd_{0.9}Bi_{0.1}Fe_{1.9}O_4$
	600	16.5	8.376	588	3.627	2.961	5.393	6.749	20.0	100% $Zn_{0.1}Cd_{0.9}Bi_{0.1}Fe_{1.9}O_4$
	900	34.8	8.423	598	3.648	2.978	5.431	6.636	18.2	73% $Zn_{0.1}Cd_{0.9}Bi_{0.1}Fe_{1.9}O_4$/ 22% Fe_2SiO_4/5%Fe_2O_3
	1200	52.4	8.468	607	3.667	2.994	5.489	6.538	16.0	34% $Zn_{0.1}Cd_{0.9}Bi_{0.1}Fe_{1.9}O_4$/35% Fe_2O_3/31% SiO_2
$Ca_{0.1}Cd_{0.9}Bi_{0.1}Fe_{1.9}O_4$@$SiO_2$	300	9.6	8.366	586	3.623	2.958	5.359	6.714	20.2	100% $Ca_{0.1}Cd_{0.9}Bi_{0.1}Fe_{1.9}O_4$
	600	15.9	8.406	594	3.640	2.972	5.408	6.624	18.4	100% $Ca_{0.1}Cd_{0.9}Bi_{0.1}Fe_{1.9}O_4$
	900	36.2	8.446	603	3.657	2.986	5.449	6.525	16.5	72% $Ca_{0.1}Cd_{0.9}Bi_{0.1}Fe_{1.9}O_4$/18% Fe_2SiO_4
	1200	56.5	8.488	612	3.675	3.001	5.497	6.429	14.5	71% $Ca_{0.1}Cd_{0.9}Bi_{0.1}Fe_{1.9}O_4$/29% SiO_2
$Co_{0.1}Cd_{0.9}Bi_{0.1}Fe_{1.9}O_4$@$SiO_2$	300	7.9	8.353	583	3.617	2.953	5.352	6.792	21.2	100% $Co_{0.1}Cd_{0.9}Bi_{0.1}Fe_{1.9}O_4$
	600	15.7	8.389	590	3.633	2.966	5.398	6.711	19.6	100% $Co_{0.1}Cd_{0.9}Bi_{0.1}Fe_{1.9}O_4$
	900	33.0	8.435	600	3.652	2.982	5.440	6.599	17.6	100% $Co_{0.1}Cd_{0.9}Bi_{0.1}Fe_{1.9}O_4$
	1200	49.8	8.473	608	3.669	2.996	5.495	6.513	15.6	100% $Co_{0.1}Cd_{0.9}Bi_{0.1}Fe_{1.9}O_4$

Across all calcination temperatures, the undoped $CdBi_{0.1}Fe_{1.9}O_4$ exhibits the smallest D_{CS}, whereas doping with Zn^{2+}, Ca^{2+}, and Co^{2+} ions lead to a noticeable increase in D_{CS} (Table 1). The calcination temperature-driven growth of D_{CS} can be explained by (i) elevated

https://doi.org/10.3390/nano16040259

thermal energy, which enhances atomic diffusion, promotes particle coalescence, and favors the formation of larger crystallites; (ii) agglomeration of incompletely crystallized particles, which reduces the effective nucleation density and supports the formation of single, well-defined crystals; and (iii) variations in cation valence and ionic radii, which induce lattice strain, generate additional nucleation sites, and facilitate crystallite formation [14,27,28]. Additionally, peak broadening, lattice strain and the constraining effect of the SiO_2 matrix can further modulate the crystallite growth. D_{CS} is a key determinant of the magnetic behavior of the NCs, particularly when the particle size nears the crystallite size [14,23,24]. These results highlight that the synthesis strategy and calcination conditions critically influence the structural and magnetic characteristics of the NCs. The Reference Intensity Ratio (RIR) method was used to determine the quantitative phase analysis (Table 1) [45,46].

The lattice parameter (a) exhibits the highest values in undoped $CdBi_{0.1}Fe_{1.9}O_4$ across all calcination temperatures. Doping with Zn^{2+}, Ca^{2+}, or Co^{2+} ions leads to a reduction in the lattice parameter, ascribed to the smaller ionic radii of the dopants, namely Zn^{2+} (0.74 Å) [3,9,46,47], Ca^{2+} (0.95 Å) [39], Co^{2+} (0.75 Å) [19,21,25] compared to Cd^{2+} (0.97 Å) [4,5,10,12]. The lattice parameter increases with calcination temperature from 8.344 to 8.498 Å, reflecting the combined effect of ionic radii, lattice distortions, and structural defects [14,16,23,27]. Differences between experimental and theoretical lattice parameters arise from the idealized assumption that ions behave as rigid spheres within a perfectly ordered lattice. In the $CdBi_{0.1}Fe_{1.9}O_4$ structure, the Cd^{2+} ions and dopant (Zn^{2+}, Ca^{2+} and Co^{2+}) ions predominantly occupy the A- sites, whereas Fe^{3+} and Bi^{3+} ions are distributed across both A- and B- sites, depending on the degree of cation disorder and specific synthesis conditions [10,17,18]. Lattice shrinkage is primarily governed by the cation charge distribution, surface tension, and dipole interactions within the nanocrystallites, which also limit grain growth. The internal stresses associated with variations in the lattice parameter can hinder grain development further during calcination [16,23,27].

The cell volume (V) gradually increases with calcination temperature, ranging from 581 to 614 $Å^3$. Doping $CdBi_{0.1}Fe_{1.9}O_4@SiO_2$ with Zn^{2+}, Ca^{2+}, Co^{2+} ions causes a slight reduction in cell volume, likely due to the combined effects of higher molecular mass, the formation of oxygen vacancies, and decreased crystallization temperature. These factors jointly offset the expected volumetric contraction associated with the densification [14,27,28].

The hopping lengths for the tetrahedral A (d_A) and octahedral B (d_B) sites are closely correlated to the lattice parameter. Lattice expansion is primarily driven by the greater ionic radii of Cd^{2+}, Ca^{2+}, and Bi^{3+} ions compared to those of Zn^{2+}, Co^{2+}, and Fe^{3+} ions. Moreover, the Cd^{2+} ions and the dopant ions (Zn^{2+}, Ca^{2+}, Co^{2+}) show a limited preference for the A sites, whereas Fe^{3+} and Bi^{3+} ions are unequally distributed between the A- and B- sites, depending on the composition of undoped and doped $CdBi_{0.1}Fe_{1.9}O_4@SiO_2$ [14,27,28].

Higher d_A and d_B indicate greater cation separation, which increases the energy barrier for charge carrier transfer and influences the magnetic characteristics. Since the crystal field splitting at (A) sites is smaller than at the B sites, variations in cation distribution, particularly of Zn^{2+}, Ca^{2+}, Co^{2+}, Cd^{2+}, and Fe^{3+} ions, and changes in the degree of inversion can modify the 3d orbitals energy levels, resulting in band gap modifications [14,27,28,45]. Furthermore, replacing magnetic Cd^{2+} ions with Zn^{2+}, Ca^{2+}, or Co^{2+} ions adjust the cation distribution within the ferrite lattice, thereby affecting the A-B superexchange interactions, and enabling tunable magnetic behavior [45].

The d_{XRD} reflects the dislocation density per unit volume in a crystalline material [11,13] and is primarily influenced by the lattice parameter and molecular mass [15]. In this study, the d_{XRD} increases with Zn^{2+}, Ca^{2+}, Co^{2+} ions doping in $CdBi_{0.1}Fe_{1.9}O_4@SiO_2$ NCs, whereas it gradual decrease with the increasing calcination temperature (from 6.855 to 6.513 g/cm^3), which could be attributed to the pore formation during the thermal treat-

https://doi.org/10.3390/nano16040259

ment [15]. Variations in d_{XRD} associated with changes in the lattice parameter are attributed to modified cation distribution between the A- and B- sites. Replacing Cd^{2+} ions with Zn^{2+}, Ca^{2+}, or Co^{2+} ions reduces grains compactness and promotes grain growth, likely due to the pore development through calcination and subsequent sample processing. Additionally, disparities in ionic radii introduce lattice strain and structural disorder, thereby perturbing the spinel framework of the ferrite [11,13].

The porosity (P) of undoped $CdBi_{0.1}Fe_{1.9}O_4$@SiO_2 (22.1–16.7%) was higher than that of $Zn_{0.1}Cd_{0.9}Bi_{0.1}Fe_{1.9}O_4$@$SiO_2$ (21.7–16%), $Ca_{0.1}Cd_{0.9}Bi_{0.1}Fe_{1.9}O_4$@$SiO_2$ (20.2–14.5%), $Co_{0.1}Cd_{0.9}Bi_{0.1}Fe_{1.9}O_4$@$SiO_2$ (21.2–15.6%) NCs. This trend is attributed to enhanced vacancy generation and pore formation during synthesis, whereas higher calcination temperatures facilitate densification, resulting in a progressive reduction in porosity [48,49]. Low porosity, as observed for $Co_{0.1}Cd_{0.9}Bi_{0.1}Fe_{1.9}O_4$@$SiO_2$ calcined at 1200 °C, is essential for achieving superior magnetic performance, whereas higher porosity levels can negatively affect magnetic behavior [14,27,28,48,49]. Additionally, doping with Zn^{2+}, Ca^{2+}, Co^{2+} ions influences porosity by promoting the formation of oxygen vacancies, which slightly expand the lattice while maintaining the structural symmetry of the ferrite framework [14,27,28,48,49].

3.3. *AFM Analysis*

XRD analysis indicates that the calcination temperatures above 600 °C promotes the development of well-defined ferrite crystallites, while the samples calcined at 300 °C remain amorphous. High calcination temperatures also enhance the tendency of adjacent particles to coalesce. Such behavior has been reported for fine clay particles [50], where physical attraction without chemical bonds governs aggregation; dispersion in an aqueous medium ensures proper particle separation and individualization [51]. Thus, the NCs calcined at 600, 900 and 1200 °C were dispersed into the ultra-pure water by magnetic agitation at 3000 rpm for 15 min. Afterward, glass slides were vertically immersed into the dispersions, allowing the Brownian motion of nanoparticles to promote adsorption onto the slide surface. The literature indicates that an adsorption time of 10–30 s is sufficient to ensure a uniform a thin film of nanoparticles transferred onto the solid substrate [51]. Following this guideline, the ferrite nanoparticles were adsorbed for 15 s, followed by the gentle extraction of the glass slide followed by natural drying. This procedure ensures a proper individualization of the adsorbed nanoparticles and results in thin films with topographical features that depend on the sample's composition.

The $CdBi_{0.1}Fe_{1.9}O_4$@SiO_2 consists of very small, rounded nanoparticles after calcination at 600 °C, with a diameter of 18 nm (Figure 3a). This diameter is slightly larger than the corresponding crystallite size of 14.9 nm, which can be attributed to the amorphous SiO_2 shell embedding the ferrite crystallite as the composite core. Ferrite core increases with the calcination temperature developing the rectangle shape characteristics of its crystal structure. Thus, the particles calcined at 900 °C exhibit partial alteration of the rounded shape due to the ferrite core growth alongside the amorphous SiO_2 coating, leading to a slight increase in diameter to about 35 nm (Figure 3b). This trend is further accentuated after calcination at 1200 °C (Figure 3c) where the nanoparticles become slightly elongated due to the extensive development of the ferrite crystallite core, resulting in diameters up to 52 nm. These observations are consistent with literature reports on $CdFe_2O_4$ [52], where nanoparticle elongation is evident in SEM images [53]. Moreover, the perovskite structure of $BiFeO_3$ has a square-to-elongated crystal lattice, in agreement with the AFM observations [54], with particle sizes increasing with the calcination temperature [55].

https://doi.org/10.3390/nano16040259

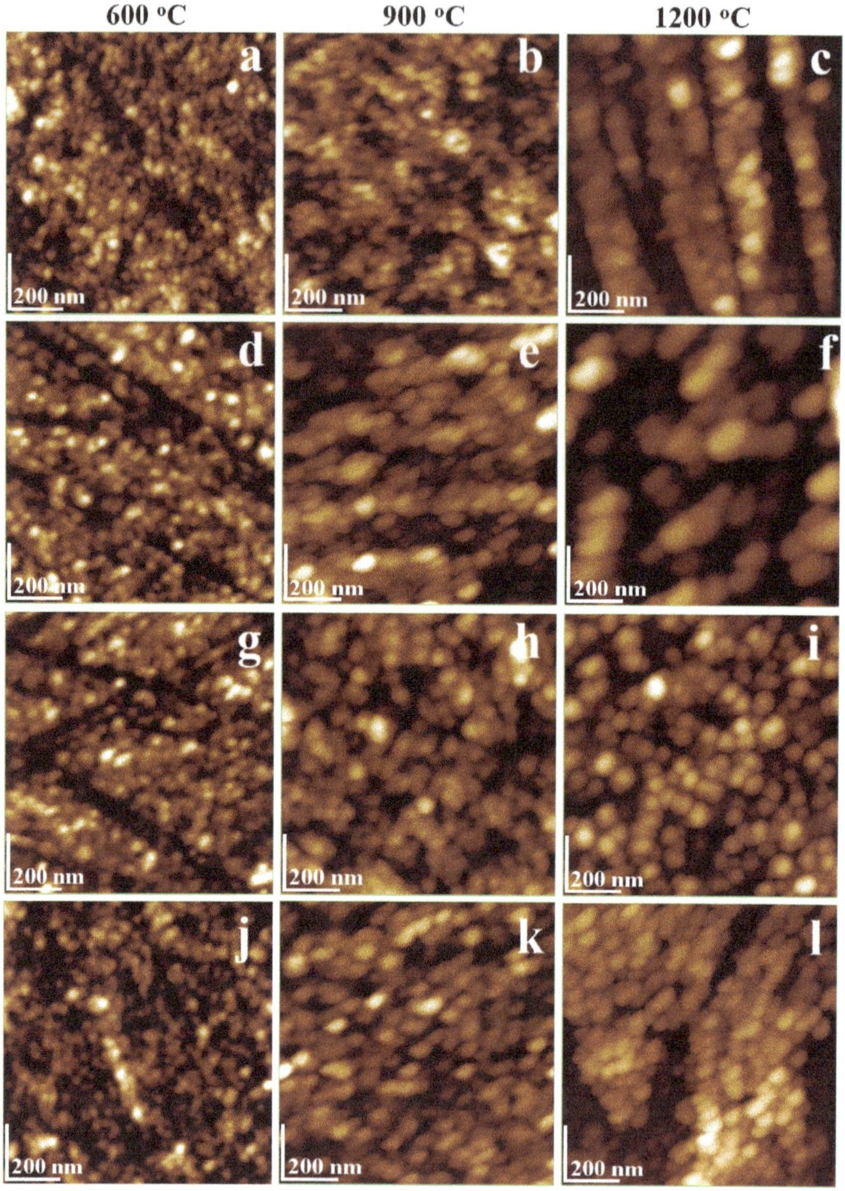

Figure 3. AFM topographic images of the powders' thin films transferred onto solid substrate: $CdBi_{0.1}Fe_{1.9}O_4@SiO_2$ NC calcined at (**a**) 600 °C, (**b**) 900 °C, (**c**) 1200 °C; $Zn_{0.1}Cd_{0.9}Bi_{0.1}Fe_{1.9}O_4@SiO_2$ NC calcined at (**d**) 600 °C, (**e**) 900 °C, (**f**) 1200 °C; $Ca_{0.1}Cd_{0.9}Bi_{0.1}Fe_{1.9}O_4@SiO_2$ NC calcined at (**g**) 600 °C, (**h**) 900 °C, (**i**) 1200 °C; $Co_{0.1}Cd_{0.9}Bi_{0.1}Fe_{1.9}O_4@SiO_2$ NC calcined at (**j**) 600 °C, (**k**) 900 °C, (**l**) 1200 °C.

Previous studies have shown that doping $BiFeO_3$ with Cd^{2+} and Mg^{2+} ions significantly improves its microstructural and magnetic properties by stabilizing the spinel phase, thereby enhancing both dielectric and magnetic performance [56]. The AFM images indicate a synergistic effect of Cd and Bi substitution on the growth and development of the ferrite core. Nayak et.al emphasizes that structural changes within $BiFeO_3$ play a key role

https://doi.org/10.3390/nano16040259

in tailoring its magnetic behavior [57]. Building on this, we propose that additional tuning of the ferrite characteristics can be achieved through the strategic incorporation of minor dopants. This hypothesis is supported by the observed improvements in the crystal lattice, as observed in XRD patterns presented in Figure 2.

Zn^{2+} doping of $CdBi_{0.1}Fe_{1.9}O_4$ exhibits a pronounced effect, promoting well-developed particles even at a relatively low calcination temperature (600 °C, Figure 3d). The nanoparticles display a diameter of approximately 22 nm and a predominantly rounded shape. The increase in calcination temperature facilitates optimal growth of the ferrite core, which alters the nanoparticle morphology: the particles become slightly elongated after calcination at 900 °C and develop a blockier, boulder-like shape with softened square corners after calcination at 1200 °C, partially moderated by the amorphous SiO_2 shell (Figure 3e,f). A similar evolution in morphology is observed on the NCs doped with Ca^{2+} (Figure 3g–i).

Co^{2+} doping of $CdBi_{0.1}Fe_{1.9}O_4$ exhibit hybrid behavior, revealing small, rounded nanoparticles after calcination at 600 °C (Figure 3j) with a diameter of approximatively 19 nm, closely resembling the undoped $CdBi_{0.1}Fe_{1.9}O_4$. The doping effect on the nanoparticle structure becomes apparent after calcination at 900 °C (Figure 3k), inducing a significant increase of the particle size to about 38 nm and pronounced elongation due to the enhanced growth of the ferrite core. Further increasing the calcination temperature increases to 1200 °C leads to only a minor increase in particle diameter to 54 nm, but evidence a strong consolidation of the ferrite core, ensuring optimal stability and well-defined individualization of the NC structure.

Considering the approximate rounded shape of the nanoparticles and ferrite crystallite cores, the thickness of the amorphous SiO_2 shell is estimated to about 2 nm. This value is consistent with the expectations of the synthesis procedure. It should be noted, however, that AFM cannot reliably differentiate between the ferrite core and the amorphous SiO_2 shell, whereas XRD analysis allows for this differentiation. This represents a limitation of the present study.

Amorphous SiO_2 shell provides an optimal insulation of the ferrite nuclei during the ferrite crystallization, preventing the coalescence between adjacent crystallites. This facilitates the optimal encapsulation of the ferrite core into well individualized composite nanoparticles [58–60]. The optimal encapsulation promotes uniform dispersion of the nanoparticles into the aqueous environment which favors the uniform adsorption onto the solid substrate (e.g., glass slide) (Figure 4). The roughness of the surface strongly depends on the nanoparticle's diameter and the uniformity of adsorbed layer. For example, the samples calcined at 600 °C exhibit low roughness and minimal local height variations ensuring a thin and compact film (Figure 4a,d,g,j). The continuous growth of the ferrite core partly alters the nanoparticle shapes (Figure 4b,e,h,k) leading to the relative increase of the surface roughness (Table 2). The slightly roughness enhancement facilitates the nanoparticle functionalization with biomedical agents (e.g., anticancer drug for targeted delivery onto the tumor sites) according to the literature data [61,62]. Moreover, the higher calcination temperature promotes optimal crystallization of the ferrite core and further increases surface roughness due to the adsorption of larger and more irregularly shaped nanoparticles. Despite this, the resulting thin films remain relatively uniform owing to the well-individualized nanoparticles, which prevent coalescence. The literature highlights the versatility and potential of such ferrites due to their structural stability in composite materials, while also noting associated environmental concerns [63]. In our study, the well-dispersed nanoparticles observed in the adsorbed thin films enable efficient recovery from environmental media via magnetic separation, allowing applications such as oil spill remediation [64] or degradation of dye pollutants [65].

https://doi.org/10.3390/nano16040259

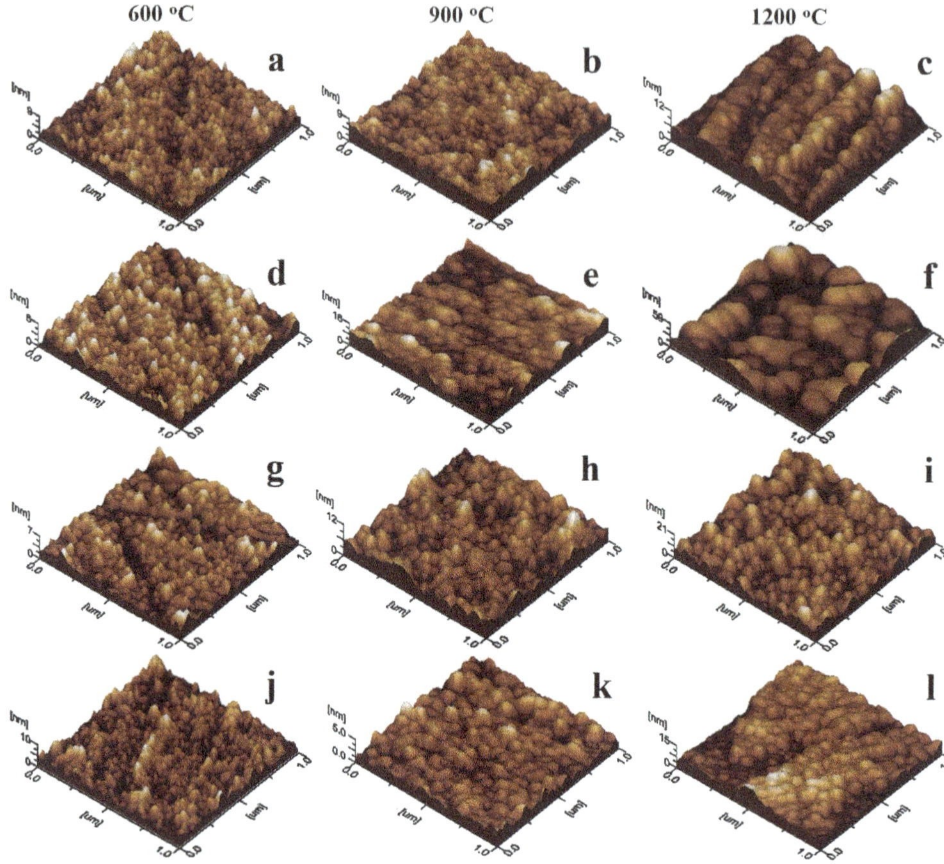

Figure 4. AFM tridimensional profiles of the powders' thin films transferred onto solid substrate: $CdBi_{0.1}Fe_{1.9}O_4@SiO_2$ NC calcined at (**a**) 600 °C, (**b**) 900 °C, (**c**) 1200 °C; $Zn_{0.1}Cd_{0.9}Bi_{0.1}Fe_{1.9}O_4@SiO_2$ NC calcined at (**d**) 600 °C, (**e**) 900 °C, (**f**) 1200 °C; $Ca_{0.1}Cd_{0.9}Bi_{0.1}Fe_{1.9}O_4@SiO_2$ NC calcined at (**g**) 600 °C, (**h**) 900 °C, (**i**) 1200 °C; $Co_{0.1}Cd_{0.9}Bi_{0.1}Fe_{1.9}O_4@SiO_2$ NC calcined at (**j**) 600 °C, (**k**) 900 °C, (**l**) 1200 °C.

Table 2. Topographic parameters and particle diameter measured with AFM.

Sample	Temp, °C	Height, nm	Ra, nm	D, nm
$CdBi_{0.1}Fe_{1.9}O_4@SiO_2$	600	9	0.89	18
	900	9	1.14	35
	1200	12	1.87	52
$Zn_{0.1}Cd_{0.9}Bi_{0.1}Fe_{1.9}O_4@SiO_2$	600	8	0.86	22
	900	16	2.25	43
	1200	59	9.06	72
$Ca_{0.1}Cd_{0.9}Bi_{0.1}Fe_{1.9}O_4@SiO_2$	600	7	0.90	20
	900	12	1.54	41
	1200	21	3.40	64
$Co_{0.1}Cd_{0.9}Bi_{0.1}Fe_{1.9}O_4@SiO_2$	600	10	1.10	19
	900	5	0.63	38
	1200	15	2.44	54

https://doi.org/10.3390/nano16040259

3.4. Magnetic Behavior

The magnetic hysteresis loops $M(\mu_0 H)$ and their first derivatives $(dM/d(\mu_0 H)$ of NCs calcined at 1200 °C exhibit typical soft ferrimagnetic behavior (Figure 5). At the nanoscale, the high surface-to-volume ratio positions many atoms at or near the particle surface, where broken bonds, defects, spin disorder, and spin canting generate surface spin effects that reduce magnetization and broaden the H_c distribution [66]. As surface contributions increase, surface-dominated interactions outweigh core effects, causing deviations from bulk behavior. Thus, the magnetic characteristics are strongly governed by particle size, shape, and morphology, which depend on synthesis and thermal treatment conditions [27,28,66,67].

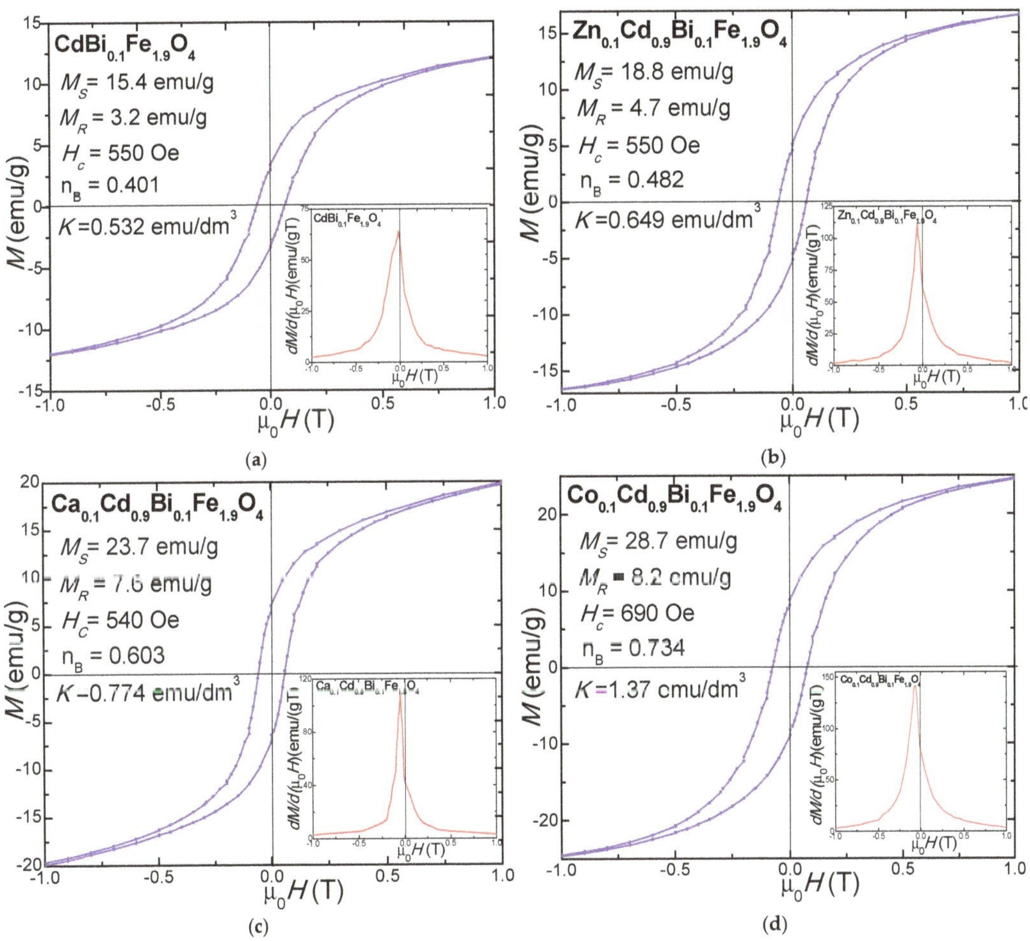

Figure 5. Magnetic hysteresis loops and magnetization derivative of (**a**) $CdBi_{0.1}Fe_{1.9}O_4@SiO_2$, (**b**) $Zn_{0.1}Cd_{0.9}Bi_{0.1}Fe_{1.9}O_4@SiO_2$, (**c**) $Ca_{0.1}Cd_{0.9}Bi_{0.1}Fe_{1.9}O_4@SiO_2$, and (**d**) $Co_{0.1}Cd_{0.9}Bi_{0.1}Fe_{1.9}O_4@SiO_2$ NCs calcined at 1200 °C.

All samples exhibit characteristic *S*-shaped hysteresis loops (Figure 5), which indicated that relatively high magnetocrystalline anisotropy (K) requires stronger fields for saturation. The moderate H_c observed is attributed to partial particle coalescence, which enhances interparticle magnetic coupling and overall magnetization. However, diminished magnetic

https://doi.org/10.3390/nano16040259

performance may result from a magnetically inactive surface layer associated with defects, altered cation distribution, incomplete saturation, and lattice defects [14,16,27].

The derivative curves $dM/d(\mu_0H)$ vs. $\mu 0H$ reveal magnetization reversal dynamics. For the NCs calcined at 1200 °C, a dominant peak near H_c indicates a single magnetic phase [14,16,23], while a broad peak reflects a distribution of particle sizes and H_c. The minimal peak shift from the origin indicates a narrow H_c distribution due to strong interparticle coupling [68,69]. Sharp peaks are associated with well-crystallized, magnetically homogeneous samples, whereas broader features indicate increased size dispersion and internal defects [14,27]. These results confirm the formation of uniform, well-crystallized nanocrystals.

Doping $CdBi_{0.1}Fe_{1.9}O_4@SiO_2$ with non-magnetic Zn^{2+}, Ca^{2+}, and Co^{2+} ions modifies the magnetocrystalline anisotropy and particle size, leading to enhanced M_S, remnant magnetization (M_R), magnetic moments per unit cell (n_B), H_C and K compared to undoped $CdBi_{0.1}Fe_{1.9}O_4$ [14,27,28]. This enhancement is attributed to the dopant ions which promote soft ferrimagnetic behavior [14,16,23,27,67].

M_s of ferrites is primarily governed by the superexchange interactions between the A- and B- sites. An increase in D_c enhances M_s by reducing surface-related defects and improving magnetic ordering [67]. In this study, the highest M_s value (28.7 emu/g) was observed for $Co_{0.1}Cd_{0.9}Bi_{0.1}Fe_{1.9}O_4@SiO_2$ NC, whereas the undoped $CdBi_{0.1}Fe_{1.9}O_4@SiO_2$ NC exhibited lower M_s values (15.4 emu/g). The surface phenomena such as broken chemical bonds, deviations from the ideal cation distribution, lattice defects, and randomly oriented magnetic moments, contribute to reducing the overall magnetic performance. Additionally, the structural disorder at the particle surface can disrupt the alignment of magnetic moments, a phenomenon typically associated with single-domain configurations [14,16,23,27,67]. Partial substitution of Cd^{2+} with Zn^{2+}, Ca^{2+}, Co^{2+} ions may also reduce M_s due to the spin canting arising from the triangular spin configuration within the sublattice, as described by the Yaffet–Kittel [14,16,23,27]. Overall, the enhanced characteristics of $Co_{0.1}Cd_{0.9}Bi_{0.1}Fe_{1.9}O_4@SiO_2$ NC, including its relatively high Ms and favorable combination of structural and morphological characteristics, make it a promising candidate for technological applications in communication devices [22,26,32].

The remanent magnetization (M_R) of undoped $CdBi_{0.1}Fe_{1.9}O_4@SiO_2$ is 3.2 emu/g. The doping with Zn^{2+} ions increases M_R to 4.7 emu/g, while Co^{2+} ions doping results in an even greater increase, reaching 8.2 emu/g, demonstrating the significant effect of the dopant type on the M_R of ferrites.

The K value was calculated under the assumption that the spinel ferrite nanoparticles are approximatively spherical. The K is governed by crystal symmetry, intrinsic magneto crystalline anisotropy, as well as particle size and shape [14,16,23,27]. Among the studied NCs calcined at 1200 °C, $Co_{0.1}Cd_{0.9}Bi_{0.1}Fe_{1.9}O_4@SiO_2$ NC exhibits the highest K value of 1.34 erg/dm^3. The increase in K for the $CdBi_{0.1}Fe_{1.9}O_4@SiO_2$ NC with the addition of dopant ions, reaching the highest value for $Co_{0.1}Cd_{0.9}Bi_{0.1}Fe_{1.9}O_4@SiO_2$ NC, follows a trend similar to M_S, although no clear linear or colinear correlation between M_S and K is observed. Generally, K values increase with doping, which can be attributed to several factors: (i) partial pinning of surface spins in magnetically disordered surface layers, requiring stronger magnetic fields to achieve saturation, and (ii) the presence of randomly oriented grains and lattice vacancies, which further contribute to magnetic disorder. Additionally, crystalline K is affected by the lattice strain arising from the partial substitution of Cd^{2+} by Zn^{2+}, Ca^{2+}, or Co^{2+} ions, whichalters the local structural environment and contributes to K enhancement [14,16,23,27].

The lowest n_B value is observed in undoped $CdBi_{0.1}Fe_{1.9}O_4$ (0.401) increasing upon doping with Zn^{2+} (0.482), Ca^{2+} (0.603) or Co^{2+} (0.734) doping, which indicates an enhancement of non-collinear spin structures as described by the Yafet–Kittel model. The low M_s

https://doi.org/10.3390/nano16040259

in undoped $CdBi_{0.1}Fe_{1.9}O_4$ is attributed to spin canting in the B- sublattice, Jahn-Teller cations, surface magnetic dead layers, cation distribution deviations, incomplete saturation, random moment orientation, and lattice defects. Magnetization in larger particles is dominated by domain wall motion, while smaller particles form single domains aligned with the applied field [69].

The modest increase in H_c values for the doped NCs can be attributed to the contributions from magnetocrystalline anisotropy, microstrain effects, particle size distribution, and reduction in magnetic domain dimensions [31]. The undoped $CdBi_{0.1}Fe_{1.9}O_4@SiO_2$ NC exhibits an H_c value of 530 Oe, whereas $Co_{0.1}Cd_{0.9}Bi_{0.1}Fe_{1.9}O_4@SiO_2$ NC exhibit the highest H_C of 690 Oe. This enhancement is likely related to the variations in D_c, K and the formation of particle agglomerates that increase the particle size beyond the single-domain threshold, resulting in multi-domain structures and reduced domain wall pinning at grain boundaries [14,16,23,27,67]. Conversely, lower H_c values suggest surface spin distortion arising from variations in the magnetocrystalline K. In addition, an increased density of surface lattice defects, such as the atom displacement in the surface layers, could develop local energy barriers that further contribute to increased H_C [14,16,23,27,67].

Embedding of magnetic ferrites within the SiO_2 matrix stabilizes and organizes the magnetic domains, leading to well-defined magnetic nano-crystallites within the NC. This structural reinforcement improves the mechanical integrity and processability, enabling potential applications in magnetic coatings, thin films, and other nanostructured devices [2,3,6,32].

4. Conclusions

The influence of Zn^{2+}, Ca^{2+}, Co^{2+} ions doping on the structural, morphological, surface, and magnetic characteristics of $CdBi_{0.1}Fe_{1.9}O_4$ was systematically investigated. At 300 °C, the ferrites exhibited poorly crystallinity, whereas $Co_{0.1}Cd_{0.9}Bi_{0.1}Fe_{1.9}O_4@SiO_2$ calcined at 1200 °C displayed a well-defined single-phase ferrite structure. Doping not only enhanced ferrite crystallinity but also promoted the development of crystalline SiO_2 phases (silicate, cristobalite, and tridymite). Although all NCs maintained a cubic spinel structure, XRD analysis revealed that the dopant incorporation induced variations in the structural parameters. AFM imaging confirmed that the ferrite crystallites observed by XRD are encapsulated within amorphous SiO_2, forming well-individualized nanoparticles. At lower calcination temperatures, the nanoparticles are predominantly rounded, whereas higher calcination temperatures enhance ferrite core growth, leading to distorted particle shapes and increased sizes. The surface roughness of the adsorbed thin films strongly correlates with nanoparticle size and morphology: small, rounded particles yield low roughness, while larger, irregular nanoparticles significantly increase surface roughness, making them suitable for targeted functionalization. The M_S (15.4–28.7 emu/g), M_R (3.2–8.2 emu/g), H_c (530–690 Oe) and K (0.532–1.24 erg/dm^3) diminished with the incorporation of dopant ions, approaching soft ferrimagnetic-like behavior. The analysis using Néel's collinear and Yafet–Kittel spin-canting models, considering the surface spin effects, indicates that doping and thermal treatment are critical factors in governing the magnetic behavior and particle size distribution of the nanocomposites.

Author Contributions: Conceptualization, T.D.; methodology, T.D., I.P. and O.C.; software, T.D.; validation, T.D., I.P. and O.C.; formal analysis, T.D., I.P. and O.C.; investigation, T.D.; resources, T.D., I.P. and O.C.; data curation, T.D.; writing—original draft preparation, T.D., I.P. and O.C.; writing—review and editing, T.D. and O.C.; visualization, T.D., I.P. and O.C.; supervision, T.D. All authors have read and agreed to the published version of the manuscript.

Funding: This research received no external funding.

Data Availability Statement: Data are available on request from the corresponding author.

https://doi.org/10.3390/nano16040259

Acknowledgments: O.C. acknowledges the support of Innovation Plan 2022–2027 (PNCDI IV), European and International Cooperation Program, contr. no. 3BMTR/2025.

Conflicts of Interest: The authors declare no conflicts of interest.

References

1. Bagade, A.V.; Nagwade, P.A.; Nagawade, A.V.; Thopate, S.R.; Pund, S.N. A review on synthesis, characterization and applications of cadmium ferrite and its doped variants. *Orient. J. Chem.* **2022**, *38*, 1–5. [CrossRef]
2. Shakil, M.; Inayat, U.; Arshad, M.I.; Nabi, G.; Khalid, N.R.; Tariq, N.H.; Shah, A.; Iqbal, M.Z. Influence of zinc and cadmium co-doping on optical and magnetic properties of cobalt ferrites. *Ceram. Int.* **2020**, *46*, 7767–7773. [CrossRef]
3. Nandanwar, A.K.; Chaodhary, D.L.; Kamde, S.N.; Choudhary, D.S.; Rewatkar, K.G. Study of structural and magnetic properties of zinc-substituted cadmium ferrite nanocrystals. *Mater. Todays Proc.* **2020**, *29*, 951–955. [CrossRef]
4. Siddique, M.A.; Irfan, M.; Rasool, R.T.; Arshad, M.; Ashraf, G.A.; Akhtar, M.N.; Manzoor, S.; Raza, M.R.; Khan, M.A. Role of Zn cations substitution on structural, vibrational, elastic, dielectric, bandwidth, and microwave absorbance of Co–Cd ferrites at high frequencies. *Mater. Chem. Phys.* **2024**, *315*, 128972. [CrossRef]
5. Kaur, H.; Singh, A.; Kumar, V.; Ahlawat, D.S. Structural, thermal and magnetic investigations of cobalt ferrite doped with Zn^{2+} and Cd^{2+} synthesised by auto combustion method. *J. Magn. Magn. Mater.* **2019**, *474*, 505–511. [CrossRef]
6. Ghorbani, H.; Eshraghi, M.; Dodaran, A.A.S. Structural and magnetic properties of cobalt ferrite nanoparticles doped with cadmium. *Phys. B Condens. Matter* **2022**, *634*, 413816. [CrossRef]
7. Aakash, P.; Nordblad, R.; Mohan, S. Mukherjee, Structural, magnetic and hyperfine characterisations of nanocrystalline Zn-Cd doped nickel ferrites. *J. Magn. Magn. Mater.* **2017**, *441*, 710–717. [CrossRef]
8. Sanad, M.M.S.; Azab, A.A.; Taha, T.A. Introduced oxygen vacancies in cadmium ferrite anode materials via Zn^{2+} incorporation for high performance lithium-ion batteries. *Mater. Sci. Semicond. Proc.* **2022**, *143*, 106567. [CrossRef]
9. Ghatak, S.; Chakraborty, G.; Sinha, M.; Pradhan, S.K.; Meikap, A.K. Dielectric relaxation and magnetic field dependent alternating current conductivity of nanocrystalline cadmium–zinc ferrite below room temperature. *Phys. B* **2011**, *406*, 3261–3266. [CrossRef]
10. Sinha, M.; Pradhan, S.K. Synthesis of nanocrystalline Cd–Zn ferrite by ball milling and its stability at elevated temperatures. *J. Alloys Compd.* **2010**, *489*, 91–98. [CrossRef]
11. Parajuli, D.; Taddesse, P.; Murali, N.; Samatha, K. Study of structural, electromagnetic and dielectric properties of cadmium substituted Ni–Zn nanosized ferrites. *J. Indian Chem. Soc.* **2022**, *99*, 100380. [CrossRef]
12. Gupta, M.; Gupta, M.; Anu Mudsainiyan, R.K.; Randhawa, B.S. Physico-chemical analysis of pure and Zn doped Cd ferrites ($Cd_{1-x}Zn_xFe_2O_4$) nanofabricated by Pechini sol-gel method. *J. Anal. Appl. Pyrol.* **2015**, *116*, 75–85. [CrossRef]
13. Stefănescu, M.; Stoia, M.; Caizer, C.; Dippong, T.; Barvinschi, P. Preparation of $Co_xFe_{3-x}O_4$ nanoparticles by thermal decomposition of some organo-metallic precursors. *J. Therm. Anal. Calorim.* **2009**, *97*, 245–250. [CrossRef]
14. Dippong, T.; Cadar, O.; Deac, I.G.; Lazar, M.; Borodi, G.; Levei, E.A. Influence of ferrite to silica ratio and thermal treatment on porosity, surface, microstructure and magnetic properties of $Zn_{0.5}Ni_{0.5}Fe_2O_4/SiO_2$ nanocomposites. *J. Alloys Compd.* **2020**, *828*, 154409. [CrossRef]
15. Dippong, T.; Levei, E.A.; Cadar, O.; Goga, F.; Barbu Todoran, L.; Borodi, G. Size and shape-controlled synthesis and characterization of $CoFe_2O_4$ nanoparticles embedded in a PVA-SiO_2 hybrid matrix. *J. Anal. Appl. Sci.* **2017**, *128*, 121–130. [CrossRef]
16. Dippong, T.; Levei, E.A.; Diamandescu, L.; Bibicu, I.; Leostean, C.; Borodi, G.; Barbu-Tudoran, L. Structural and magnetic properties of $Co_xFe_{3-x}O_4$ versus Co/Fe molar ratio. *J. Magn. Magn. Mater.* **2015**, *394*, 111–116. [CrossRef]
17. Kulkarni, A.B.; Mathad, S.N. Variation in structural and mechanical properties of Cd-doped Co-Zn ferrites. *Mater. Sci. Energy Technol.* **2019**, *2*, 455–462. [CrossRef]
18. Hcini, S.; Kouki, N.; Omri, A.; Dhahri, A.; Bouazizi, M.L. Effect of sintering temperature on structural, magnetic, magnetocaloric and critical behaviors of Ni-Cd-Zn ferrites prepared using sol-gel method. *J. Magn. Magn. Mater.* **2018**, *464*, 91–102. [CrossRef]
19. Chinchay-Espino, H.A.; Montes-Albino, G.M.; Morales-Cruz, C.M.; Dobbertin-Sanchez, S.E.; Rojas-Flores, S. Effect of cobalt substitution on the structural and magnetic properties of bismuth ferrite powders. *Crystals* **2022**, *12*, 1058. [CrossRef]
20. Panwar, M.N.; Khan, H.M.; Bano, N.; Zahid, M.; Mazhar, M.E.; Bilal, M.; Saleem, M.I.; Solre, G.F.B.; Gomez, P.H.; Li, Z. Influence of bismuth and cobalt doping on structural, dielectric, and magnetic properties of M-type calcium hexagonal ferrites. *J. Alloys Compd.* **2024**, *994*, 174623. [CrossRef]
21. Patil, S.; Meti, S.; Kanavi, P.S.; Bhajantri, R.F.; Anandalli, M.; Mondal, R.; Karmakar, S.; Muhiuddin, M.; Rahman, M.R.; Kumar, B.C.; et al. A study on solubility of bismuth cations in nickel cobalt ferrite nanoparticles and their influence on dielectric and magnetic properties. *Mater. Sci. Eng. B* **2023**, *296*, 116570. [CrossRef]
22. Vijaya, K.; Kumar, N.; Kumar, S. Electrical properties of bismuth doped $CoBi_xFe_{2-x}O_4$ nano ferrites. *Mater. Today Proc.* **2016**, *3*, 4193–4198. [CrossRef]

https://doi.org/10.3390/nano16040259

23. Dippong, T.; Cadar, O.; Levei, E.A.; Deac, I.G. Microstructure, porosity and magnetic properties of $Zn_{0.5}Co_{0.5}Fe_2O_4/SiO_2$ nanocomposites prepared by sol-gel method using different polyols. *J. Magn. Magn. Mater.* **2020**, *498*, 166168. [CrossRef]
24. Zahir, R.; Chowdhury, F.U.Z.; Uddin, M.M.; Hakim, M.A. Structural, magnetic and electrical characterisation of Cd-substituted Mg ferrites synthesised by double sintering technique. *J. Magn. Magn. Mater.* **2016**, *410*, 55–62. [CrossRef]
25. Ahmad, R.; Hussain Gul, I.; Zarrar, M.; Anwar, H.; Niazi, M.B.K.; Khan, A. Improved electrical properties of cadmium substituted cobalt ferrites nano-particles for microwave application. *J. Magn. Magn. Mater.* **2016**, *405*, 28–35. [CrossRef]
26. Valan, M.F.; Manikandan, A.; Arul Antony, S. Microwave combustion synthesis and characterization studies of magnetic $Zn_{1-x}Cd_xFe_2O_4$ ($0 \leq x \leq 0.5$) nanoparticles. *J. Nanosci. Nanotechnol.* **2014**, *15*, 4543–4551. [CrossRef]
27. Dippong, T.; Levei, E.A.; Cadar, O.; Mesaros, A.; Borodi, G. Sol-gel synthesis of $CoFe_2O_4$:SiO_2 nanocomposites—Insights into the thermal decomposition process of precursors. *J. Anal. Appl. Pyrol.* **2017**, *125*, 169–177. [CrossRef]
28. Dippong, T.; Levei, E.A.; Cadar, O.; Goga, F.; Toloman, D.; Borodi, G. Thermal behavior of Ni, Co and Fe succinates embedded in silica Matrix. *J. Therm. Anal. Calorim.* **2019**, *136*, 1587–1596. [CrossRef]
29. Divakara, S.G.; Mahesh, B. A comprehensive review on current trends in greener and sustainable synthesis of ferrite nanoparticles and their promising applications. *Results Eng.* **2024**, *21*, 101702.
30. Ali, S.S.L.; Selvaraj, S.; Batoo, K.M.; Chauhan, A.; Rana, G.; Kalaichelvan, S.; Radhakrishnan, A. Green synthesis of cubic spi-nel ferrites and their potential biomedical applications. *Ceram. Int.* **2024**, *50*, 52159–52189. [CrossRef]
31. Abidin, M.Z.U.; Ikram, M.; Moeen, S.; Nazir, G.; Kanoun, M.B.; Goumri-Said, S. A comprehensive review on the synthesis of ferrite nanomaterials via bottom-up and top-down approaches advantages, disadvantages, characterizations and computa-tional insights. *Coord. Chem. Rev.* **2024**, *520*, 216158. [CrossRef]
32. Badr, M.H.; Kudrevatykh, N.V.; Hassan, M.A.; Moustafa, M.; Rammah, Y.S.; Abouhaswa, A.S.; EL-Hamalawy, A.A. Structural and magnetic analysis of Cd-Zn spinel ferrite nanoparticles. *Phys. Scr.* **2023**, *98*, 025823. [CrossRef]
33. Rao, P.A.; Raghavendra, V.; Suryanarayana, B.; Paulos, T.; Murali, N.; Varma, P.V.S.K.P.; Prasad, R.G.; Ramakrishna, Y.; Chan-dramouli, K. Cadmium substitution effect on structural, electrical and magnetic properties of Ni-Zn nano ferrites. *Res. Phys.* **2020**, *19*, 103487.
34. Amer, M.A.; Meaz, T.M.; El-Kestawy, M.; Ghoneim, A.I. Magnetic and structural studies of trivalent Co-substituted Cd–Mn ferrites. *J. Magn. Magn. Mater.* **2016**, *405*, 137–144. [CrossRef]
35. Gonçalves, M.J.; Faria, V.L.; Nascimento, A.B.; Germscheidt, R.L.; Patra, S.; Saravaria, L.P.H.; Bonacin, A.J.; Munoz, R.A.A.; Angnes, L. Sensing performances of spinel ferrites MFe_2O_4 (M = Mg, Ni, Co, Mn, Cu and Zn) based electrochemical sensors: A review. *Anal. Chim. Acta* **2022**, *1233*, 340362. [CrossRef]
36. Panda, R.K.; Muduli, R.; Behera, D. Electric and magnetic properties of Bi substituted cobalt ferrite nanoparticles: Evolution of grain effect. *J. Alloys Compd.* **2015**, *634*, 239–245. [CrossRef]
37. Kumar, H.; Singh, J.P.; Srivastava, R.C.; Negi, P.; Agrawal, H.M.; Asokan, K. FTIR and electrical study of dysprosium doped cobalt ferrite nanoparticles. *J. Nanosci.* **2014**, *2014*, 862415. [CrossRef]
38. Rodriguez-Carvajal, J. Recent advances in magnetic structure determination by neutron powder diffraction. *Phys. B Condens. Matter* **1993**, *192*, 55–69. [CrossRef]
39. Dhankhar, S.; Tiwari, P.; Baskar, K.; Basu, S.; Singh, S. Effect of low cobalt doping on morphology and properties of calcium ferrite and its application as cathode in solid oxide fuel cell. *J. Curr. Appl. Phys.* **2017**, *17*, 467–473. [CrossRef]
40. Huili, H.; Grindi, B.; Viau, G.; Tahar, L.B. Influence of the stoichiometry and grain morphology on the magnetic properties of Co substituted Ni–Zn nanoferrites. *Ceram. Int.* **2016**, *42*, 17594–17604. [CrossRef]
41. Ramesh, S.; Sekhar, B.C.; Rao, P.S.V.S.; Rao, B.P. Microstructural and magnetic behavior of mixed Ni–Zn–Co and Ni–Zn–Mn ferrites. *Ceram. Int.* **2014**, *40*, 8729–8735. [CrossRef]
42. Arulmurugan, R.; Vaidyanathan, G.; Sendhilnathan, S.; Jeyadevan, B. Mn-Zn ferrite nanoparticles for ferrofluid preparation: Study on thermalemagnetic properties. *J. Magn. Magn. Mater.* **2006**, *298*, 83–94. [CrossRef]
43. Zheng, Z.G.; Zhong, X.C.; Zhang, Y.H.; Yu, J.Y.; Zeng, D.C. Synthesis, structure and magnetic properties of nanocrystalline $Zn_xMn_{1-x}Fe_2O_4$ prepared by ball milling. *J. Alloys Compd.* **2008**, *466*, 377–382. [CrossRef]
44. Raji, P.; Balachandra Kumar, K. Structural, elastic and magnetic properties of Ca doped copper ferrite nanoparticles. *Phys. B Condens. Matter.* **2022**, *632*, 413759. [CrossRef]
45. Mohamed, W.S.; Alzaid, M.; Abdelbaky, M.S.M.; Amghouz, Z.; García-Granda, S.; Abu-Dief, A.M. Impact of Co^{2+} substitution on microstructure and magnetic properties of $Co_xZn_{1-x}Fe_2O_4$ nanoparticles. *Nanomat* **2019**, *9*, 1602. [CrossRef] [PubMed]
46. Hubbard, C.R.; Snyder, R.L. RIR—Measurement and use in quantitative XRD. *Powder Diffr.* **1988**, *3*, 74–77. [CrossRef]
47. Mahmood, A.; Maqsood, A.; Tayyab, S.; Ali, M.; Ul Abideen, Z. Effect of cadmium on the fabrication and thermally tailoring electrical transport properties of zinc ferrite nanoparticles. *Ceram. Int.* **2026**. [CrossRef]
48. Ge, Y.; Li, C.; Waterhouse, G.I.N.; Zhang, Z.; Yu, L. $ZnFe_2O_4$@SiO_2@Polypyrrole nanocomposites with efficient electromagnetic wave absorption properties in the K and Ka band regions. *Ceram. Int.* **2021**, *47*, 1728–1739. [CrossRef]

https://doi.org/10.3390/nano16040259

49. Powar, R.R.; Phadtare, V.D.; Parale, V.G.; Park, H.H.; Pathak, S.; Kamble, P.R.; Piste, P.B.; Zambare, D.N. Structural, morphological, and magnetic properties of $Zn_xCo_{1-x}Fe_2O_4$ (0 < x < 1) prepared using a chemical co-precipitation method. *Ceram. Int.* **2018**, *44*, 20782–20789.
50. Avram, S.E.; Birle, B.V.; Cosma, C.; Tudoran, L.B.; Moldovan, M.; Cuc, S.; Borodi, G.; Petean, I. Ecological building material obtained through the moderate thermal consolidation of ceramic slurry collected from industrial waste waters. *Materials* **2025**, *18*, 1715. [CrossRef]
51. Avram, S.E.; Tudoran, L.B.; Borodi, G.; Petean, I. Microstructural Characterization of the Mn Lepidolite Distribution in Dark Red Clay Soils. *Appl. Sci.* **2025**, *15*, 6445. [CrossRef]
52. Jdidi, A.R.; Nouira, W.; Selmi, A.; Drissi, N.; Aissa, M.; Hcini, S.; Gassoumi, M. Impact of Calcination Temperature on the Properties and Photocatalytic Efficiency of $Cd_{0.6}Mg_{0.2}Cu_{0.2}Fe_2O_4$ Spinel Ferrites Synthesized via the Sol–Gel Method. *Crystals* **2025**, *15*, 457. [CrossRef]
53. Al-Kadhi, N.S.; Basha, M.T. Enhanced removal of Cd(II) ions from aqueous media via adsorption on facilely synthesized copper ferrite nanoparticles. *Molecules* **2024**, *29*, 3711. [CrossRef] [PubMed]
54. Mallek-Zouari, I.; Kaddar, Y.; Ben Taazayet, W.; Mounkachi, O.; Hlil, E.-K.; Thabet Mliki, N.; El Moutaouakil, A. first-principles investigation of structural, electronic, and magnetic properties of $BiFeO_3$ and $Bi_2Fe_4O_9$ nanostructures. *Int. J. Mol. Sci.* **2025**, *26*, 4671. [CrossRef] [PubMed]
55. Petrukhin, D.; Salnikov, V.; Nikitin, A.; Sidane, I.; Slimani, S.; Alberti, S.; Peddis, D.; Omelyanchik, A.; Rodionova, V. Effect of bismuth ferrite nanoparticles on physicochemical properties of polyvinylidene fluoride-based nanocomposites. *J. Compos. Sci.* **2024**, *8*, 329. [CrossRef]
56. Gan, G.; Zhang, H.; Li, Q.; Li, J.; Huang, X.; Xie, F.; Xu, F.; Zhang, Q.; Li, M.; Liang, T.; et al. Low loss, enhanced magneto-dielectric properties of Bi_2O_3 doped Mg-Cd ferrites for high frequency antennas. *J. Alloys Compd.* **2018**, *735*, 2634–2639. [CrossRef]
57. Nayak, S.; Agnihotri, P.; Nayak, J.P.; Joshi, C.; Rai, R. Exploring bismuth ferrite: Structural changes and future applications. *Curr. Appl. Phys.* **2025**, *73*, 49–76. [CrossRef]
58. Melis, N.; Loche, D.; Thakkar, S.V.; Cutrufello, M.G.; Sini, M.F.; Sedda, G.; Pilia, L.; Frongia, A.; Casula, M.F. Magnetic aerogels for room-temperature catalytic production of bis(indolyl)methane derivatives. *Molecules* **2024**, *29*, 2223. [CrossRef]
59. Omelyanchik, A.; Salvador, M.; D'Orazio, F.; Mameli, V.; Cannas, C.; Fiorani, D.; Musinu, A.; Rivas, M.; Rodionova, V.; Varvaro, G.; et al. magnetocrystalline and surface anisotropy in $CoFe_2O_4$ nanoparticles. *Nanomaterials* **2020**, *10*, 1288. [CrossRef]
60. Virumbrales, M.; Saez-Puche, R.; Torralvo, M.J.; Blanco-Gutierrez, V. Mesoporous silica matrix as a tool for minimizing dipolar interactions in $NiFe_2O_4$ and $ZnFe_2O_4$ Nanoparticles. *Nanomaterials* **2017**, *7*, 151. [CrossRef]
61. Kharat, P.B.; Somvanshi, S.B.; Dawi, E.A.; Ibrahim, R.M.; Mopari, A.M. Surface-functionalized nanoscale nickel-ferrites ($NiFe_2O_4$) nanoparticles as heating agents for cancer theranostics. *Inorg. Chem. Commun.* **2025**, *182*, 115569. [CrossRef]
62. Giannousi, K.; Toursidis, P.D.; Chatzatoglou, E.; Dendrinou-Samara, C. Manganese ferrite nanoparticles functionalized with gallic acid for inhibition of insulin amyloids formation and mitigation of oxidative stress. *Colloids Surf. A Physicochem. Eng. Asp.* **2025**, *727*, 138114. [CrossRef]
63. Chakraborty, S.; Roy, S.; Ahmaruzzaman, M.d. Advancing the frontier of spinel ferrites in sustainable remediation: Fundamentals, synthetic strategies, and mechanistic insights. *J. Water Process Eng.* **2026**, *81*, 109276. [CrossRef]
64. Košak, A.; Hadela, A.; Poberžnik, M.; Lobnik, A. magnetic separation of oil spills from water using cobalt ferrite nanoparticles with fluorocarbon functionalization. *Int. J. Mol. Sci.* **2025**, *26*, 6562. [CrossRef]
65. Fernandes, R.J.C.; Cardoso, B.D.; Rodrigues, A.R.O.; Pires, A.; Pereira, A.M.; Araújo, J.P.; Pereira, L.; Coutinho, P.J.G. Zinc/magnesium ferrite nanoparticles functionalized with silver for optimized photocatalytic removal of malachite green. *Materials* **2024**, *17*, 3158. [CrossRef]
66. Rao, S.; Choudary, G.S.V.R.K.; Rao, R.H.; Sujatha, C. Structural and magnetic properties of ultrafine $CoFe_2O_4$ nanoparticles. *Proc. Mat. Sci.* **2015**, *10*, 19–27. [CrossRef]
67. Haque, S.U.; Saikia, K.K.; Murugesan, G.; Kalainathan, S. A study on dielectric and magnetic properties of lanthanum substituted cobalt ferrite. *J. Alloys Compd.* **2017**, *701*, 612–618. [CrossRef]
68. Ghosh, M.P.; Murkherjee, S. Microstructural, magnetic and hyperfine characterizations of Cu-doped cobalt ferrite nanoparticles. *J. Am. Chem. Soc.* **2019**, *102*, 7509–7520. [CrossRef]
69. Ansari, S.M.; Suryawanshi, S.R.; More, M.A.; Sen, D.; Kolekar, Y.D.; Ramana, C.V. Field emission properties of nano-structured cobalt ferrite ($CoFe_2O_4$) synthesized by low-temperature chemical method. *Chem. Phys. Lett.* **2018**, *701*, 151–156. [CrossRef]

https://doi.org/10.3390/nano16040259

nanomaterials

MDPI

Article

Theoretical and Experimental Analysis of Hydroxyl and Epoxy Group Effects on Graphene Oxide Properties

Ximena Jaramillo-Fierro [1,*] and Guisella Cuenca [2]

1 Departamento de Química, Facultad de Ciencias Exactas y Naturales, Universidad Técnica Particular de Loja, San Cayetano Alto, Loja 1101608, Ecuador

2 Ingeniería Química, Facultad de Ciencias Exactas y Naturales, Universidad Técnica Particular de Loja, San Cayetano Alto, Loja 1101608, Ecuador; gpcuenca@utpl.edu.ec

* Correspondence: xvjaramillo@utpl.edu.ec; Tel.: +593-7-3701444

Abstract: In this study, we analyzed the impact of hydroxyl and epoxy groups on the properties of graphene oxide (GO) for the adsorption of methylene blue (MB) dye from water, addressing the urgent need for effective water purification methods due to industrial pollution. Employing a dual approach, we integrated experimental techniques with theoretical modeling via density functional theory (DFT) to examine the atomic structure of GO and its adsorption capabilities. The methodology encompasses a series of experiments to evaluate the performance of GO in MB dye adsorption under different conditions, including differences in pH, dye concentration, reaction temperature, and contact time, providing a comprehensive view of its effectiveness. Theoretical DFT calculations provide insights into how hydroxyl and epoxy modifications alter the electronic properties of GO, improving adsorption efficiency. The results demonstrate a significant improvement in the dye adsorption capacity of GO, attributed to the interaction between the functional groups and MB molecules. This study not only confirms the potential of GO as a superior adsorbent for water treatment, but also contributes to the optimization of GO-based materials for environmental remediation, highlighting the synergy between experimental observations and theoretical predictions in advances in materials science to improve sustainability.

Keywords: adsorbent materials; graphene oxide; methylene blue; density functional theory; water purification; environmental remediation

1. Introduction

Water pollution caused by industrial dyes represents a significant environmental issue [1]. During the dyeing process in the textile industry, between 10% and 25% of dyes are lost, and from 2% to 20% are directly discharged as aqueous effluents into the environment [2]. This results in the release of substantial quantities of dyes into water bodies, adversely affecting water quality, aquatic organisms, and human health. [3]. Beyond the visual impact, many dyes that are released into the water are toxic, carcinogenic, or mutagenic to living organisms, primarily due to the presence of aromatic compounds such as benzidine and naphthalene [4]. Methylene blue (MB), a widely used dye in the textile industry, has been classified as a high-risk compound due to its high toxicity and persistence in the environment [5].

For the removal of methylene blue and other dyes from water bodies, various physicochemical and biological methods have been developed [6,7] Adsorption, a physicochemical method, stands out for its low cost, simple operation and high efficiency, as well as the use of technological advancements to improve the process, including the development of new materials with a simple design [8–10]. This process relies on the electrostatic attraction between the contaminant molecules and the adsorption sites on the surface of the adsorbent [11], which occurs through various mechanisms, including electrostatic interactions, hydrogen bonding, and π–π interactions [12–14].

The search for effective adsorbents for dye removal, such as methylene blue from water, has led to the exploration of various materials, including graphene and its derivatives [15–20]. Graphene (G), composed of a single layer of carbon atoms arranged in a 2D hexagonal lattice [21,22], exhibits high electrical and thermal conductivity, granting it unique properties for contaminant adsorption [23,24]. Its high hydrophobic surface area and theoretical specific surface area make it an effective adsorbent [25]. Graphene oxide (GO), produced from graphene through oxidative processes [26–30], introduces functional groups that enhance adsorption capacity through chemical interactions [31–35]. Reduced graphene oxide (rGO) is derived from GO by reduction, partially restoring the graphene structure and its adsorption properties [36,37].

Recently, interest in the use of graphene-based materials such as methylene blue for water decontamination has increased, leveraging their large surface area and porous structure for efficient dye molecule adsorption, with removal efficiencies exceeding 95% [38–40]. Experimental and computational studies, including density functional theory (DFT) analyses, have been conducted to elucidate the adsorption capacity of MB on graphene, GO, and rGO, providing insights into the molecular adsorption mechanisms and predicting material adsorption capacities [41–46].

Over the last century, the atomic structure of graphene oxide has been investigated, with various models proposed for its functional group composition and arrangement. Experimental confirmations have identified epoxy and hydroxyl groups, and sp^2 coordinated carbons in carbon layers. Likewise, previous studies have identified carbonyl and carboxyl functional groups at the edges of carbon layers. These experimental findings also suggest the presence of non-oxidized aromatic "islands", separated by aliphatic rings with hydroxyl groups, epoxy, and double bonds. Furthermore, theoretical research highlights the role of epoxy and hydroxyl groups in structural stability, suggesting a model where carbon layers exhibit bulges and a random distribution of oxygen-containing groups [47,48].

Despite numerous models, the precise structure of graphene oxide remains unresolved. Therefore, this study aims to understand the atomic structure of graphene oxide due to its significant impact on electronic properties and, consequently, on methylene blue adsorption capacity. The experimental and theoretical results presented here shed light on the molecular adsorption of MB dye by graphene oxide. Furthermore, DFT calculations of pristine graphene (G), hydroxyl-decorated graphene (GOL), epoxy-decorated graphene (GOXI), and graphene decorated with both functional groups (GOL/GOXI) are explored to determine their influence on the electronic properties of the material. Using the Vienna Ab Initio Simulation Package (VASP), the molecular interactions and adsorption mechanisms on G, GOL, and GOXI surfaces are investigated, comparing the effects of hydroxyl and epoxy groups on MB adsorption capacity. Additionally, the experimental results aim to establish the influence of solution pH, dye concentration, reaction temperature, and contact time on graphene oxide's ability to adsorb methylene blue dye.

The novelty of this study lies in its comprehensive approach that combines detailed experimentation and theoretical analysis to clarify the adsorption mechanism of methylene blue (MB) on graphene oxide (GO). This research highlights, for the first time, the synergisms between experimentation and theory to achieve an in-depth and detailed understanding of the molecular-level interactions between GO and MB. By exploring the effect of critical operational variables and detailing the vital role of functional groups in enhancing adsorption, this study not only advances the scientific knowledge of adsorbent materials but also opens new perspectives aimed at the development of more efficient and sustainable technologies for wastewater treatment.

2. Materials and Methods

2.1. Materials

All chemicals were used as received, without further purification: graphite powder (<150 μm, 99.99%, Sigma-Aldrich, Burlington, MA, USA); hydrochloric acid (HCl, 37%, Sigma-Aldrich, Burlington, MA, USA); hydrogen peroxide (H_2O_2, 30%, Merk, Darmstadt,

Germany); methylene blue (MB, $C_{16}H_{18}N_3ClS$, Sigma-Aldrich, Burlington, MA, USA); potassium permanganate ($KMnO_4$, ≥99.0%, Sigma-Aldrich, Burlington, MA, USA); sodium hydroxide (NaOH, 1310-73-2, 40.00 g/mol, Merk, Darmstadt, Germany); sulfuric acid (H_2SO_4, 95.0–98.0%, Sigma-Aldrich, Burlington, MA, USA), Methanol (CH_3OH, ≥99.8%, Sigma-Aldrich, Burlington, MA, USA).

2.2. Graphene Oxide (GO) Synthesis

GO was obtained using the modified Hummers method [49]. A total of 3.0 g of powdered graphite was added to 70 mL of sulfuric acid (H_2SO_4) with constant stirring. Then, 9.0 g of potassium permanganate ($KMnO_4$) was added with moderate stirring and placed in an ice bath, where it was kept for 30 min. Next, the solution was transferred to a water bath at a temperature of 50 °C with constant stirring for 30 min. Then, 150 mL of distilled water (H_2O) was added and stirred for 20 min, maintaining a temperature below 90 °C. A total of 500 mL of H_2O and 15 mL of 30% hydrogen peroxide (H_2O_2) were added. The resulting solution was kept at rest for 24 h at room temperature. Subsequently, the precipitate was centrifuged at 1000 rpm for 12 min and washed with 15 mL of hydrochloric acid (HCl) (1:10); this was carried out three times. Again, it was centrifuged at 1000 rpm for 12 min and washed with distilled H_2O; this continued until a pH of around 7.00 was reached. Finally, the sample was transferred to a beaker and dried at 100 °C in the oven for 24 h. After the oxidation process, 100 mg of the sample was dispersed in 1 L of distilled water via sonication for 30 min. The resulting solution was centrifuged to separate graphene oxide (GO) from non-exfoliated graphite oxide particles.

2.3. Characterization of Graphene Oxide (GO)

The characterization of samples was completed through applying the methodology described in our previous study [50]. X-ray diffraction (XRD) measurements were made using a Bruker-AXS D8-Discover diffractometer (Bruker AXS, Karlsruhe, Germany) equipped with Cu Kα radiation (λ = 1.5406 Å). The identification of crystalline phases was conducted via the Crystallography Open Database (COD, version 2023). For microstructural analysis, micrographs and energy-dispersive X-ray (EDX) spectra were acquired with a JEOL JSM 6400 scanning electron microscope (SEM) (JEOL, Peabody, MA, USA), which was fitted with an in-house-developed JEOL dispersive X-ray spectrometer (EDS). Fourier-transform infrared (FTIR) spectra were captured on a PerkinElmer GX2000-FTIR Spectrometer (PerkinElmer, Inc., Waltham, MA, USA). The specific surface area (SSA) of graphene oxide (GO) was determined by liquid nitrogen physisorption at −196 °C using a ChemiSorb 2720 instrument (Micromeritics, Norcross, GA, USA). The gas mixture used for the analysis consisted of 30% nitrogen (N_2) in helium (He). The SSA calculation was based on the Brunauer–Emmet–Teller (BET) equation, utilizing the Chemisoft TPx system (version 1.03; Micromeritics, 2011) for data analysis via the single-point method.

The point of zero charge (PZC) of GO was assessed at room temperature (20 ± 2 °C). The methodology involved adding 0.1 g of GO powder to a 50 mL tube containing 25 mL of a 0.1 M NaCl solution. The pH was adjusted to values between 3 and 10 using 0.1 M HCl or NaOH, recorded as pH_i. Following 24 h of agitation at 250 rpm, the pH of the supernatant (pH_f) was measured. The pH_{PZC} was determined at the intersection of the experimental curve (initial pH versus final pH) with the line defined by x = y (initial pH = final pH). This procedure was replicated three times to calculate the average pH_{PZC} value for GO [51,52]. Lastly, the quantification of residual methylene blue (MB) in solutions was performed using a Jenway 7350 spectrophotometer (Cole-Parmer, Staffordshire, UK) at a wavelength of 623 nm.

2.4. Experimental Methodology

This investigation conducted a series of adsorption experiments utilizing graphene oxide (GO) with aqueous solutions of methylene blue (MB) to evaluate the effects of various parameters, including solution pH, initial adsorbate concentration, reaction system temperature, and adsorbate–adsorbent contact time. The acquired data were analyzed through fitting with isotherm and kinetic models using the least-squares non-linear regression method [53].

The adsorption capacity of GO (100 mg L^{-1}) for an MB solution (10 mg L^{-1}) was assessed across a pH range from 3 to 12, adjusting the pH with 0.1 M HCl and 0.1 M NaOH solutions. These solutions were stirred continuously for 24 h at ambient temperature before measuring the residual MB concentration. Further adsorption experiments were performed in a batch reactor at a controlled pH of 7.0 ± 0.1 and at room temperature (20 ± 2.0 °C), using 200 mg L^{-1} of GO. To explore the maximum adsorption capacity for MB, the concentration of a 500 mL MB solution was varied from 2.5 to 50 mg L^{-1}. Adsorption thermodynamics and kinetics for MB were investigated using a 20 mg L^{-1} MB solution in a 500 mL volume, with absorbance measurements at 623 nm taken using a UV-visible spectrophotometer and a pre-established calibration curve (R^2 = 0.9987) according to the Lambert–Beer Law. Experiments were performed in triplicate, and results were reported as mean values. The amount of dye adsorbed onto GO was calculated using Equation (2) [54], as presented in Table 1.

Table 1. Summary of the mathematical equations used in this study for data analysis.

Denomination	Equation		Parameters
Adsorbate adsorbed	$q_e = (C_0 - C_e) \times \frac{v}{w}$	(1)	C_0 = Initial concentration (mg L^{-1}) Ce = Equilibrium concentration (mg L^{-1}) w = Mass of the adsorbent (g) v = Volume of the solution (L)
Langmuir	$\frac{C_e}{q_e} = \frac{1}{K_L.q_{max}} + \frac{C_e}{q_{max}}$	(2)	q_{max} = Maximum monolayer adsorption (mg g^{-1}) K_L = Equilibrium Langmuir constant related to the adsorption energy (L mg^{-1}) C_e = Concentration of adsorbate in solution at equilibrium (mg L^{-1})
Freundlich	$q_e = K_F C_e^{\frac{1}{n}}$	(3)	K_F = Freundlich constant (L mg^{-1}) 1/n = Adsorption intensity constant. Note: For favorable adsorption, the value of n should be between 1 and 10
Temkin	$q_e = B\ \ln(AC_e)$	(4)	q_e = Adsorbate adsorbed per unit weight (mg g^{-1}) at equilibrium A = Temkin isotherm constant (L g^{-1}) C_e = Concentration of adsorbate in solution at equilibrium (mg L^{-1}) B = Constant related to the heat adsorption
Constant of heat adsorption	$B = \frac{RT}{b}$	(5)	b = Temkin constant (J mol^{-1}) T = Absolute temperature (K) R = Gas constant (8.314 J mol^{-1} K^{-1})
Separation factor	$R_L = \frac{1}{(1+K_L C_e)}$	(6)	K_L = Equilibrium Langmuir constant related to the adsorption energy (L mg^{-1}) C_e = Concentration of adsorbate in solution at equilibrium (mg L^{-1}) Note: $0 < R_L < 1$, suitable adsorption, $R_L > 1$ suitable adsorption, $R_L = 0$ irreversible adsorption, $R_L = 1$ linear adsorption.
Gibbs free energy	$\Delta G^0 = -RT\ \ln k_C$	(7)	ΔG^0 = Gibbs free energy (kJ mol^{-1}), ΔH^0 = Enthalpy (kJ mol^{-1}) ΔS^0 = Entropy (kJ mol^{-1} K^{-1})

Table 1. *Cont.*

Denomination	Equation		Parameters
Van't Hoff equation	$\ln k_C = \frac{-\Delta H^0}{R} \times \frac{1}{T} + \frac{\Delta S^0}{R}$ $k_C = k_L \times M_w \times 1000 \times 55.5$	(8) (9)	k_C = Dimensionless parameter T = Absolute temperature (K) R = Universal gas constant (8.314 J mol^{-1} K^{-1}) k_L = Langmuir constant (L mg^{-1}) M_w = Adsorbate weight (g mol^{-1})
Pseudo first order	$\ln(q_e - q_t) = \ln(q_e) - k_1 t$	(10)	k_1 = Rate constant (min^{-1}) q_e = Adsorbate adsorbed per unit weight (mg g^{-1}) at equilibrium q_t = Adsorbate adsorbed per unit weight (mg g^{-1}) at any time (t)
Pseudo second order	$\frac{t}{q_t} = \frac{1}{k_2 q_e^2} + \frac{1}{q_e} t$	(11)	k_2 = Rate constant (g mg^{-1} min^{-1}) q_e = Adsorbate adsorbed per unit weight (mg g^{-1}) at equilibrium q_t = Adsorbate adsorbed per unit weight (mg g^{-1}) at any time (t)
Elovich	$q_t = \frac{1}{\beta} \ln(\alpha\beta) + \frac{1}{\beta} \ln(t)$	(12)	q_t = Adsorbate adsorbed per unit weight (mg g^{-1}) at any time (t) α = Constant related to chemisorption rate β = Constant which depicts the extent of surface coverage
Intraparticle-diffusion	$q_t = k_3 t^{\frac{1}{2}} + A$	(13)	k_3 = Intraparticle diffusion rate constant (mg g^{-1} $min^{-1/2}$) A = constant indicating the width of the boundary layer (mg g^{-1}). The larger the value of A, the greater the boundary layer effect.
Particle-diffusion	$-\ln\left(1 - \left(\frac{q_t}{q_e}\right)^2\right) = \frac{2\pi^2 D_p}{r^2} t$	(14)	q_e = Adsorbate adsorbed per unit weight (mg g^{-1}) at equilibrium q_t = Adsorbate adsorbed per unit weight (mg g^{-1}) at any time (t) C_z = Ion concentration o the adsorbent (mg kg^{-1}). D_p = Diffusion coefficient in the adsorbent phase (m^2 min^{-1}) r = Average radius of the adsorbent particles (1×10^{-7} m) t = Contact time (min)
External-film-diffusion	$-\ln\left(1 - \left(\frac{q_t}{q_e}\right)\right) = \frac{D_f C_s}{h\, r\, C_z} t$	(15)	q_e = Adsorbate adsorbed per unit weight (mg g^{-1}) at equilibrium q_t = Adsorbate adsorbed per unit weight (mg g^{-1}) at any time (t) D_f = Diffusion in the film phase surrounding the adsorbent particles (m^2 min^{-1}) C_s = Ion concentration in the solution (mg L^{-1}) h = Film thickness around the adsorbent particles (10^{-6} m in poorly stirred solutions) r = Average radius of the adsorbent particles (1×10^{-7} m) t = Contact time (min)
Adsorption energy	$E_{ads} = E_{sorb/surf} - E_{surf} - E_{sorb}$	(16)	$E_{sorb/surf}$ = Energy of the supersystem produced by the adsorbed molecule on the surface (eV) E_{surf} = Energy of the surface (eV) E_{sorb} = Energy of the isolated molecule in vacuum (eV)

Equilibrium adsorption of MB was analyzed using the Langmuir [55], Freundlich [56], and Temkin [57] isotherm models, represented by Equations (3)–(5), respectively, as detailed in Table 1. Additionally, the adsorption heat (B) and separation factor (RL) constants were calculated using Equations (6) and (7) [55], providing insight into the adsorption characteristics.

Thermodynamic parameters, including Gibbs free energy (ΔG^0), enthalpy (ΔH^0), and entropy (ΔS^0), were determined using Equation (8) in Table 1 [58]. The relationship between these parameters was established using the van't Hoff equation, represented by Equation (9) in Table 1. The dimensionless parameter k_C was calculated by multiplying k_L by the molecular weight of the adsorbate (M_w) and adjusting for the presence of 1000 moles of pure water per liter, as detailed in Equation (10) [59] in Table 1.

Adsorption kinetics were examined using pseudo-first-order, pseudo-second-order, Elovich, intraparticle diffusion, external-film diffusion, and internal-pore diffusion models [56,57], each described by mathematical expressions (11) through (16), as outlined in Table 1. Finally, GO was desorbed after one treatment cycle to evaluate its recyclability. MB desorption from the loaded adsorbent was achieved using pure methanol. The adsorbent was then dried and reused under identical conditions for subsequent cycles. This recycling process was repeated over three cycles [60]. In each cycle, a new 500 mL MB solution (20 mg L^{-1}) was used with the GO concentration maintained at 200 mg L^{-1}.

2.5. Theoretical Methodology

The Density Functional Theory (DFT) investigation was conducted following the methodology outlined in our previous study [50]. Theoretical calculations were carried out using the Vienna Ab Initio Simulation Package (VASP) version 6.0, developed by VASP Software GmbH in Vienna, Austria [61,62]. Molecular modeling and visualization were performed using BioVia Materials Studio, version 5.5, provided by BioVia in San Diego, CA, USA.

The ionic potential of both inner nuclei and electrons was represented using pseudopotentials, which utilized the projector augmented wave (PAW) methodology [63]. All calculations employed the Perdew–Burke–Ernzerhof (PBE) generalized gradient approximation (GGA) function to describe electronic exchange–correlation interactions, which is known for its effectiveness in characterizing interactions between adsorbates and various surfaces, including sp^2 carbon-based materials [64]. The plane wave cutoff energy was established at 500 eV. Kohn–Sham equations were solved in a self-consistent manner, ensuring that energy variations between computational cycles remained below 10^{-5} eV [65].

Recognizing the limitation of GGA functionals, including hybrids, in describing the long-range electron correlations crucial for van der Waals (vdW) forces, this study incorporated the Grimme dispersion correction (DFT-D2) approach. This was particularly relevant for accurately modeling interactions in systems involving large atoms/molecules, such as those with aromatic rings [66–68]. The dispersion correction was applied to simulate both the surface structure and the adsorption of MB molecules onto the GO surface.

Given the complex nature of the structure and chemical composition of GO, it is generally accepted that epoxy (–C–O–C–) and hydroxyl (–OH) functional groups are randomly distributed on the GO sheet surface, while carboxyl (–COOH) and carbonyl (–C=O) groups are predominantly found at the edges [69]. The theoretical adsorption study thus considered two configurations of functionalized graphene oxide, focusing on models with either epoxy or hydroxyl groups.

The simulation of methylene blue (MB) molecule adsorption onto GO surfaces utilized optimized lattice parameters: a = 22.22 Å, b = 38.48 Å, and c = 16.00 Å, with the angles between the lattice vectors set to $<90° \times 90° \times 120°>$ [70]. The dynamic configuration adopted point symmetry P1, and the Brillouin zone was sampled using Γ-centered (2 × 1 × 2) Monkhorst–Pack grids [71]. Simulations were performed under non-spin-polarized conditions, with atomic positions relaxed until forces on the atoms were reduced to below 0.01 eV/Å. The GO model was represented by a p(5 × 5) supercell containing 370 carbon atoms, two oxygen-containing groups, and a 15 Å vacuum region in the out-of-plane direction. Surface energies (γs) of the GO structures were calculated following Equation (17), as outlined in Table 1 [72,73]. The adsorption energy (E_{ads}) of MB molecules on the surfaces of graphene oxide was determined using Equation (18) in Table 1 [74,75].

For a deeper understanding of the chemical interactions between MB and GO, a population analysis was conducted using Bader's method, a valuable approach for assessing bond ionicity by quantifying the charge transfer between atoms [76,77]. Charge difference analysis further quantified the redistribution of charge on the GO surface due to MB adsorption [76]. Additionally, the electron localization function (ELF) was employed to gain insights into the nature of MB-GO interactions [78], where the distribution of maximum electron density (Region of Maximum Density (RMD)) around atomic nuclei provides information on the interaction type. Symmetric RMD distribution suggests ionic or van der Waals interactions, while the migration of RMD between atomic centers indicates increasing covalent character, achieving ideal covalent symmetry [79].

3. Results

3.1. Characterization of GO

3.1.1. XRD Analysis

Figure 1 presents the X-ray diffraction pattern obtained for the synthesized material. According to the literature, graphite shows a sharp peak at approximately $2\theta = 26.00°$ (d = 0.34 nm), which corresponds to the diffraction peak (0 0 2), while, due to the increase in the distance between layers, graphite oxide graphite and graphene show a sharp peak at approximately $2\theta = 11.15°$ (d = 0.81 nm) that corresponds to the diffraction peak (0 0 1) [80,81]. As can be seen in Figure 1, the diffraction pattern agrees with the literature, since it shows a peak at $2\theta = 12.12°$ (d = 0.73 nm) that corresponds to the (0 0 1) diffraction peak of graphene oxide. Furthermore, another peak is observed at $2\theta = 22.87°$, characteristic of reduced graphene oxide, indicating that part of the graphene oxide was reduced during the synthesis process.

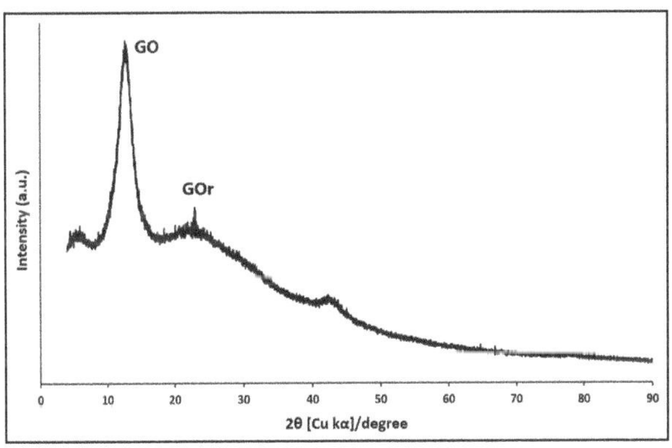

Figure 1. X-ray diffraction pattern of GO synthesized using the Hummers method.

3.1.2. SEM and EDS Analysis

To study the morphology of the synthesized material surface, an SEM analysis was performed. Figure 2 shows a somewhat rough structure of the material, with sheets folded over each other. The GO morphology shown at different magnifications in Figure 2a,b indicates that the exfoliation process was carried out successfully and some GO layers were synthesized.

To investigate the composition of the synthesized material, EDS measurements were performed. The results shown in Figure 3 reveal that there was only elements C (52.6%) and O (47.4%) were present in the synthesized material, suggesting high purity or no source of significant contamination for the obtained product.

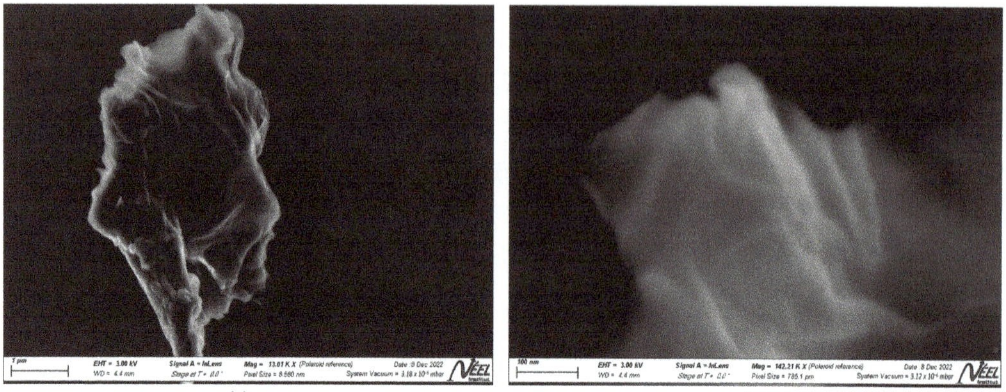

(a) (b)

Figure 2. SEM micrographs of GO at magnifications of (a) 1 μm and (b) 100 nm.

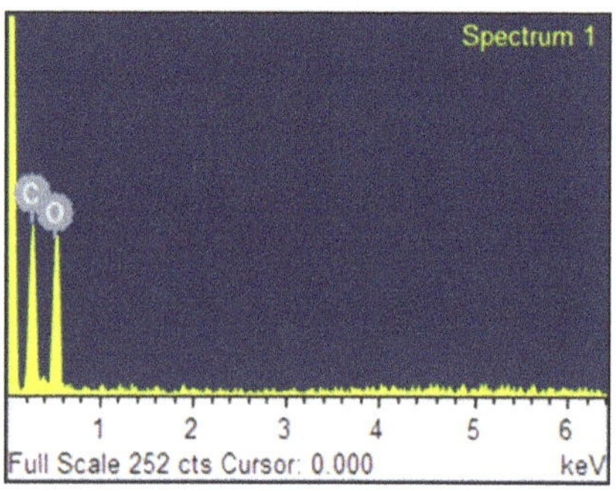

Figure 3. EDS spectra of synthesized GO.

3.1.3. FTIR Analysis

The presence of oxygen functional groups, especially hydroxyl, epoxy and carboxyl, which are key to the functionality of GO [82], was confirmed by FTIR analysis. The existence of these groups demonstrates the good oxidation of graphite. Figure 4 shows the spectrum of GO and GOr to compare and analyze their structural and functional differences.

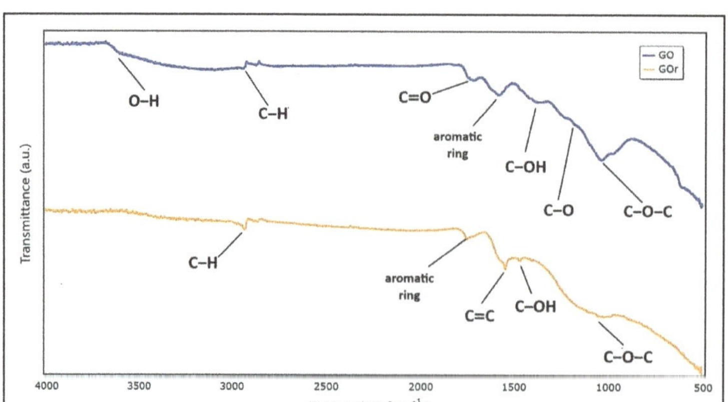

Figure 4. Fourier transform infrared spectra of GO and GOr.

Likewise, the adsorption of MB dye on GO was confirmed by FTIR analysis. Figure 5 shows the spectrum of MB, GO-MB, and GO, illustrating the changes in functional groups and molecular interactions upon dye adsorption.

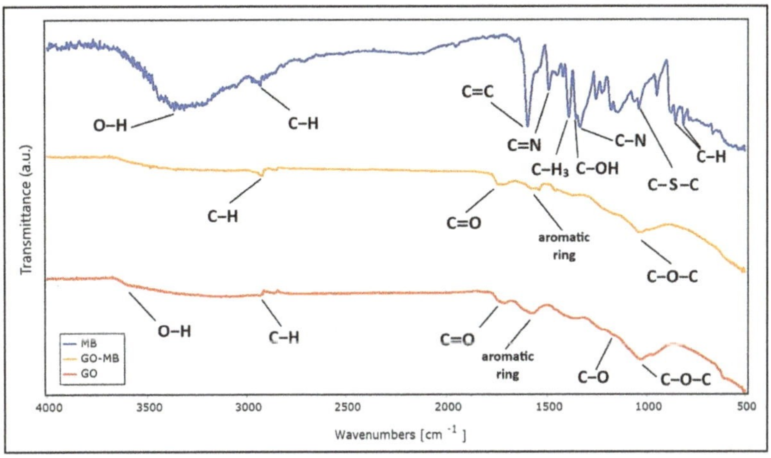

Figure 5. Fourier transform infrared spectra of MB, GO-MB and GO.

3.2. Experimental Adsorption Studies

3.2.1. Effect of pH on MB Adsorption

Figure 6 illustrates the ability of graphene oxide (GO) to remove MB dye as a function of the solution pH. This figure reveals that the dye can be adsorbed at acidic pHs below PZC (pH_{PZC} = 5.5). As the pH of the dye solution increases, the adsorption of cationic MB dye molecules on the GO surface is enhanced. However, Figure 6 also shows a decrease in adsorption capacity starting at pH > 10, which could be due to the hydrolysis of GO [83]. According to the literature, GO is stable at pH values between 4 and 10 [84].

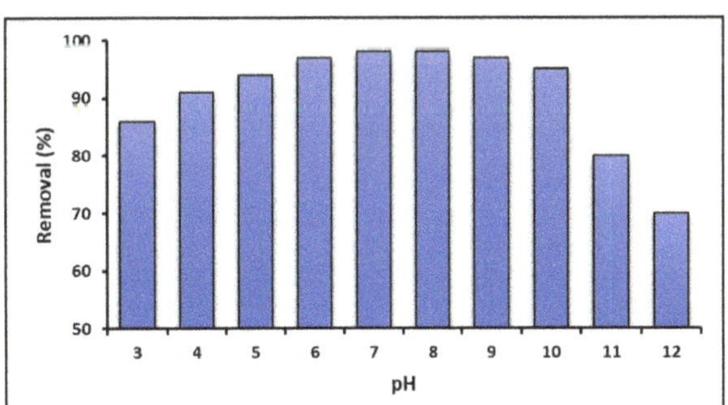

Figure 6. Effect of solution pH on MB adsorption (GO concentration = 100 mg L^{-1}; MB concentration = 10 mg L^{-1}; temperature = 20 ± 2.0 °C; time = 24 h).

3.2.2. Effect of Initial Concentration of MB

The effect of the initial concentration of the dye is closely related to its concentration and the sites present on the adsorbent surface [85]. The effect of the initial concentration of MB on the adsorption capacity of GO was determined by measuring the amount of residual MB in the solution via UV-Vis spectrophotometry at λ = 623 nm and applying Equation (1), as shown in Table 1. The adsorption maximum of GO was calculated using equilibrium isotherms at different initial concentrations of MB, from 1.5 to 50 mg L^{-1}. The fit of the experimental data is presented in Figure 7 for the Langmuir, Freundlich and Temkin isotherm models, according to Equations (2)–(4), as shown in Table 1, respectively. According to Figure 7, a better fit was demonstrated with the Langmuir isotherm model.

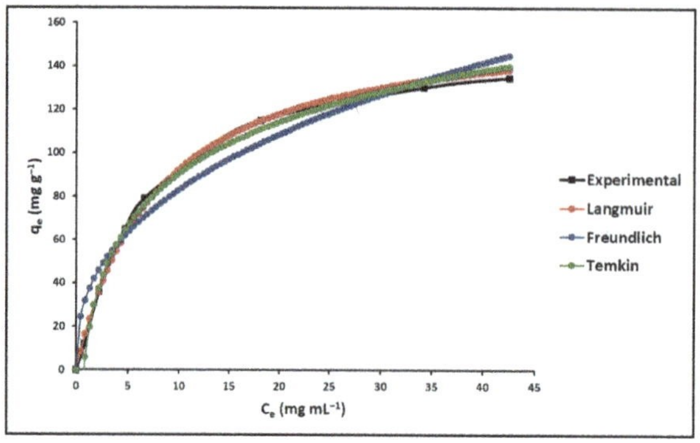

Figure 7. Effect of concentration on MB adsorption (GO concentration = 200 mg L^{-1}; solution pH = 7.0 ± 0.1; temperature = 20 ± 2.0 °C).

The parameters calculated from the isotherms are presented in Table 2. The estimated constants for the Langmuir, Freundlich, and Temkin models at absolute temperatures of 293.15 K, 303.15 K, and 313.15 K are also displayed in Table 2. In this table, the values of the separation factor or equilibrium parameter (R_L) are in the range of 0–1. Similarly, the coefficient "n", which reports the adsorption intensity, is within the range of 1–10.

Table 2. Isotherm parameters for MB adsorption on GO at different temperatures.

Isotherm Parameters		293.15 K	303.15 K	313.15 K
Langmuir	q_{max} (mg g^{-1})	163.30 (±5.90)	197.94 (±6.56)	214.27 (±6.95)
	K_L (L mg^{-1})	0.13 (±0.02)	0.17 (±0.03)	0.21 (±0.03)
	R_L	0.28	0.23	0.19
	χ^2	6.05	6.33	6.79
	R^2	0.98	0.97	1.00
Freundlich	K_F (L mg^{-1})	34.09 (±2.98)	38.22 (±2.94)	44.72 (±2.74)
	N	2.59 (±0.45)	2.90 (±0.49)	3.40 (±0.53)
	1/n	0.39	0.34	0.29
	χ^2	2.42	3.12	3.45
	R^2	0.90	0.92	0.95
Temkin	B	34.39 (±2.87)	38.55 (±2.24)	45.12 (±2.98)
	A	1.39 (±0.31)	1.55 (±0.42)	1.82 (±0.45)
	χ^2	2.92	3.06	3.86
	R^2	0.96	0.93	0.96

3.2.3. Effect of Temperature on MB Adsorption

It is widely known that the adsorption process depends on the temperature of the solution. In this context, thermodynamic parameters provide insight into the viability and spontaneity of a process [86]. To determine these thermodynamic parameters, in particular the Gibbs free energy change (ΔG°), enthalpy change (ΔH°), and surface entropy change (ΔS°), the equilibrium constant was established at various temperatures, according to Equations (7)–(9), as shown in Table 1. Figure 8 displays the results of this analysis.

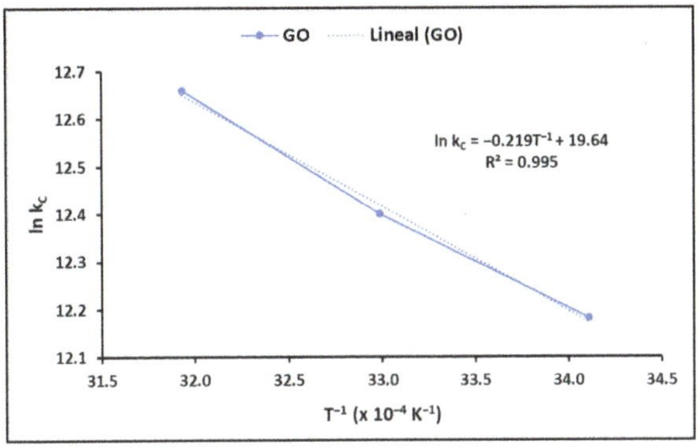

Figure 8. Temperature effect on MB adsorption (GO concentration = 200 mg L^{-1}; solution pH = 7.0 ± 0.1; temperature = 20 ± 2.0 °C).

The values of the thermodynamic parameters obtained in this study are described in Table 3. ΔG° indicates the degree of spontaneity in the process, and the more negative the values are, the more favorable the adsorption. On the other hand, positive values of ΔH° suggest that the process is endothermic, while positive values of ΔS° suggest an increase in randomness at the solution–solid interface during the adsorption process.

Table 3. Thermodynamic parameters of the MB adsorption onto GO.

Temperature (K)	ln k_C	ΔG° (kJ mol^{-1})	ΔH° (kJ mol^{-1})	ΔS° (kJ mol^{-1} K^{-1})
293.15	12.18	−29.69		
303.15	12.40	−31.25	18.21	0.16
313.15	12.66	−32.96		

3.2.4. Effect of Contact Time on MB Adsorption

In this study, the kinetics of the adsorption process of MB on GO were investigated. For this, three models were used: the pseudo-first-order model (Lagergren), the pseudo-second-order model (Ho), and the Elovich Model. Figure 9 shows the fit of the experimental data based on Equations (10)–(12), as presented in Table 1.

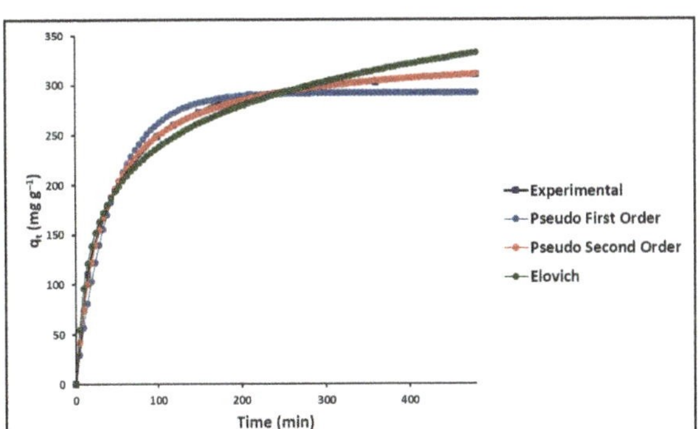

Figure 9. Time effect on MB adsorption (GO concentration = 200 mg L^{-1}; MB concentration = 20 mg L^{-1}; solution pH = 7.0 ± 0.1; temperature = 20 ± 2.0 °C).

As shown in Figure 8, all models exhibit an initial phase of rapid adsorption followed by a second phase of slow adsorption. Table 4 shows that the pseudo-second-order model best describes the kinetic behavior of the dye adsorption process on the GO surface, which agrees with the reports of other authors [87].

Table 4. Kinetic parameters for MB adsorption onto GO.

Kinetic Parameters		293.15 K
Pseudo-first-order	q_{max} (mg g^{-1})	292.39 (±5.57)
	k_1 (L mg^{-1})	0.02 (±1.77 × 10^{-3})
	χ^2	6.84
	R^2	0.97
Pseudo-second-order	q_{max} (mg g^{-1})	332.29 (±2.30)
	k_2 (L mg^{-1})	9.01 × 10^{-5} (±3.33 × 10^{-6})
	χ^2	11.21
	R^2	1.00
Elovich	A	30.82 (±5.61)
	B	0.02 (±9.39 × 10^{-4})
	χ^2	6.34
	R^2	0.98
Intraparticle diffusion	k (mg g^{-1} min$^{-1/2}$)	145.18 (±8.54)
	A	44.53 (±7.12)
	R^2	0.91
External-film diffusion	D_f (m^2 min^{-1})	6.12 × 10^{-12}
	R^2	0.89
Internal-pore diffusion	D_p (m^2 min^{-1})	1.44 × 10^{-18}
	R^2	0.91

On the other hand, the adjustment of the intraparticle diffusion model (Figure 10) shows, according to Equation 13 of Table 1, that the adsorption of MB on GO takes place in three stages: an initial stage of fast speed (k_1 = 231.45 mg g^{-1} min$^{-1/2}$), another, slower, intermediate stage (k_2 = 66.85 mg g^{-1} min$^{-1/2}$), and a final stage that tends to equilibrium (k_3 = 9.28 mg g^{-1} min$^{-1/2}$). The values of the kinetic constants for each stage, as well as the values of the effective diffusion coefficients D_f and D_p, calculated using Equations (14) and (15) of Table 1, respectively, are shown in Table 4.

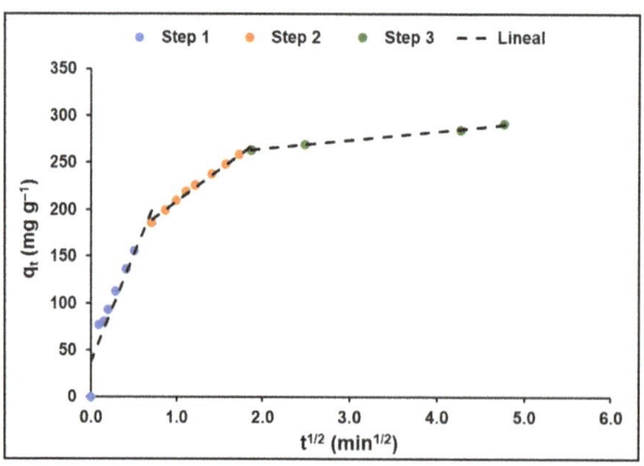

Figure 10. Diffusional effects on MB adsorption (GO concentration = 200 mg L^{-1}; MB concentration = 20 mg L^{-1}; solution pH = 7.0 ± 0.1; temperature = 20 ± 2.0 °C).

From Figure 10, it is evident that in the first, higher-speed, stage, MB molecules diffuse from the liquid film toward the adsorbent surface. In the second stage, the diffusion rate was lower, probably because the adsorption process occurred. In the last phase, the equilibrium trend was characterized by a much lower rate due to the decrease in MB concentration in the aqueous solution and the reduction in active sites in the adsorbent.

3.2.5. Desorption and Reusability of GO for MB Removal

Since the recycling and regeneration of the adsorbent is very important in practical applications, GO recycling experiments were carried out for five consecutive cycles. After each cycle, the materials were washed with methanol to desorb the adsorbed dye and were again subjected to the adsorption process, maintaining the same operational conditions as the previous cycle. Figure 11 shows that the recovery efficiency of GO is generally high, and the adsorption capacity is little affected over five consecutive adsorption–desorption cycles. In fact, after the fifth cycle, the MB adsorption capacity of GO decreased by only 12%. From these results showing a high recycling efficiency, it is suggested that the GO synthesized in this study is suitable for practical application.

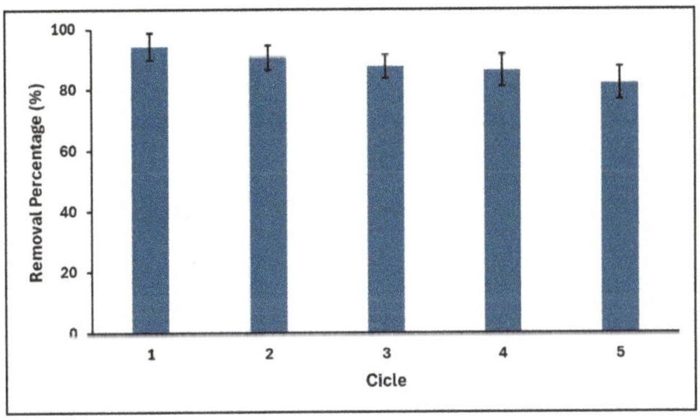

Figure 11. Recovery efficiency of GO after MB adsorption (GO concentration = 200 mg L^{-1}; MB concentration = 20 mg L^{-1}; solution pH = 7.0 ± 0.1; temperature = 20 ± 2.0 °C).

3.3. *Computational Studies*

3.3.1. Optimization of Adsorbent Structures

In this study, the effect of the hydroxyl and epoxy functional groups in graphene was determined using 3 × 3 supercells. These groups are common in oxidized graphene and are considered impurities in its structure. Several experimental studies have shown that oxidized graphene sheets have a thickness that is an integer multiple of approximately 6.7 Å, suggesting the likely presence of hydroxyl and epoxy groups on both sides of the graphene sheet [69]. However, the focus of this work is on possible configurations on a single side of the graphene oxide sheet.

Figure 12 shows the simulation models for pristine graphene (G), graphene oxide decorated with only one hydroxyl group (GOL), graphene oxide decorated with only one epoxy group (GOXI), and graphene oxide decorated with both hydroxyl and epoxy groups (GOL/GOXI). It is worth mentioning that analyzing the hydroxyl and epoxy functional groups, in isolation and in combination, is essential to understand the effects of impurity dispersion and doping in graphene [88].

As can be seen in Figure 12, the hydroxyl group forms a bond perpendicular to the plane of graphene, causing changes in bond lengths and angles. On the other hand, the epoxy group establishes a bond with two adjacent carbon atoms, which generates distortions in the graphene structure. The structures obtained in our calculations agree well with those reported in other theoretical studies [47,48].

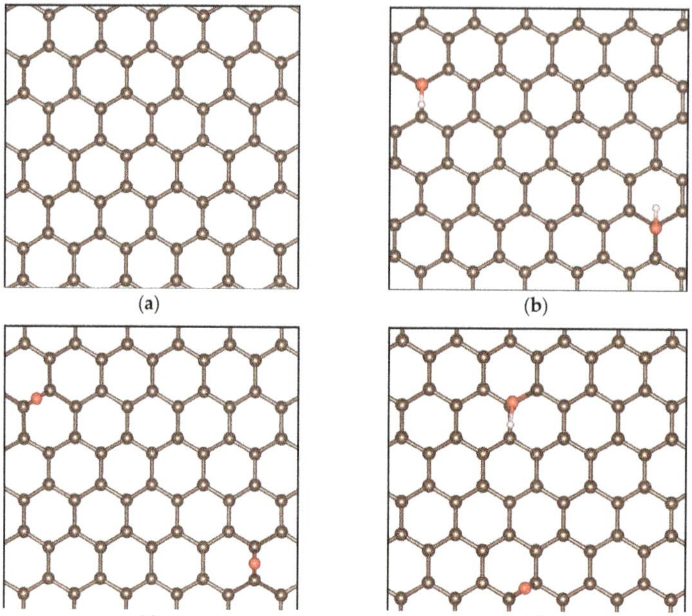

(a) (b) (c) (d)

Figure 12. Simulation models of (**a**) pristine graphene (G), (**b**) GO-hydroxyl (GOL), (**c**) GO-epoxy (GOXI), and (**d**) GO-hydroxyl/epoxy (GOL/GOXI). The brown balls are C atoms, the red balls are O atoms, and the white balls are H atoms.

3.3.2. Electronic Properties of Adsorbent Structures

In this study, the electronic properties of pristine graphene and graphene oxides were determined by analyzing the band structure (SB) and the density of states (DOS). Figure 13 shows the band structures (SB) for the configurations of (a) pristine graphene (G) and the graphene oxides—(b) GOL, (c) GOXI, and (d) GOL/GOXI—along the high-symmetry directions A—L—M—Γ—A—H—K [89]. This figure shows that the presence of oxygenated functional groups in the graphene monolayers does not shift the Dirac point regarding the Fermi level.

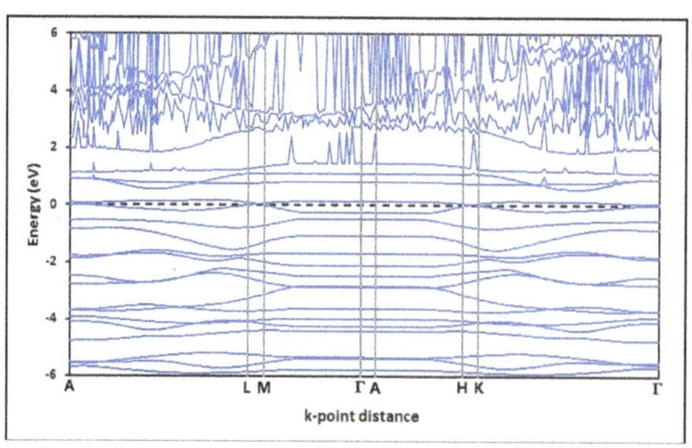

(a)

Figure 13. *Cont.*

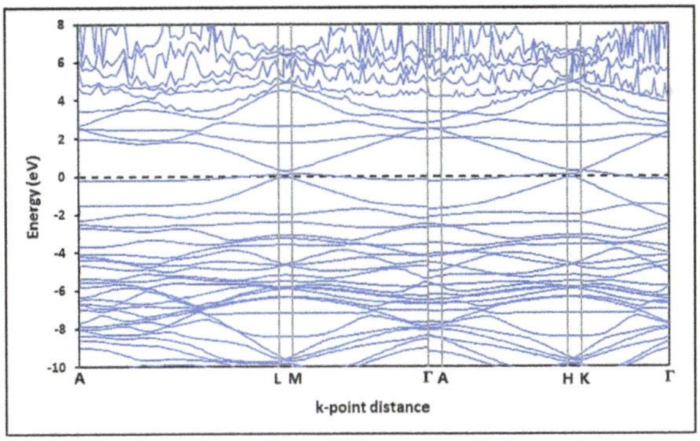

(**b**)

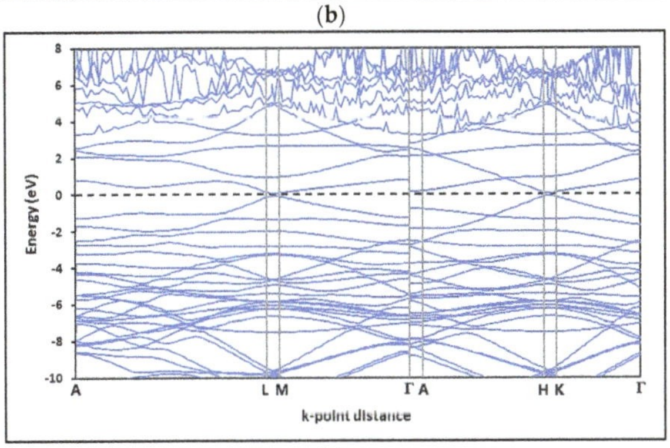

(**c**)

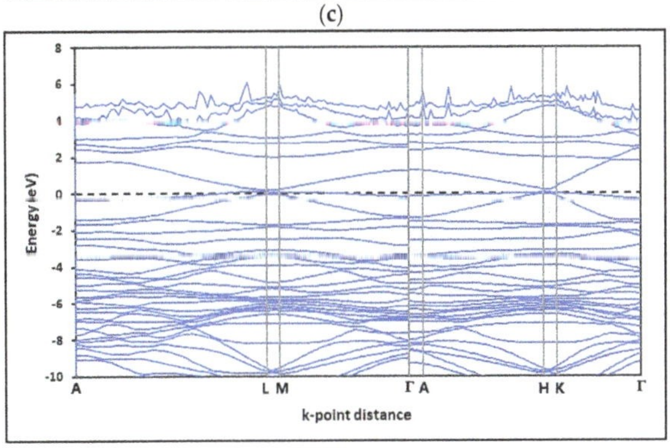

(**d**)

Figure 13. Band structure (BS) of (**a**) pristine graphene (G), (**b**) GO-hydroxyl (GOL), (**c**) GO-epoxy (GOXI), and (**d**) GO-hydroxyl/epoxy (GOL/GOXI).

On the other hand, the DOS of pristine graphene (G) and graphene oxides GOL, GOXI, and GOL/GOXI are shown in Figure 14a–d, respectively. Figure 14b,c show that the GOL and GOL/GOXI oxides present a sharp peak in the density of states near the Fermi level, which indicates the presence of highly localized electronic states in this region. Furthermore, the comparison between the density of states of pristine graphene and graphene oxides highlights the differences in the electronic properties of these materials, which has important implications for the design and development of electronic devices and advanced materials.

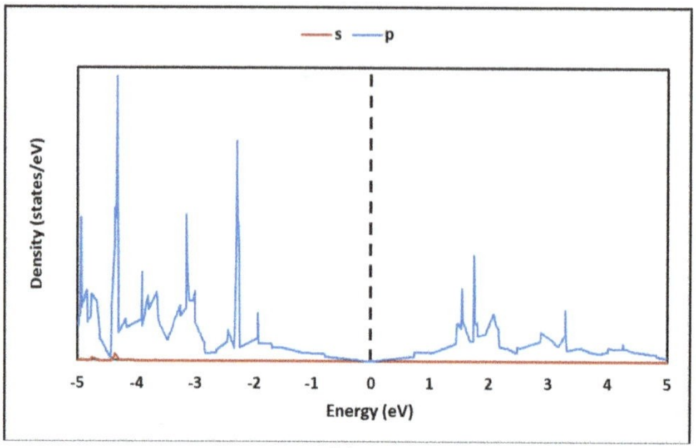

(**a**)

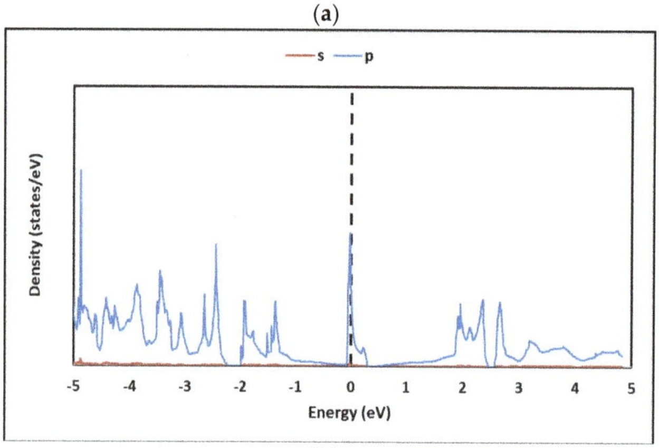

(**b**)

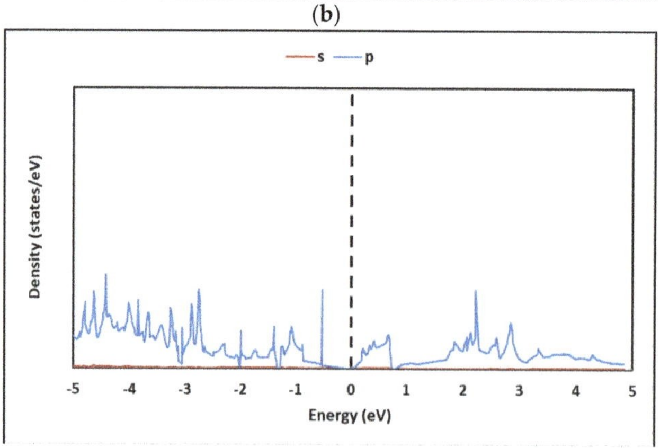

(**c**)

Figure 14. *Cont.*

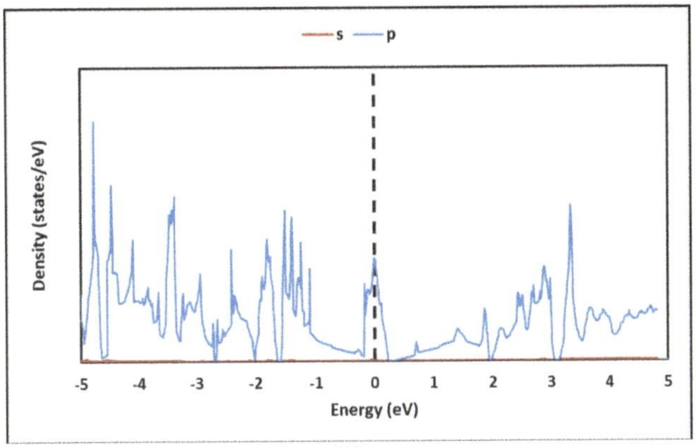

(**d**)

Figure 14. Density of states (DOS) of (**a**) pristine graphene (G), (**b**) GO-hydroxyl (GOL), (**c**) GO-epoxy (GOXI), and y (**d**) GO-hydroxyl/epoxy (GOL/GOXI).

Figure 14a shows the density of states (DOS) of pristine graphene (G), evidencing the location of the Fermi level at zero. The DOS of Figure 14b shows, for the GOL oxide, the existence of a distinctive peak at the Fermi level, which could be induced by the neutral OH group with an odd number of electrons in the supercell. In Figure 14c, the epoxy group of the GOXI oxide eliminates the localized state at the Fermi level, generating a small band gap (approximately 0.1 eV) in the DOS of the system. Finally, in Figure 14d, the DOS of GOL/GOXI oxide shows a similar trend to GOL oxide, with a distinct peak at the Fermi level.

3.3.3. Computational Adsorption Studies

Figure 15 shows the simulation models for MB adsorption on the surfaces of pristine G (graphene), GOL (graphene oxide decorated with a hydroxyl group), and GOXI (graphene oxide decorated with an epoxy group). This figure shows that the hydroxyl and epoxy functional groups are randomly distributed in the respective GO monolayers. As can be seen from Figure 15b,c, the presence of hydroxyl and epoxy groups can cause significant local distortion in the structure of pristine graphene.

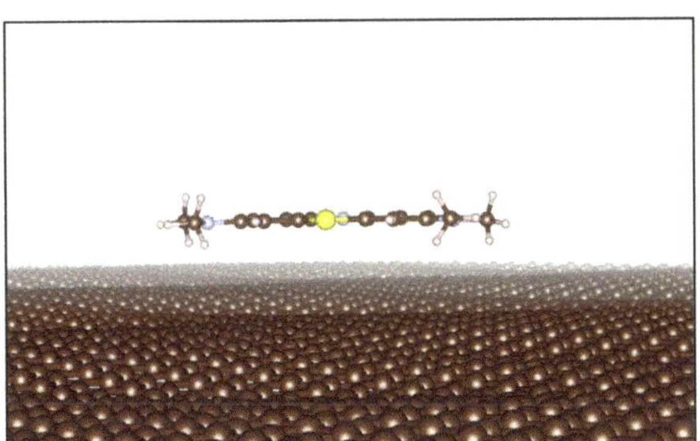

(**a**)

Figure 15. *Cont.*

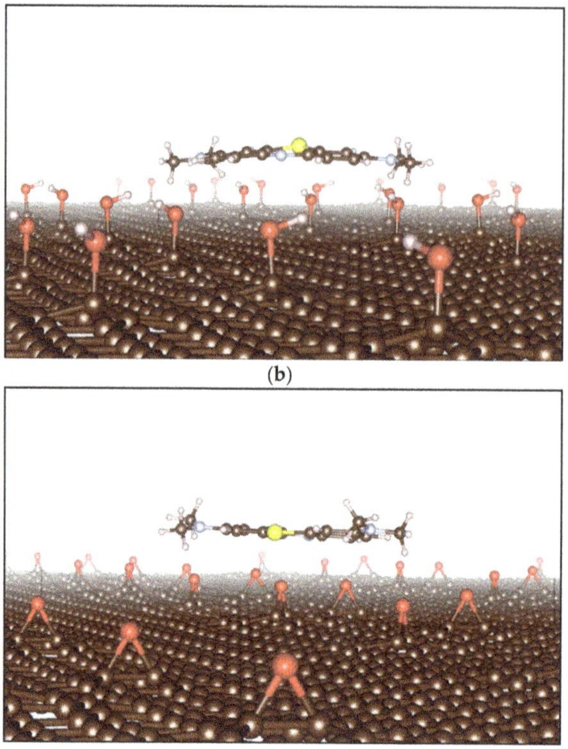

Figure 15. Optimized structures of (a) MB adsorbed on Graphene (G), (b) MB adsorbed on GOL oxide, and (c) MB adsorbed on GOXI oxide. The brown balls are C atoms, the red balls are O atoms, the white balls are H atoms, the gray balls are N atoms, and the yellow balls are S atoms.

Table 5 summarizes the values of adsorption energy, interfacial distance, and transferred charge for the optimized reaction systems. By convention, transferred load values show a ($^+$) sign when the structure gains load (e) and a ($^-$) sign when it loses load (e).

Table 5. Calculated values of absorption energy, interfacial distance, and transferred charge.

Optimized System	Adsorption Energy (kJ mol^{-1})	Interfacial Distance (Å)	Charge Transfer (e)	
			Surface	MB
G-MB	−25.96	C-H$_{MB}$ = 3.05	+0.39	−0.39
GOL-MB	−67.27	H-S$_{MB}$ = 2.47	+0.68	−0.68
GOXI-MB	−53.53	O-H$_{MB}$ = 2.26	+0.57	−0.57

The most energetically stable adsorption for the MB molecule occurred on graphene oxide (GO) structures, and was higher for GOL than for GOXI. These results were confirmed in the Bader analysis, which demonstrated a greater charge transfer for the GOL-MB system, as seen in Table 5.

Likewise, to verify the chemical interaction between the adsorbent surfaces (G, GOL, and GOXI) and the MB molecule, a population analysis of the G-MB, GOL-MB, and GOXI-MB systems was performed using the Bader method [77,90]. This analysis is valuable as it allows for a description of the ionicity of chemical bonds by evaluating the charge transfer between the bonded atoms [76]. Table 6 shows, for each system, the average charge per atom type, before (BA) and after (AA) MB adsorption. The atoms belonging to the adsorbent surface were designated with the subscript ($_{surf}$) while those belonging to the methylene blue molecule were designated with the subscript ($_{MB}$).

Table 6. Calculated values of transferred charge.

Atoms	G-MB			GOL-MB			GOXI-MB		
	BA (e)	AA (e)	ΔCharge	BA (e)	AA (e)	Δcharge	BA (e)	AA (e)	ΔCharge
C_{surf}	0.0000	−0.0012	0.0012	0.0060	0.0040	0.0020	0.0141	0.0099	0.0042
O_{surf}	-	-	-	−1.4837	−1.4913	0.0076	−1.1389	−0.9428	−0.1961
H_{surf}	-	-	-	0.9999	0.9999	0.0000	-	-	-
C_{MB}	0.0802	0.4812	−0.4009	0.0802	0.4609	−6.0916	0.0802	0.4670	−0.3868
H_{MB}	0.2756	0.0186	0.2498	0.2756	0.0463	4.1264	0.4870	0.2005	0.2865
N_{MB}	−2.2436	−2.6266	0.3830	−2.2436	−2.5924	0.2421	−2.2436	−2.6007	0.3395
S_{MB}	0.4870	0.2372	0.2570	0.4870	0.2449	1.0465	0.2756	0.0386	0.2370

Furthermore, the electron localization function (ELF) was used in this study to gain a deeper understanding of the G-MB, GOL-MB, and GOXI-MB interactions [78]. In the ELF analysis, when the region of maximum density (RMD) is more symmetrically distributed around the nucleus, it means that a more ionic or Van der Waals interaction is taking place. In contrast, as the covalent character of a bond strengthens, RMD migration between centers becomes more pronounced until a perfectly symmetric geometry is achieved in the ideal covalent scenario. The ELF section for the G-MB, GOL-MB, and GOXI-MB interactions is shown in Figure 16. As can be seen in this figure, there is a slightly stronger interaction between the MB molecule and the hydroxyl functional group of graphene oxide. The figure shows that the RMDs are located along the line connecting the nuclei and can be separated from the nuclei themselves by a path. However, RMDs do not surround the nucleus in either system [79].

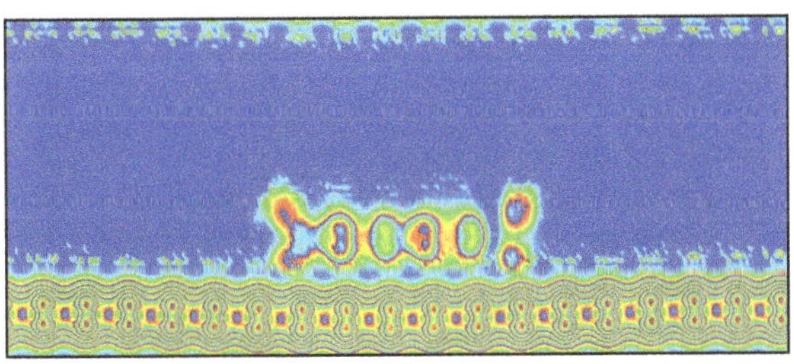

(a)

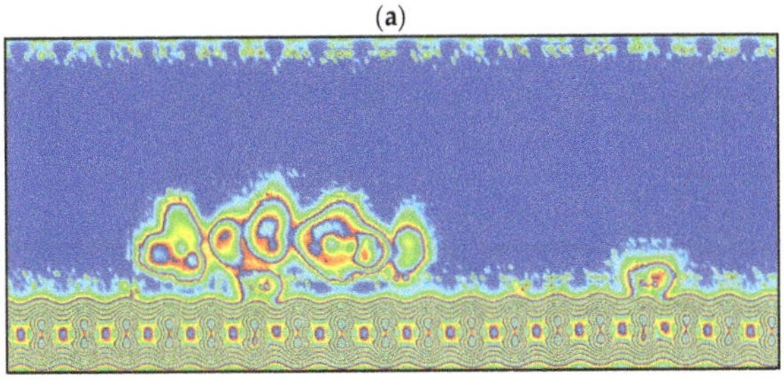

(b)

Figure 16. *Cont.*

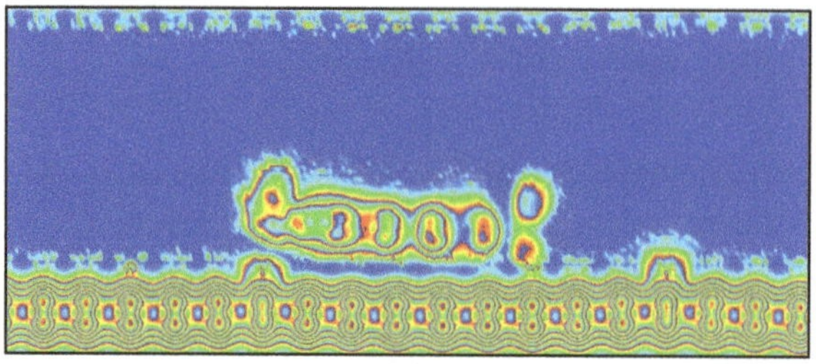

(c)

Figure 16. ELF analysis of (**a**) MB adsorbed on graphene (G), (**b**) MB adsorbed on GOL graphene oxide (GOL), and (**c**) MB adsorbed on GOL graphene oxide (GOXI). The red, green, and blue (RGB) color scale for ELF analysis represents electronic localization (dark red) to electronic delocalization (light blue).

4. Discussion

4.1. Characterization of GO

In the XRD analysis, the peak in GO is greater than that of GOr, due to the incorporation of oxygen functional groups and intercalated water. This result agrees with that reported in other studies [80], where the increase in interlayer spacing facilitates exfoliation and improves dispersibility in aqueous solutions, a desirable property for applications in composites and adsorption materials.

The rough and folded morphology observed in SEM analysis is related to surface defects due to the deviation from the sp^2 to sp^3 character as a consequence of the high density of oxygen functional groups [91]. According to the literature, the use of ultrasound during the exfoliation of graphite oxide sheets to generate layers of graphene oxide, as well as the thermal treatment used during the drying process, could cause deformations in the structure of the sheets and their folding [80]. Moreover, the elemental composition, with a C/O ratio close to 1.25, is comparable to that reported in the literature [92], reflecting significant oxidation but preserving the essential carbonaceous structure for conductivity and mechanical strength.

Regarding the FTIR analysis, it is evident that hydroxyl, epoxy, carboxyl, and carbonyl are the dominant functional groups in GO. Thus, the band at around 2900 cm^{-1} represents the C-H stretching of heterocyclic compounds. The bands observed at 1714 cm^{-1} and 1575 cm^{-1} depict the vibrations of the stretching carboxylic acid group (C=O) and C=C aromatic stretching, respectively. GO also showed strong characteristic peaks at 1377 cm^{-1}, 1219 cm^{-1}, and 1031 cm^{-1}, corresponding to the bending vibrations of C–OH, and stretching vibrations of both C–O in C–OH and C–O–C (epoxy), respectively. The band at around 500 cm^{-1} may correspond to C–O stretching vibrations. These results agree with those reported by other authors [93]. The broad peak between 3000 and 3600 cm^{-1} is due to the vibration of O–H stretching for carboxyl and hydroxyl groups and the water molecules adsorbed between the GO sheets. From these results, it can be suggested that the most likely configuration of the graphene oxide sheet consists of a combination of clean graphene regions and fully oxidized regions where the functional groups would be locally arranged in a random manner. However, due to the highly unstable nature of the oxidation process, several intermediate phases may still exist in the experimental samples. In the same context of FTIR analysis, GO was evaluated after adsorbing MB. The results indicate that, in addition to the characteristic peaks of GO, there are new peaks that would be related to the structure of MB. The new peaks were observed at 1540 cm^{-1}, 1458 cm^{-1}, 1397 cm^{-1}, and 1335 cm^{-1}, corresponding to the C=C stretching, C=N stretching, C–H_3 bending, and C–N stretching vibrations of the terminal saturated dimethyl amino group, respectively [94]. Furthermore, the adsorption of methylene blue (MB) on graphene oxide (GO) induces characteristic changes in the FTIR bands of GO, providing detailed information on the underlying adsorption mechanisms. The decrease in intensity in the O-H band at around 3400 cm^{-1} suggests that the hydroxyl groups of GO are involved in the formation of hydrogen bonds with MB, while the changes in the C=O stretching band at around 1720 cm^{-1} could indicate interactions between the carboxylic groups of GO and the positive charges of MB, consistent with electrostatic interactions. Furthermore, the alteration in the bands associated with C=C bonds at around 1620 cm^{-1} reflects possible π–π interactions between the aromatic rings of GO and MB [94]. These spectral changes not only confirm the adsorption of MB on the GO surface, but also highlight the synergistic role of hydrogen bonds, electrostatic forces and π–π interactions in the adsorption process, which is important for the efficient removal of organic contaminants in aqueous solutions.

In this study, the specific surface area (SSA = 178.4 m^2 g^{-1}), monolayer volume (MV = 40.9 cm^3 g^{-1}) and point of zero charge (PZC = 5.5) of the synthesized GO were also evaluated. SSA is an important indicator of the adsorbent capacity of a material, with higher values generally indicating a better adsorption performance due to the larger surface area available for interaction with the adsorbate molecules. Compared to other graphene-based materials reported in the literature, with SSA in the range of 30–880 m^2 g^{-1} [95], the GO synthesized here shows good potential, especially considering that its surface functionalities offer specific sites for the adsorption processes. This balance between the surface area and chemical functionality of GO is crucial for applications where a specific interaction with the adsorbate is desired, such as in the selective removal of contaminants. On the other hand, the results of PZC shown in Figure S1, suggest that the surface of the adsorbent will be positively charged when the pH of the solution is lower than 5.5, and negatively charged when the pH of the solution is superior to this value. Therefore, a $pH > pH_{PZC}$ would promote favorable electrostatic interactions between the adsorbent surface and cationic molecules, such as MB dye [96]. The combination of PZC and SSA in the characterization of GO offers a comprehensive view of how this material can be optimized for specific applications. For instance, in contaminant adsorption processes, understanding the PZC allows for pH adjustment to maximize electrostatic interactions, while a high SSA ensures the wide availability of adsorption sites. However, it is important to note that the presence of oxygen functional groups, as evidenced in the FTIR analysis, can impact both the PZC and SSA by introducing porosity and increasing the dispensability of the material, which could affect the accessibility of the pores and the charge distribution on the GO surface. Additionally, the interaction between PZC and SSA under different environmental conditions can significantly influence the stability and reusability of GO. For example, in highly acidic or basic environments, modifications to the surface charge can not only affect the adsorption efficiency but also the structural stability of the material, which is critical for repetitive adsorption applications.

4.2. Experimental Adsorption Studies

4.2.1. Effect of pH on MB Adsorption

The effect of pH on the adsorption of MB dye on graphene oxide (GO) is a fundamental aspect of understanding the interaction between the adsorbate and the adsorbent, as well as optimizing the adsorption efficiency in practical applications. The adsorption process depends on the pH of the solution, since this influences both the distribution of surface charges on the adsorbent and the speciation of the dye in the solution [97]. According to the literature, molecular MB can exist in an aqueous solution as an undissociated molecule (MB°) and as a cationic species (MB^{+}). At pH = 3, the MB° form predominates, at pH = pKa = 3.8, MB° and MB^{+} coexist in equal proportions, and at pH > 6, the cationic form MB^{+} is practically the only species present [98].

In this study, it was observed that GO exhibits a higher MB adsorption capacity as the pH of the solution increases, up to an optimum close to pH = 7.0. However, at extremely basic pH values, the adsorption capacity drastically decreases, which can be attributed to the destabilization of GO and the creation of positively charged sites on its surface, leading to electrostatic repulsion with MB molecules [99,100]. Furthermore, according to the literature, the molecular structure of MB could be modified at a high pH due to gradual demethylation [101]. The variation in the adsorption capacity of MB with pH can be explained by the modification of the surface charge of GO and the ionization of MB molecules. At a low pH, the acidic environment facilitates the protonation of oxygen functional groups on the GO surface, generating a positive surface charge that can repel cationic MB. As the pH increases, these groups are deprotonated, giving rise to a more negatively charged surface that favors the adsorption of MB through electrostatic attraction.

The high adsorption capacity observed at alkaline pH values can be attributed to the increase in hydroxyl ions, resulting in greater electrostatic attraction between the MB^+ cationic species and the negatively charged surface of graphene oxide [102]. However, at extremely high alkaline pH levels, OH ions can form complexes with other ions, such as MB^+, potentially influencing dye adsorption on the adsorbent surface [103]. This phenomenon could result in the precipitation of dye molecules on the GO surface. Consequently, the adsorption mechanism at an alkaline pH is likely to be a combination of electrostatic attraction and precipitation [104]. On the other hand, Figure 6 also shows that graphene oxide has a reasonably good MB+ adsorption capacity at $pH < pH_{PZC}$, where electrostatic interactions do not favor adsorption. Under these experimental conditions ($pH < 5.5$), it is suggested that the adsorption of the dye could occur via ion exchange since the cationic species MB^+ would be competing with H^+ for the active sites on the GO surface [18].

4.2.2. Effect of Initial Concentration of MB

The adsorption capacity of GO was evaluated using Langmuir, Freundlich, and Temkin adsorption isotherms, and a favorable fit was found with the Langmuir model, suggesting monolayer adsorption on homogeneous surfaces of GO. This behavior indicates a specific and uniform interaction between MB and the active sites on GO, which is crucial for purification applications where high selectivity and efficiency are required. From the Langmuir model, the maximum adsorption capacity (q_{max}) was estimated, reflecting the maximum amount of MB that can be adsorbed per unit mass of GO under optimal conditions. Furthermore, in this study, it was found that the value of the Langmuir separation factor (R_L) was in the range of $0 < R_L < 1$ for the tested initial MB concentrations, indicating that adsorption is favorable under the experimental conditions that were used. Values of R_L between 0 and 1 suggest favorable adsorption, while $R_L > 1$ would indicate unfavorable adsorption [105]. The fit of the data to both the Langmuir model, and the R_L values indicates not only the feasibility and favorability of MB adsorption on GO but also that surface saturation occurs in a monolayer. This is indicative of a high affinity between MB and GO, possibly due to π–π interactions between the aromatic rings of MB and the graphene structure of GO, as well as possible electrostatic interactions depending on the pH of the solutions. The coefficients obtained from the Freundlich model, especially the value of n, which falls between 1 and 10, suggest that the adsorption is physical and favorable. The physical nature of adsorption is beneficial from the perspective of adsorbent reusability, as physical interactions are generally easier to reverse than chemical ones, facilitating the regeneration of GO.

4.2.3. Effect of Temperature on MB Adsorption

The effect of temperature on the adsorption of methylene blue (MB) on graphene oxide (GO) is a key factor in understanding the thermodynamic nature of the adsorption process. In this study, it was observed that, with an increase in temperature, the adsorption capacity of MB on GO also increases, indicating that adsorption is an endothermic process. This behavior suggests that at higher temperatures, there is a greater affinity between MB and GO, probably due to the increased mobility of the MB molecules and the expansion of GO pores, which facilitates access to more adsorption sites. Additionally, the obtained thermodynamic parameters, such as the change in Gibbs free energy (ΔG°), enthalpy (ΔH°), and entropy (ΔS°), confirmed the endothermic nature of the process. Negative values of ΔG° at different temperatures indicate that adsorption is spontaneous, while a positive value of ΔH° reinforces the idea that the process is endothermic. The positive increase in

ΔS° suggests an increase in disorder at the solid–liquid interface during adsorption, which is typical in physical adsorption processes.

4.2.4. Effect of Contact Time on MB Adsorption

This study reveals that the adsorption kinetics of MB on GO follow the pseudo-second-order model, indicating that the rate of adsorption depends on the amount of MB adsorbed at equilibrium and the amount of MB adsorbed at any given time. According to the pseudo-second-order model, rate of MB adsorption increases over time until equilibrium is reached, which fits well with the model, suggesting that the rate-limiting step is the chemical interaction between MB and the active sites on the GO surface. The high correlation coefficient R^2 values for this model indicate a good fit of the experimental data, reflecting the applicability of the model in describing the adsorption kinetics of the GO-MB system. The fit with the pseudo-second-order model underscores the characteristics of GO as a material with active sites capable of chemically interacting with MB molecules, facilitating efficient adsorption. This observation is in line with the implications of the Langmuir isotherm observed in this study, where the formation of a monolayer of MB on the GO surface indicates that adsorption occurs at specific sites with a uniform adsorption energy.

On the other hand, MB adsorption on GO showed typical intraparticle diffusion behavior, where three distinct stages were identified based on the kinetic constant of each stage. The first stage, characterized by a high kinetic constant, corresponds to the external diffusion of MB towards the GO surface. The second stage, with a lower kinetic constant, reflects the intraparticle diffusion of MB within the pores of GO. Finally, the last stage, with the lowest kinetic constant, indicates adsorption equilibrium, where the rate of adsorption slows down due to the saturation of adsorption sites. The diffusion coefficients in the external film phase (D_f) and the adsorbent phase (D_p) provide information on the resistance to mass transfer during the adsorption process. In this study, the values of D_f and D_p indicate that intraparticle diffusion and mass transfer through the external film boundary significantly contribute to the adsorption kinetics, which is consistent with the results obtained in the kinetic modeling. The relationship between these coefficients and the results reflects the complexity of the adsorption process, where both diffusion mechanisms (intraparticle and through the film boundary) play important roles in the overall rate of adsorption. This highlights the need to optimize operational conditions, such as agitation and adsorbent particle size, to minimize resistance to mass transfer and maximize adsorption efficiency.

Finally, evidence from the experimental study showed that the maximum MB adsorption capacity (q_{max}) of the synthesized GO is comparable to that of other adsorbents, as previously reported by other authors (Table 7).

Table 7. Comparison of adsorption capacity with other studies.

Adsorbent	Adsorption Capacity (mg g^{-1})	Reference
Graphene Oxide	332	[106]
Graphene Oxide	287	[107]
Graphene Oxide	244	[108]
3D Graphene Oxide sponge	397	[109]
Graphite Oxide	351	[110]
Graphene	154	[111]
Reduced Graphene Oxide	68	[112]
Coconut-Shell-Activated Carbon	200	[113]
Multi-Wall Carbon Nanotube	48	[114]
Graphene Oxide	145	[115]
Magnetic Graphene Oxide	25	[116]
Fe_3O_4/Chitosan/Graphene	48	[117]
Graphene Oxide	163	This study

A comparison of the methylene blue (MB) adsorption capacity of graphene oxide (GO) synthesized in this study, with a capacity of 163 mg g^{-1}, against other adsorbents reported in the literature, offers a detailed perspective on the performance of different materials in the adsorption of contaminants. The GO in this study shows an intermediate performance compared to other forms of GO, with capacities ranging from 145 mg g^{-1} to 397 mg g^{-1} for the 3D GO sponge, suggesting that specific structural modifications, such as increasing the three-dimensionality, can significantly improve the adsorption capacity. This is probably due to an increase in the surface area available for adsorption and possibly the presence of more active sites. On the other hand, reduced graphene oxide (rGO) shows wide variability in its capacity, from as low as 68 mg g^{-1}, indicating how reduction treatments significantly affect its adsorption properties, possibly altering the quantity and nature of the functional groups on the surface of the material. Compared with other types of activated carbon and nanomaterials, such as coconut-shell-activated carbon with 200 mg g^{-1} and multi-walled carbon nanotubes with 48 mg g^{-1}, it is evident that the GO synthesized here shows a competitive ability, especially considering that nanotubes and activated carbon are highly optimized for adsorption. Differences in the adsorption capacity of the tested materials can be attributed to variations in their morphology, surface area, porosity, and surface chemistry. Therefore, to improve the adsorbent capacity of synthesized GO in future research, it would be beneficial to explore structural modifications, such as creating nanoscale pores or increasing porosity through chemical or physical activation techniques. Furthermore, functionalization of the GO surface with chemical groups that have a high affinity for MB could increase the specific adsorption capacity of this compound. It would also be interesting to investigate the synergy between GO and other materials in hybrid composites, which could combine the unique properties of various adsorbents to achieve superior adsorption capacities and specific selectivity for different contaminants.

4.3. *Computational Studies*

4.3.1. Optimization of Adsorbent Structures

This study focused on analyzing how the introduction of hydroxyl and epoxy functional groups affects the structural and electronic properties of GO, and therefore, its adsorption capacity. The findings reveal that the functionalization of GO with these groups can significantly modify both the morphology and surface chemistry of the material, which has direct implications for its performance as an adsorbent.

The introduction of hydroxyl and epoxy groups into the structure of GO was optimized to increase the efficacy in the adsorption of MB molecules. These functional groups not only improve the dispersibility of GO in water but also offer specific sites for adsorption through hydrogen bonding and van der Waals forces interactions. The optimized structure of GO showed a significant improvement in adsorption capacity compared to non-functionalized graphene, which is attributed to the increased available surface area and the presence of active sites for adsorption.

4.3.2. Electronic Properties of Adsorbent Structures

The introduction of oxygen functional groups into graphene, such as hydroxyls and epoxys, induces significant changes in the electronic structure of the material, which are reflected in the density of states (DOS) and the band structure (BS). These changes have profound implications for the hybridization of orbitals, the Fermi level, and the Dirac point, key elements for understanding and applying the material in electronic devices and sensors. Pure graphene possesses an sp^2 hybridization of carbon orbitals, which gives it a flat structure and unique electronic properties, such as high electrical conductivity. The introduction of oxygen functional groups disturbs this sp^2 hybridization, introducing regions of sp^3 hybridization. This results in a local distortion of the graphene flat structure and the formation of defective electronic states that affect the mobility of charge carriers and the optical and electrical properties of the material. The Fermi level in pure graphene is located at the Dirac point, where the conduction and valence bands meet, resulting in a null bandgap and high intrinsic conductivity.

The functionalization of graphene with oxygenated groups can shift the Fermi level due to charge transfer between graphene and the functional groups. This shift modifies the conduction properties of the materials, as the Fermi level dictates the occupancy of available electronic states and, therefore, the conductivity. The Dirac point, characteristic of graphene, is the crossing point of the conduction and valence bands in the band diagram, which are crucial for the unique electronic properties of graphene. In this study, the functionalization of graphene with oxygenated groups produced a slight perturbation at the Dirac point, as electronic states were introduced into the bandgap, potentially opening a small energy gap at the Dirac point. This change could certainly reduce the electronic mobility of the material but increases its semiconductivity, allowing for graphene to be adapted for applications that require a bandgap, such as in transistors or gas sensors.

Several authors have shown how the introduction of defects and functional groups alters the electronic properties of graphene and improves its adsorption capacity, corroborating the findings of this study. This phenomenon is probably due to the introduced functional groups, which could increase the chemical compatibility between the adsorbent and adsorbate, facilitating the adsorption process [118,119]. Therefore, the results of this study support the idea that the functionalization of graphene is an effective strategy to increase its versatility and effectiveness as an adsorbent material.

4.3.3. Computational Adsorption Studies

The computational study of the molecular adsorption of methylene blue (MB) on graphene (G) and graphene oxides (GO) sheds light on how these interactions can be optimized for applications such as water purification and sensor design. Table 7 provides detailed data on the adsorption of MB on graphene and graphene oxide (GO) surfaces, highlighting differences in adsorption energy, interfacial distance, and charge transfer. These parameters are crucial for understanding the interaction between MB and the different graphene-based materials. The adsorption energy is an indicator of the strength of the interaction between the adsorbate (MB) and the adsorbent (G or GO). The negative values achieved in this study indicate that all adsorptions are spontaneous processes. The adsorption energy of −25.96 kJ/mol, calculated for the G-MB system, indicates a moderately strong interaction, suggestive of predominant physical interactions, such as van der Waals forces and π–π stacking between the aromatic rings of MB and the graphene surface. In the GOL-MB system, the adsorption energy significantly increases to −67.27 kJ/mol, indicating a much stronger interaction compared to pure graphene. This suggests that the presence of hydroxyl groups on GO facilitates additional interactions, possibly through hydrogen bonding, increasing the adsorbent affinity for MB. Finally, the adsorption energy of −53.53 kJ/mol for the GOXI-MB system also shows a strong interaction, although slightly less than that of GO with hydroxyl groups. This could be due to the specific characteristics of the epoxy groups, which also allow for strong interactions, but with a different bonding dynamic compared to hydroxyls.

The interfacial distance between MB and the adsorbent surface provides an idea of how close the molecules are to each other, which directly affects the nature of the interaction. The interfacial distance of 3.05 Å in the G-MB system is consistent with noncovalent interactions, such as van der Waals forces and π–π stacking. In contrast, the shorter distances of 2.47 Å and 2.26 Å calculated for the GOL-MB and GOXI-MB systems, respectively, suggest stronger interactions, likely due to the formation of hydrogen bonds between the functional groups of GO and MB, in addition to π–π interactions.

The charge transfer from the adsorbent to the adsorbate (or vice versa) is an indicator of the formation of a chemical bond or significant electrostatic interactions. The charge transfer of ±0.39 e in the G-MB system indicates a moderate interaction, consistent with van der Waals forces and π–π stacking. Meanwhile, the more significant charge transfers of ±0.68 e and ±0.57 e from the GOL-MB and GOXI-MB systems, respectively, suggest a stronger chemical interaction, possibly due to the formation of hydrogen bonds and increased polarization at the interface.

It is important to mention that the horizontal orientation of the MB molecule on the G and GO surfaces adopted in this study is justified by maximizing the π–π interactions between the aromatic systems of MB and the graphene-based structures. This adsorption mechanism is consistent with the findings of other studies, which highlight how the horizontal orientation facilitates a larger contact surface, increasing the interaction between the adsorbate and the adsorbent [120,121]. The choice of this orientation is supported by molecular dynamics simulations and total energy calculations that show an energetic preference for the horizontal arrangement over vertical or inclined configurations. The ELF analysis reinforces this interpretation, showing a redistribution of electronic density that favors the formation of weak bonds and potentially hydrogen bonds between MB and the functional groups of GO. The ELF analysis provides a detailed view of the electronic distribution at the adsorbate–adsorbent interface, revealing areas of high electronic density indicative of possible chemical interaction sites. In the case of the horizontal orientation of MB, the ELF analysis suggests that π–π interactions between the aromatic rings of MB and graphene/GO are prominent, with the possible participation of hydrogen bonds between the oxygenated groups of GO and the methyl groups of MB. This type of interaction contributes to a more stable and stronger adsorption, which is crucial for adsorption efficiency.

The experimental and theoretical results of this study demonstrate that the functionalization of graphene with hydroxyl and epoxy groups significantly increases its efficiency as an adsorbent material, not only by improving the existing physical interactions (π–π stacking), but also by introducing new chemical interactions (hydrogen bonds). From the experimental results, it can be inferred that the modified GO exhibits optimal adsorption capacity under alkaline conditions and at elevated temperatures, suggesting that adsorption is an endothermic process and that the interactions between GO and MB are strengthened with increasing temperature. The influence of pH can be attributed to changes in the surface charge of GO and the ionization of MB, favoring adsorption through electrostatic interactions in alkaline environments. Furthermore, it was observed that the adsorption capacity increases with the initial concentration of MB until reaching equilibrium, indicating the saturation of the available adsorption sites in the GO. The contact time also plays a crucial role, as the amount of adsorbed MB was found to increase with time until equilibrium is reached, reflecting the dynamics of the interaction between the adsorbate and the adsorbent. From the theoretical perspective, the use of density functional theory (DFT) allowed for GO-MB interactions to be explored at the atomic and molecular levels, providing a deep understanding of the underlying mechanisms. DFT calculations revealed that oxygen functional groups on GO facilitate the more effective adsorption of MB by creating specific adsorption sites and enhancing electrostatic and hydrogen bonding interactions. A simulation of the modified GO structures showed how surface modifications alter the electronic properties of the material, directly influencing its interaction with MB. These theoretical results not only corroborate the experimental findings but also provide a framework to understand how the structural and functional characteristics of GO contribute to its effectiveness as an adsorbent material.

Figure 17 illustrates the proposed mechanism for the adsorption of methylene blue on graphene oxide. The figure represents the three key interactions: electrostatic interactions, hydrogen bond formation, and π–π interactions.

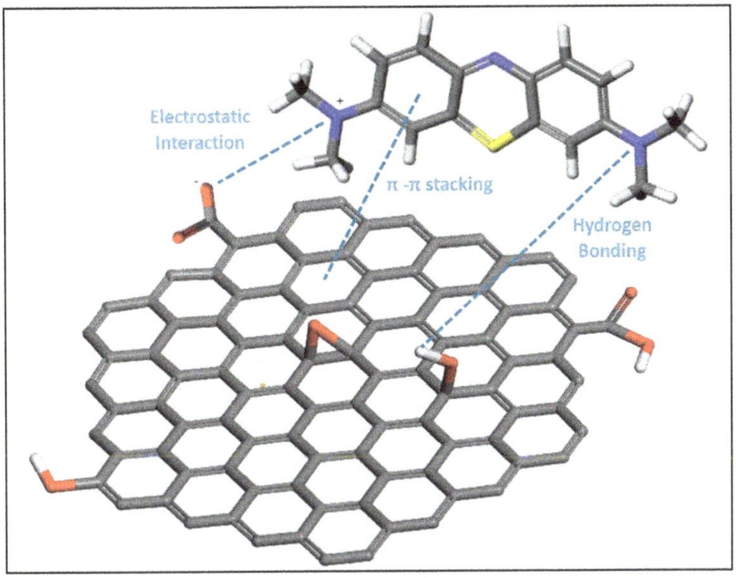

Figure 17. Proposed mechanism for MB adsorption on GO. The gray sticks are C bonds, the red sticks are O bonds, the white sticks are H bonds, the blue sticks are N bonds, and the yellow sticks are S bonds.

5. Conclusions

This study furthers the understanding of the adsorption capabilities of modified graphene oxide (GO) for the removal of methylene blue (MB) dye from aqueous solutions, focusing on the implications for water purification and environmental remediation. The successful modification of GO via the incorporation of oxygen functional groups, such as hydroxyls and epoxides, transforms both the structure and surface chemistry of the material. This transformation, evidenced through characterization techniques such as XRD and SEM/EDS/FTIR, results in an increase in the interlaminar distance and the introduction of surface defects, which are crucial to improve the adsorption capacity of GO.

The research reveals that the underlying adsorption mechanisms benefit significantly from the presence of these functional groups, which not only increase the hydrophilicity of GO, facilitating its dispersion in aqueous media, but also provide specific sites for the adsorption of MB molecules through hydrogen bonds and electrostatic interactions. Furthermore, the influence of factors such as pH, temperature, the initial concentration of MB, and contact time on the adsorption efficiency was highlighted, showing that alkaline conditions and elevated temperatures favor the interaction between GO and MB, highlighting the endothermic character of the adsorption process.

From a theoretical perspective, density functional theory (DFT) calculations provide a detailed understanding of the molecular interactions between GO and MB, corroborating and complementing the experimental findings. These calculations suggest that surface modifications to GO not only favor the creation of specific adsorption sites but also alter the electronic properties of the material, thus improving its interaction with MB. This detailed understanding of GO-MB interactions at the molecular level lays the foundation for the rational design of carbon-based materials with optimized adsorption properties for specific applications in water treatment and environmental remediation.

Finally, this study demonstrates the potential of modified GO as an efficient adsorbent material for the removal of organic contaminants from aqueous solutions, marking a significant advance in the search for sustainable and effective solutions to the global challenges of water purification and environmental remediation. The findings underline the importance of careful chemical modification and the optimization of operating conditions to maximize adsorption efficiency, opening new avenues for the future development of more advanced water treatment technologies.

Supplementary Materials: The following supporting information can be downloaded at: https://www.mdpi.com/article/10.3390/nano14080714/s1, Figure S1. Point of zero charge (PZC) of GO.

Author Contributions: Conceptualization, X.J.-F.; methodology, X.J.-F. and G.C.; software, X.J.-F.; validation, X.J.-F. and G.C.; formal analysis, X.J.-F.; investigation, X.J.-F. and G.C.; resources, X.J.-F.; data curation, X.J.-F. and G.C.; writing—original draft preparation, X.J.-F. and G.C.; writing—review and editing, X.J.-F.; visualization, X.J.-F. and G.C.; supervision, X.J.-F.; project administration, X.J.-F.; funding acquisition, X.J.-F. All authors have read and agreed to the published version of the manuscript.

Funding: This research was funded by Universidad Técnica Particular de Loja.

Data Availability Statement: Data are available from the authors upon reasonable request.

Acknowledgments: The authors would like to thank Mónica Vega, Mayra Montalván and Xavier Cattoen for the XRD, FTIR and SEM/EDS measurements, respectively.

Conflicts of Interest: The authors declare no conflicts of interest.

References

1. Xiao, G.N.; Man, S.Q. Surface-enhanced Raman scattering of methylene blue adsorbed on cap-shaped silver nanoparticles. *Chem. Phys. Lett.* **2007**, *447*, 305–309. [CrossRef]
2. Carmen, Z.; Daniel, S. Textile Organic Dyes—Characteristics, Polluting Effects and Separation/Elimination Procedures from Industrial Effluents—A Critical Overview. In *Organic Pollutants Ten Years After the Stockholm Convention—Environmental and Analytical Update*; IntechOpen: London, UK, 2012; ISBN 978-953-307-917-2.
3. Dutta, A.K.; Ghorai, U.K.; Chattopadhyay, K.K.; Banerjee, D. Removal of textile dyes by carbon nanotubes: A comparison between adsorption and UV assisted photocatalysis. *Phys. E Low-Dimensional Syst. Nanostruct.* **2018**, *99*, 6–15. [CrossRef]
4. Samsami, S.; Mohamadi, M.; Sarrafzadeh, M.H.; Rene, E.R.; Firoozbahr, M. Recent advances in the treatment of dye-containing wastewater from textile industries: Overview and perspectives. *Process Saf. Environ. Prot.* **2020**, *143*, 138–163. [CrossRef]
5. Manappadan, Z.; Kumar, S.; Joshi, K.; Govindaraja, T.; Krishnamurty, S.; Selvaraj, K. Unravelling the distinct surface interactions of modified graphene nanostructures with methylene blue dye through experimental and computational approaches. *J. Hazard. Mater.* **2020**, *388*, 121755. [CrossRef]
6. Al-Tohamy, R.; Ali, S.S.; Li, F.; Okasha, K.M.; Mahmoud, Y.A.G.; Elsamahy, T.; Jiao, H.; Fu, Y.; Sun, J. A critical review on the treatment of dye-containing wastewater: Ecotoxicological and health concerns of textile dyes and possible remediation approaches for environmental safety. *Ecotoxicol. Environ. Saf.* **2022**, *231*, 113160. [CrossRef]
7. Sophia, A.C.; Lima, E.C. Removal of emerging contaminants from the environment by adsorption. *Ecotoxicol. Environ. Saf.* **2018**, *150*, 1–17. [CrossRef] [PubMed]
8. Burakov, A.E.; Galunin, E.V.; Burakova, I.V.; Kucherova, A.E.; Agarwal, S.; Tkachev, A.G.; Gupta, V.K. Adsorption of heavy metals on conventional and nanostructured materials for wastewater treatment purposes: A review. *Ecotoxicol. Environ. Saf.* **2018**, *148*, 702–712. [CrossRef] [PubMed]
9. Sahoo, T.R.; Prelot, B. Adsorption processes for the removal of contaminants from wastewater: The perspective role of nanomaterials and nanotechnology. In *Nanomaterials for the Detection and Removal of Wastewater Pollutants*; Elsevier: Amsterdam, The Netherlands, 2020; pp. 161–222, ISBN 9780128184899.
10. Liu, T.; Wang, Z.; Wang, X.; Yang, G.; Liu, Y. Adsorption-photocatalysis performance of polyaniline/dicarboxyl acid cellulose@graphene oxide for dye removal. *Int. J. Biol. Macromol.* **2021**, *182*, 492–501. [CrossRef]
11. Cossu, R.; Ehrig, H.-J.; Muntoni, A. Physical–Chemical Leachate Treatment. In *Solid Waste Landfilling*; Elsevier: Amsterdam, The Netherlands, 2018; pp. 575–632, ISBN 978-0-12-818336-6.
12. Tan, K.L.; Hameed, B.H. Insight into the adsorption kinetics models for the removal of contaminants from aqueous solutions. *J. Taiwan Inst. Chem. Eng.* **2017**, *74*, 25–48. [CrossRef]
13. Ray, S.S.; Gusain, R.; Kumar, N. Adsorption in the context of water purification. In *Carbon Nanomaterial-Based Adsorbents for Water Purification*; Elsevier: Amsterdam, The Netherlands, 2020; pp. 67–100, ISBN 978-0-12-821959-1.
14. Kecili, R.; Hussain, C.M. Mechanism of adsorption on nanomaterials. In *Nanomaterials in Chromatography: Current Trends in Chromatographic Research Technology and Techniques*; Elsevier: Amsterdam, The Netherlands, 2018; pp. 89–115, ISBN 9780128127926.
15. Rout, D.R.; Jena, H.M.; Baigenzhenov, O.; Hosseini-Bandegharaei, A. Graphene-based materials for effective adsorption of organic and inorganic pollutants: A critical and comprehensive review. *Sci. Total Environ.* **2023**, *863*, 160871. [CrossRef]

16. Atkovska, K.; Paunovik, P.; Dimitrov, A.; Lisichkov, K.; Alghuthaymi, M.; Grozdanov, A. Graphene and activated graphene as adsorbents for removal of heavy metals from water resources. In *Carbon Nanomaterials for Agri-Food and Environmental Applications*; Elsevier: Amsterdam, The Netherlands, 2019; pp. 177–191, ISBN 9780128197868.
17. Minitha, C.R.; Lalitha, M.; Jeyachandran, Y.L.; Senthilkumar, L.; Rajendra Kumar, R.T. Adsorption behaviour of reduced graphene oxide towards cationic and anionic dyes: Co-action of electrostatic and π–π interactions. *Mater. Chem. Phys.* **2017**, *194*, 243–252. [CrossRef]
18. Wang, G.; Li, G.; Huan, Y.; Hao, C.; Chen, W. Acrylic acid functionalized graphene oxide: High-efficient removal of cationic dyes from wastewater and exploration on adsorption mechanism. *Chemosphere* **2020**, *261*, 127736. [CrossRef] [PubMed]
19. Nissanka, B.; Kottegoda, N.; Jayasundara, D.R. Probing structural variations of graphene oxide and reduced graphene oxide using methylene blue adsorption method. *J. Mater. Sci.* **2020**, *55*, 1996–2005. [CrossRef]
20. Wang, H.; Yi, L.; Huang, F.; Huang, Q.; Zhou, T. Facile synthesis of graphene nanosheets on wastewater sediments for high efficient adsorption of methylene blue. *Sep. Purif. Technol.* **2024**, *337*, 126366. [CrossRef]
21. Verma, D.; Goh, K.L. Functionalized Graphene-Based Nanocomposites for Energy Applications. In *Functionalized Graphene Nanocomposites and Their Derivatives: Synthesis, Processing and Applications*; Elsevier: Amsterdam, The Netherlands, 2019; pp. 219–243. [CrossRef]
22. Hu, H.; Xu, K. Physicochemical technologies for HRPs and risk control. In *High-Risk Pollutants in Wastewater*; Elsevier: Amsterdam, The Netherlands, 2020; pp. 169–207, ISBN 9780128164488.
23. Ben Gouider Trabelsi, A.; Kusmartsev, F.V.; Kusmartseva, A.; Alkallas, F.H.; Alfaify, S.; Shkir, M. Raman Spectroscopy Imaging of Exceptional Electronic Properties in Epitaxial Graphene Grown on SiC. *Nanomaterials* **2020**, *10*, 2234. [CrossRef] [PubMed]
24. Baig, N.; Ihsanullah; Sajid, M.; Saleh, T.A. Graphene-based adsorbents for the removal of toxic organic pollutants: A review. *J. Environ. Manag.* **2019**, *244*, 370–382. [CrossRef] [PubMed]
25. Kushwaha, A.; Goswami, S.; Hans, N.; Singh, A.; Vishwakarma, H.S.; Devi, G.; Mishra, P.; Bhan, U.; Hussain, C.M. Sorption of pharmaceutical and personal care products from the wastewater by carbonaceous materials. In *Emerging Trends to Approaching Zero Waste: Environmental and Social Perspectives*; Elsevier: Amsterdam, The Netherlands, 2021; pp. 175–196, ISBN 9780323854030.
26. Kyzas, G.Z.; Deliyanni, E.A.; Matis, K.A. Graphene oxide and its application as an adsorbent for wastewater treatment. *J. Chem. Technol. Biotechnol.* **2014**, *89*, 196–205. [CrossRef]
27. Rivera Tito, H.A.; Hernandez-Sosa, G.; Cucinotta, F.; Huang, X.; Quintana Caceda, M.E. Photoluminescent graphene oxide porous particles in solution under environmental conditions produced by hydrothermal treatment. *Mater. Today Commun.* **2019**, *20*, 100621. [CrossRef]
28. Dideikin, A.T.; Vul', A.Y. Graphene oxide and derivatives: The place in graphene family. *Front. Phys.* **2019**, *6*, 427630. [CrossRef]
29. Kumar, N.; Verma, S.; Kumar, P.; Khan, A.A.; Park, J.; Srivastava, V.C. Efficient one-pot two-step electrochemical synthesis of highly oxidized graphene oxide for enhanced energy and environmental applications: Structure elucidation through DFT simulations. *Carbon* **2024**, *218*, 118722. [CrossRef]
30. Gascho, J.L.S.; Costa, S.F.; Recco, A.A.C.; Pezzin, S.H. Graphene oxide films obtained by vacuum filtration: X-ray diffraction evidence of crystalline reorganization. *J. Nanomater.* **2019**, *2019*, 12–16. [CrossRef]
31. Zhu, J.; Wen, K.; Zhang, P.; Wang, Y.; Ma, L.; Xi, Y. Microporous and Mesoporous Materials. *Microporous Mesoporous Mater.* **2017**, *242*, 256–263. [CrossRef]
32. Fraga, T.J.M.; Carvalho, M.N.; Ghislandi, M.G.; Da Motta Sobrinho, M.A. Functionalized graphene-based materials as innovative adsorbents of organic pollutants: A concise overview. *Braz. J. Chem. Eng.* **2019**, *36*, 1–31. [CrossRef]
33. Esteban-Arranz, A.; Compte-Tordesillas, D.; Muñoz-Andrés, V.; Pérez-Cadenas, M.; Guerrero-Ruiz, A. Effect of surface, structural and textural properties of graphenic materials over cooperative and synergetic adsorptions of two chloroaromatic compounds from aqueous solution. *Catal. Today* **2018**, *301*, 104–111. [CrossRef]
34. Vacchi, I.A.; Ménard-Moyon, C.; Bianco, A. Chemical Functionalization of Graphene Family Members. *Phys. Sci. Rev.* **2017**. [CrossRef]
35. Yang, X.; Chen, C.; Li, J.; Zhao, G.; Ren, X.; Wang, X. Erratum: Graphene oxide-iron oxide and reduced graphene oxide-iron oxide hybrid materials for the removal of organic and inorganic pollutants. *RSC Adv.* **2022**, *12*, 28944. [CrossRef]
36. Dreyer, D.R.; Murali, S.; Zhu, Y.; Ruoff, R.S.; Bielawski, C.W. Reduction of graphite oxide using alcohols. *J. Mater. Chem.* **2011**, *21*, 3443–3447. [CrossRef]
37. Luo, D.; Zhang, G.; Liu, J.; Sun, X. Evaluation criteria for reduced graphene oxide. *J. Phys. Chem. C* **2011**, *115*, 11327–11335. [CrossRef]
38. Meza, C.L.; Sun Kou, R.; Castro Arroyo, T. Biosorción del colorante azul de metileno usando los cladodios de la tuna (*Opuntia ficus indica*). *Rev. la Soc. Química del Perú* **2020**, *86*, 231–245. [CrossRef]
39. Justino, D.D.; Alves, M.O.; Galvão, B.R.L.; Santamaría, R.; De Sousa, F.B.; Ortega, P.F.R. The effects of functionalization on graphene oxide for organic dye adsorption: An experimental-theoretical study using electronic structure calculations and statistical mechanical modeling. *J. Mol. Liq.* **2023**, *387*, 122612. [CrossRef]
40. Dayana Priyadharshini, S.; Manikandan, S.; Kiruthiga, R.; Rednam, U.; Babu, P.S.; Subbaiya, R.; Karmegam, N.; Kim, W.; Govarthanan, M. Graphene oxide-based nanomaterials for the treatment of pollutants in the aquatic environment: Recent trends and perspectives—A review. *Environ. Pollut.* **2022**, *306*, 119377. [CrossRef] [PubMed]

41. Mandeep; Gulati, A.; Kakkar, R. Graphene-based adsorbents for water remediation by removal of organic pollutants: Theoretical and experimental insights. *Chem. Eng. Res. Des.* **2020**, *153*, 21–36. [CrossRef]
42. Fraga, T.J.M.; Ghislandi, M.G.; Carvalho, M.N.; da Motta Sobrinho, M.A. One step forward: How can functionalization enhance the adsorptive properties of graphene towards metallic ions and dyes? *Environ. Res.* **2020**, *184*, 109362. [CrossRef] [PubMed]
43. Malloum, A.; Adegoke, K.A.; Ighalo, J.O.; Conradie, J.; Ohoro, C.R.; Amaku, J.F.; Oyedotun, K.O.; Maxakato, N.W.; Akpomie, K.G.; Okeke, E.S.; et al. Computational methods for adsorption study in wastewater treatment. *J. Mol. Liq.* **2023**, *390*, 123008. [CrossRef]
44. Aguayo-Villarreal, I.A.; Cortes-Arriagada, D.; Rojas-Mayorga, C.K.; Pineda-Urbina, K.; Muñiz-Valencia, R.; González, J. Importance of the interaction adsorbent –adsorbate in the dyes adsorption process and DFT modeling. *J. Mol. Struct.* **2020**, *1203*, 127398. [CrossRef]
45. Landin-Sandoval, V.J.; Mendoza-Castillo, D.I.; Seliem, M.K.; Mobarak, M.; Villanueva-Mejia, F.; Bonilla-Petriciolet, A.; Navarro-Santos, P.; Reynel-Ávila, H.E. Physicochemical analysis of multilayer adsorption mechanism of anionic dyes on lignocellulosic biomasses via statistical physics and density functional theory. *J. Mol. Liq.* **2021**, *322*, 114511. [CrossRef]
46. Tee, W.T.; Loh, N.Y.L.; Lai, K.C.; Hiew, B.Y.Z.; Gan, S.; Lee, L.Y. Application of 3D heteroatom-doped graphene in adsorptive removal of water pollutants: Review on hydrothermal synthesis and its influencing factors. *Sep. Purif. Technol.* **2023**, *320*, 124072. [CrossRef]
47. Yan, J.A.; Chou, M.Y. Oxidation functional groups on graphene: Structural and electronic properties. *Phys. Rev. B Condens. Matter Mater. Phys.* **2010**, *82*, 125403. [CrossRef]
48. Laref, A.; Ahmed, A.; Bin-Omran, S.; Luo, S.J. First-principle analysis of the electronic and optical properties of boron and nitrogen doped carbon mono-layer graphenes. *Carbon* **2015**, *81*, 179–192. [CrossRef]
49. Tene, T.; Usca, G.T.; Guevara, M.; Molina, R.; Veltri, F.; Arias, M.; Caputi, L.S.; Gomez, C.V. Toward large-scale production of oxidized graphene. *Nanomaterials* **2020**, *10*, 279. [CrossRef]
50. Jaramillo-Fierro, X.; Ramón, J.; Valarezo, E. Cyanide Removal by $ZnTiO_3/TiO_2/H_2O_2$/UVB System: A Theoretical-Experimental Approach. *Int. J. Mol. Sci.* **2023**, *24*, 16446. [CrossRef]
51. Lamaiphan, N.; Sakaew, C.; Sricharoen, P.; Nuengmatcha, P.; Chanthai, S.; Limchoowong, N. Highly efficient ultrasonic-assisted preconcentration of trace amounts of Ag(I), Pb(II), and Cd(II) ions using 3-mercaptopropyl trimethoxysilane-functionalized graphene oxide–magnetic nanoparticles. *J. Korean Ceram. Soc.* **2021**, *58*, 314–329. [CrossRef]
52. Bakatula, E.N.; Richard, D.; Neculita, C.M.; Zagury, G.J. Determination of point of zero charge of natural organic materials. *Environ. Sci. Pollut. Res.* **2018**, *25*, 7823–7833. [CrossRef] [PubMed]
53. Benmessaoud, A.; Nibou, D.; Mekatel, E.H.; Amokrane, S. A Comparative Study of the Linear and Non-Linear Methods for Determination of the Optimum Equilibrium Isotherm for Adsorption of Pb2+ Ions onto Algerian Treated Clay. *Iran. J. Chem. Chem. Eng.* **2020**, *39*, 153–171. [CrossRef]
54. Eke-emezie, N.; Etuk, B.R.; Akpan, O.P.; Chinweoke, O.C. Cyanide removal from cassava wastewater onto H_3PO_4 activated periwinkle shell carbon. *Appl. Water Sci.* **2022**, *12*, 157. [CrossRef]
55. Pirmoradi, M.; Hashemian, S.; Shayesteh, M.R. Kinetics and thermodynamics of cyanide removal by ZnO@NiO nanocrystals. *Trans. Nonferrous Met. Soc. China* **2017**, *27*, 1394–1403. [CrossRef]
56. Noroozi, R.; Al-Musawi, T.J.; Kazemian, H.; Kalhori, E.M.; Zarrabi, M. Removal of cyanide using surface-modified Linde Type-A zeolite nanoparticles as an efficient and eco-friendly material. *J. Water Process Eng.* **2018**, *21*, 44–51. [CrossRef]
57. Inyinbor, A.A.; Adekola, F.A.; Olatunji, G.A. Kinetics, isotherms and thermodynamic modeling of liquid phase adsorption of Rhodamine B dye onto Raphia hookerie fruit epicarp. *Water Resour. Ind.* **2016**, *15*, 14–27. [CrossRef]
58. Tran, H.N.; You, S.J.; Hosseini-Bandegharaei, A.; Chao, H.P. Mistakes and inconsistencies regarding adsorption of contaminants from aqueous solutions: A critical review. *Water Res.* **2017**, *120*, 88–116. [CrossRef]
59. Zhou, X.; Zhou, X. The Unit Problem in the Thermodynamic Calculation of Adsorption Using the Langmuir Equation. *Chem. Eng. Commun.* **2014**, *201*, 1459–1467. [CrossRef]
60. Jaramillo-Fierro, X.; González, S.; Montesdeoca-Mendoza, F.; Medina, F. Structuring of $ZnTiO_3/TiO_2$ adsorbents for the removal of methylene blue, using zeolite precursor clays as natural additives. *Nanomaterials* **2021**, *11*, 898. [CrossRef] [PubMed]
61. Kresse, G.; Furthmüller, J. Efficient iterative schemes for ab initio total-energy calculations using a plane-wave basis set. *Phys. Rev. B Condens. Matter Mater. Phys.* **1996**, *54*, 11169–11186. [CrossRef]
62. Wang, V.; Xu, N.; Liu, J.C.; Tang, G.; Geng, W.T. VASPKIT: A user-friendly interface facilitating high-throughput computing and analysis using VASP code. *Comput. Phys. Commun.* **2021**, *267*, 108033. [CrossRef]
63. Sujith, C.P.; Joseph, S.; Mathew, T.; Mathew, V. First-principles investigation of structural, electronic and optical properties of quasi-one-dimensional barium cadmium chalcogenides Ba2CdX3 (X = S, Se, Te) using HSE06 and GGA-PBE functionals. *J. Phys. Chem. Solids* **2022**, *161*, 110488. [CrossRef]
64. Perdew, J.P.; Burke, K.; Ernzerhof, M. Generalized gradient approximation made simple. *Phys. Rev. Lett.* **1996**, *77*, 3865–3868. [CrossRef]
65. Kohn, W.; Sham, L.J. Quantum density oscillations in an inhomogeneous electron gas. *Phys. Rev.* **1965**, *137*, A1697. [CrossRef]

66. Rasoolidanesh, M.; Astaraki, M.; Mostafavi, M.; Rezvani, M.; Darvish Ganji, M. Toward efficient enantioseparation of ibuprofen isomers using chiral BNNTs: Dispersion corrected DFT calculations and DFTB molecular dynamic simulations. *Diam. Relat. Mater.* **2021**, *119*, 108561. [CrossRef]
67. Rezvani, M.; Darvish Ganji, M.; Faghihnasiri, M. Encapsulation of lamivudine into single walled carbon nanotubes: A vdW-DF study. *Phys. E Low-Dimensional Syst. Nanostruct.* **2013**, *52*, 27–33. [CrossRef]
68. Tavassoli Larijani, H.; Darvish Ganji, M.; Jahanshahi, M. Trends of amino acid adsorption onto graphene and graphene oxide surfaces: A dispersion corrected DFT study. *RSC Adv.* **2015**, *5*, 92843–92857. [CrossRef]
69. Sun, Y.; Tang, X.; Bao, H.; Yang, Z.; Ma, F. The effects of hydroxide and epoxide functional groups on the mechanical properties of graphene oxide and its failure mechanism by molecular dynamics simulations. *RSC Adv.* **2020**, *10*, 29610–29617. [CrossRef]
70. Jaramillo-Fierro, X.; Cuenca, G.; Ramón, J. The Effect of La^{3+} on the Methylene Blue Dye Removal Capacity of the La/$ZnTiO_3$ Photocatalyst, a DFT Study. *Nanomaterials* **2022**, *12*, 3137. [CrossRef] [PubMed]
71. Monkhorst, H.J.; Pack, J.D. Special points for Brillouin-zone integrations. *Phys. Rev. B* **1976**, *13*, 5188–5192. [CrossRef]
72. Guo, W.; She, Z.; Yang, S.; Xue, H.; Zhang, X. Understanding the influence of Lu, La and Ga active elements on the bonding properties of Sn/SiO_2 interfaces from first principle calculations. *Ceram. Int.* **2020**, *46*, 24737–24743. [CrossRef]
73. Chang, X.; Li, X.; Xue, Q. Sensing mechanism of acetone adsorption on charged ZnO and ZnSe surfaces: Insights from DFT calculations. *Mater. Today Commun.* **2022**, *31*, 103238. [CrossRef]
74. Yang, X.; Yan, Z.; Gao, G.; Gao, R.; Zhang, T.; Su, H.; Tian, M.; Wang, S. Understanding of Photocatalytic Partial Oxidation of Methanol to Methyl Formate on Surface Doped La(Ce)-TiO_2: Experiment and Dft Calculation. *SSRN Electron. J.* **2022**, *411*, 31–40. [CrossRef]
75. Lai, W.; Zhang, K.; Shao, P.; Yang, L.; Ding, L.; Pavlostathis, S.G.; Shi, H.; Zou, L.; Liang, D.; Luo, X. Optimization of adsorption configuration by DFT calculation for design of adsorbent: A case study of palladium ion-imprinted polymers. *J. Hazard. Mater.* **2019**, *379*, 120791. [CrossRef]
76. Zhang, H.; Huang, W.; Wang, W.C.; Shi, X.Q. Ionicity of bonding in elemental solids. *J. Phys. Commun.* **2018**, *2*, 115009. [CrossRef]
77. Kumar, P.S.V.; Raghavendra, V.; Subramanian, V. Bader's Theory of Atoms in Molecules (AIM) and its Applications to Chemical Bonding. *J. Chem. Sci.* **2016**, *128*, 1527–1536. [CrossRef]
78. Savin, A.; Nesper, R.; Wengert, S.; Fässler, T.F. ELF: The Electron Localization Function. *Angew. Chem. Int.* **1997**, *36*, 1808–1832. [CrossRef]
79. Wen, C.; Zhu, Y.J.; Kanbara, T.; Zhu, H.Z.; Xiao, C.F. Effects of I and F codoped TiO2 on the photocatalytic degradation of methylene blue. *Desalination* **2009**, *249*, 621–625. [CrossRef]
80. Arabpour, A.; Dan, S.; Hashemipour, H. Preparation and optimization of novel graphene oxide and adsorption isotherm study of methylene blue. *Arab. J. Chem.* **2021**, *14*, 103003. [CrossRef]
81. Yan, H.; Tao, X.; Yang, Z.; Li, K.; Yang, H.; Li, A.; Cheng, R. Effects of the oxidation degree of graphene oxide on the adsorption of methylene blue. *J. Hazard. Mater.* **2014**, *268*, 191–198. [CrossRef] [PubMed]
82. Savazzi, F.; Risplendi, F.; Mallia, G.; Harrison, N.M.; Cicero, G. Unravelling Some of the Structure-Property Relationships in Graphene Oxide at Low Degree of Oxidation. *J. Phys. Chem. Lett.* **2018**, *9*, 1746–1749. [CrossRef] [PubMed]
83. Dan, S.; Bagheri, H.; Shahidizadeh, A.; Hashemipour, H. Performance of graphene Oxide/SiO2 Nanocomposite-based: Antibacterial Activity, dye and heavy metal removal. *Arab. J. Chem.* **2023**, *16*, 104450. [CrossRef]
84. Gao, Y.; Ren, X.; Tan, X.; Hayat, T.; Alsaedi, A.; Chen, C. Insights into key factors controlling GO stability in natural surface waters. *J. Hazard. Mater.* **2017**, *335*, 56–65. [CrossRef] [PubMed]
85. Soltani, A.; Faramarzi, M.; Parsa, S.A.M. A review on adsorbent parameters for removal of dye products from industrial wastewater. *Water Qual. Res. J.* **2021**, *56*, 181–193. [CrossRef]
86. Bhunia, K.; Chandra, M.; Kumar Sharma, S.; Pradhan, D.; Kim, S.J. A critical review on transition metal phosphide based catalyst for electrochemical hydrogen evolution reaction: Gibbs free energy, composition, stability, and true identity of active site. *Coord. Chem. Rev.* **2023**, *478*, 214956. [CrossRef]
87. De Araujo, C.M.B.; De Assis Filho, R.B.; Baptisttella, A.M.S.; Do Nascimento, G.F.O.; Da Costa, G.R.B.; Carvalho, M.N.; Ghislandi, M.G.; Sobrinho, M.A.D.M. Systematic study of graphene oxide production using factorial design techniques and its application to the adsorptive removal of methylene blue dye in aqueous medium. *Mater. Res. Express* **2018**, *5*, 065042. [CrossRef]
88. Moon, H.S.; Lee, J.H.; Kwon, S.; Kim, I.T.; Lee, S.G. Mechanisms of na adsorption on graphene and graphene oxide: Density functional theory approach. *Carbon Lett.* **2015**, *16*, 116–120. [CrossRef]
89. Hinuma, Y.; Pizzi, G.; Kumagai, Y.; Oba, F.; Tanaka, I. Band structure diagram paths based on crystallography. *Comput. Mater. Sci.* **2017**, *128*, 140–184. [CrossRef]
90. Lin, C.S.; Zhou, A.Y.; Cheng, W.D.; Ye, N.; Chai, G.L. Atom-Resolved Analysis of Birefringence of Nonlinear Optical Crystals by Bader Charge Integration. *J. Phys. Chem. C* **2019**, *123*, 31183–31189. [CrossRef]
91. Tene, T.; Guevara, M.; Valarezo, A.; Salguero, O.; Arias Arias, F.; Arias, M.; Scarcello, A.; Caputi, L.S.; Vacacela Gomez, C. Drying-time study in graphene oxide. *Nanomaterials* **2021**, *11*, 1035. [CrossRef]
92. Siong, V.L.E.; Lee, K.M.; Juan, J.C.; Lai, C.W.; Tai, X.H.; Khe, C.S. Removal of methylene blue dye by solvothermally reduced graphene oxide: A metal-free adsorption and photodegradation method. *RSC Adv.* **2019**, *9*, 37686–37695. [CrossRef]

93. Jahan, N.; Roy, H.; Reaz, A.H.; Arshi, S.; Rahman, E.; Firoz, S.H.; Islam, M.S. A comparative study on sorption behavior of graphene oxide and reduced graphene oxide towards methylene blue. *Case Stud. Chem. Environ. Eng.* **2022**, *6*, 100239. [CrossRef]
94. Wang, H.; Cao, T.; Zhou, Y.; Liu, L.; Zhang, X.; Tong, Z. A facile approach to synthesis methylene blue/reduced graphene oxide nanocomposite and simultaneous determination of dopamine and uric acid. *J. Appl. Electrochem.* **2022**, *52*, 1067–1080. [CrossRef]
95. Achawi, S.; Feneon, B.; Pourchez, J.; Forest, V. Structure–activity relationship of graphene-based materials: Impact of the surface chemistry, surface specific area and lateral size on their in vitro toxicity. *Nanomaterials* **2021**, *11*, 2963. [CrossRef]
96. Chaabane, L.; Beyou, E.; Luneau, D.; Baouab, M.H.V. Functionalization of graphene oxide sheets with magnetite nanoparticles for the adsorption of copper ions and investigation of its potential catalytic activity toward the homocoupling of alkynes under green conditions. *J. Catal.* **2020**, *388*, 91–103. [CrossRef]
97. Chen, L.; Zhu, Y.; Cui, Y.; Dai, R.; Shan, Z.; Chen, H. Fabrication of starch-based high-performance adsorptive hydrogels using a novel effective pretreatment and adsorption for cationic methylene blue dye: Behavior and mechanism. *Chem. Eng. J.* **2021**, *405*, 126953. [CrossRef]
98. Salazar-Rabago, J.J.; Leyva-Ramos, R.; Rivera-Utrilla, J.; Ocampo-Perez, R.; Cerino-Cordova, F.J. Biosorption mechanism of Methylene Blue from aqueous solution onto White Pine (*Pinus durangensis*) sawdust: Effect of operating conditions. *Sustain. Environ. Res.* **2017**, *27*, 32–40. [CrossRef]
99. Chowdhury, I.; Duch, M.C.; Mansukhani, N.D.; Hersam, M.C.; Bouchard, D. Colloidal properties and stability of graphene oxide nanomaterials in the aquatic environment. *Environ. Sci. Technol.* **2013**, *47*, 6288–6296. [CrossRef]
100. Hua, Z.; Tang, Z.; Bai, X.; Zhang, J.; Yu, L.; Cheng, H. Aggregation and resuspension of graphene oxide in simulated natural surface aquatic environments. *Environ. Pollut.* **2015**, *205*, 161–169. [CrossRef]
101. Batzias, F.A.; Sidiras, D.K. Simulation of dye adsorption by beech sawdust as affected by pH. *J. Hazard. Mater.* **2007**, *141*, 668–679. [CrossRef]
102. Fan, S.; Tang, J.; Wang, Y.; Li, H.; Zhang, H.; Tang, J.; Wang, Z.; Li, X. Biochar prepared from co-pyrolysis of municipal sewage sludge and tea waste for the adsorption of methylene blue from aqueous solutions: Kinetics, isotherm, thermodynamic and mechanism. *J. Mol. Liq.* **2016**, *220*, 432–441. [CrossRef]
103. Badeenezhad, A.; Azhdarpoor, A.; Bahrami, S.; Yousefinejad, S. Removal of methylene blue dye from aqueous solutions by natural clinoptilolite and clinoptilolite modified by iron oxide nanoparticles. *Mol. Simul.* **2019**, *45*, 564–571. [CrossRef]
104. An, F.; Liu, J.; Xu, Z.; Zheng, S. Efficient removal of three dyes using porous covalent triazine frameworks: Adsorption mechanism and role of pore distribution. *Water Sci. Technol.* **2020**, *82*, 3023–3031. [CrossRef]
105. Khatib, K.; Lahmyed, L.; El Azhari, M. Synthesis, Characterization, and Application of Geopolymer/TiO_2 Nanoparticles Composite for Efficient Removal of Cu(II) and Cd(II) Ions from Aqueous Media. *Minerals* **2022**, *12*, 1445. [CrossRef]
106. Xu, J.; Li, S.; Wang, F.; Yang, Z.; Liu, H. Efficient and Enhanced Adsorption of Methylene Blue on Triethanolamine-Modified Graphene Oxide. *J. Chem. Eng. Data* **2019**, *64*, 1816–1825. [CrossRef]
107. Sabzevari, M.; Cree, D.E.; Wilson, L.D. Graphene Oxide-Chitosan Composite Material for Treatment of a Model Dye Effluent. *ACS Omega* **2018**, *3*, 13045–13054. [CrossRef]
108. Li, Y.; Du, Q.; Liu, T.; Peng, X.; Wang, J.; Sun, J.; Wang, Y.; Wu, S.; Wang, Z.; Xia, Y.; et al. Comparative study of methylene blue dye adsorption onto activated carbon, graphene oxide, and carbon nanotubes. *Chem. Eng. Res. Des.* **2013**, *91*, 361–368. [CrossRef]
109. Liu, F.; Chung, S.; Oh, G.; Seo, T.S. Three-dimensional graphene oxide nanostructure for fast and efficient water-soluble dye removal. *ACS Appl. Mater. Interfaces* **2012**, *4*, 922–927. [CrossRef]
110. Bradder, P.; Ling, S.K.; Wang, S.; Liu, S. Dye adsorption on layered graphite oxide. *J. Chem. Eng. Data* **2011**, *56*, 138–141. [CrossRef]
111. Liu, T.; Li, Y.; Du, Q.; Sun, J.; Jiao, Y.; Yang, G.; Wang, Z.; Xia, Y.; Zhang, W.; Wang, K.; et al. Adsorption of methylene blue from aqueous solution by graphene. *Colloids Surf. B Biointerfaces* **2012**, *90*, 197–203. [CrossRef]
112. Arias, F.A.; Guevara, M.; Tene, T.; Angamarca, P.; Molina, R.; Valarezo, A.; Salguero, O.; Gomez, C.V.; Arias, M.; Caputi, L.S. The adsorption of methylene blue on eco-friendly reduced graphene oxide. *Nanomaterials* **2020**, *10*, 681. [CrossRef]
113. Islam, M.A.; Ahmed, M.J.; Khanday, W.A.; Asif, M.; Hameed, B.H. Mesoporous activated coconut shell-derived hydrochar prepared via hydrothermal carbonization-NaOH activation for methylene blue adsorption. *J. Environ. Manag.* **2017**, *203*, 237–244. [CrossRef]
114. Ai, L.; Zhang, C.; Liao, F.; Wang, Y.; Li, M.; Meng, L.; Jiang, J. Removal of methylene blue from aqueous solution with magnetite loaded multi-wall carbon nanotube: Kinetic, isotherm and mechanism analysis. *J. Hazard. Mater.* **2011**, *198*, 282–290. [CrossRef]
115. Li, Y.; Du, Q.; Liu, T.; Sun, J.; Wang, Y.; Wu, S.; Wang, Z.; Xia, Y.; Xia, L. Methylene blue adsorption on graphene oxide/calcium alginate composites. *Carbohydr. Polym.* **2013**, *95*, 501–507. [CrossRef]
116. Wang, P.; Cao, M.; Wang, C.; Ao, Y.; Hou, J.; Qian, J. Kinetics and thermodynamics of adsorption of methylene blue by a magnetic graphene-carbon nanotube composite. *Appl. Surf. Sci.* **2014**, *290*, 116–124. [CrossRef]
117. Tran, H.V.; Hoang, L.T.; Huynh, C.D. An investigation on kinetic and thermodynamic parameters of methylene blue adsorption onto graphene-based nanocomposite. *Chem. Phys.* **2020**, *535*, 110793. [CrossRef]
118. Song, K.; Min, T.; Seo, J.; Ryu, S.; Lee, H.; Wang, Z.; Choi, S.Y.; Lee, J.; Eom, C.B.; Oh, S.H. Electronic and Structural Transitions of $LaAlO_3/SrTiO_3$ Heterostructure Driven by Polar Field-Assisted Oxygen Vacancy Formation at the Surface. *Adv. Sci.* **2021**, *8*, 2002073. [CrossRef]
119. Pourzare, K.; Farhadi, S.; Mansourpanah, Y. Graphene oxide/Co_3O_4 nanocomposite: Synthesis, characterization, and its adsorption capacity for the removal of organic dye pollutants from water. *Acta Chim. Slov.* **2017**, *64*, 945–958. [CrossRef]

120. Moradi, F.; Darvish Ganji, M.; Sarrafi, Y. Remediation of phenol-contaminated water by pristine and functionalized SWCNTs: Ab initio van der Waals DFT investigation. *Diam. Relat. Mater.* **2018**, *82*, 7–18. [CrossRef]
121. Ge, S.; Geng, W.; He, X.; Zhao, J.; Zhou, B.; Duan, L.; Wu, Y.; Zhang, Q. Effect of framework structure, pore size and surface modification on the adsorption performance of methylene blue and Cu^{2+} in mesoporous silica. *Colloids Surf. A Physicochem. Eng. Asp.* **2018**, *539*, 154–162. [CrossRef]

nanomaterials

MDPI

Article

Numerical Simulation and Experimental Study of Deposition Behavior for Cold-Sprayed Nano-Structured HA/70wt.%Ti Composite Coating

Xiao Chen, Chengdi Li *, Shuangxia Zhu, Peiyun Ao and Yao Hu

Xinyu Key Laboratory of Materials Technology and Application for Intelligent Manufacturing, School of Mechanical and Electrical Engineering, XinYu University, Xinyu 338004, China; chenxiaoxyxy@126.com (X.C.); zhushuangxiaxyyx@163.com (S.Z.); aopeiyunxyxy@163.com (P.A.); yaohu1990_xyxy@163.com (Y.H.)

* Correspondence: cdl1988_xyxy@126.com; Tel.: +86-0790-6666108

Abstract: This study employs numerical simulations and experiments to examine the cold spray deposition of nanostructured hydroxyapatite ($Ca_{10}(PO4)_6(OH)_2$, HA)/70wt.%Ti composite particles under different processing conditions, based on the features of nanocomposites that strengthen interfacial adhesion and improve coating interfacial strength. Using ABAQUS/CAE combined with LS-PrePost 4.9-x64 software, the deposition behavior of the composite particles during deposition under various impact velocities was analyzed, along with the stress of the HA and Ti particles within the composite particle. The deposition behavior of both single and multiple composite particles under different gas temperatures was studied through cold spray experiments, and composite coatings were fabricated. The microstructure and phase composition were analyzed using scanning electron microscopy (SEM) and X-ray diffraction (XRD). The results showed that the numerical simulations were consistent with the experimental analyses. As the particle velocity or gas temperature increased, the degree of particle deformation upon deposition became more pronounced, accompanied by phenomena such as cracking or fragmentation and splashing rebound. At a gas temperature of 700 °C, both the bonding density of individual particles and the bonding effectiveness of multi-particle deposits were lower than those achieved at 500 °C. The coating prepared at a gas temperature of 500 °C exhibited a flatter surface, better overall bonding with the Ti interlayer, and higher internal density.

Keywords: cold spray; HA/Ti coating; numerical simulation; deposition behavior

1. Introduction

In addition to demonstrating certain biological properties, bone repair materials must also exhibit excellent mechanical properties and interfacial bonding [1,2]. Hydroxyapatite ($Ca_{10}(PO4)_6(OH)_2$, HA) is commonly employed as a biocoating material for the surface modification of titanium and its alloys due to its biological properties [3,4]. However, the application of pure HA materials in load-bearing areas of the human body is significantly limited due to poor mechanical properties, substantial physicochemical differences between the substrate and the body, and low interfacial bonding caused by issues such as insufficient crystallinity or phase transition/decomposition during the preparation process [5]. Research has demonstrated that in nanocomposites, the introduction of nanoparticles with an elevated specific surface area promotes interfacial strengthening between phases, while grain size reduction leads to corresponding enhancements in interfacial bonding strength [6]. In contrast to conventional micro-HA, nano-HA offers better mechanical performance in tissues and promotes stronger interfacial integration [6,7]. Additionally, due to the superior mechanical properties of Ti, the incorporation of nano-Ti particles into nano-HA can achieve reinforcement and toughening effects, while mitigating residual stress between the coating and substrate to enhance interfacial bonding strength [8].

The current the preparation methods for HA composite coatings primarily include magnetron sputtering [9,10], electrochemical deposition [11,12], laser cladding [13–15], and thermal spraying [3,4,16–19], among others. However, due to the challenges involved in controlling the composition and crystallinity of HA coatings or the presence of significant residual stresses, it is difficult to effectively improve the interfacial bonding strength of the coatings. As a result, the performance of HA composite coatings still struggles to meet the application requirements for bone repair materials. Thus, a novel and highly efficient technique for the fabrication of HA composite coatings must be developed to enable precise control over their microstructure, composition, and crystallinity.

Cold spraying is a coating deposition method based on the low-temperature solid-state plastic deformation of high-kinetic-energy particles upon impact with the substrate surface [20]. It has achieved applicable progress in the preparation of metal [20–23], alloy [24], and ceramic [25] coatings. Moreover, owing to its low-temperature deposition characteristics, cold spraying has emerged as a novel technology for fabricating coatings from thermally sensitive materials, such as WC-Co cermets [26–29], HA bioceramics [30–32], and nano-structured materials [26,29,30,32,33]. Compared with thermal spraying technology, cold spraying can transfer the original microstructure of the powder to the coating, which not only preserves the inherent structure of the sprayed material but also avoids issues such as decomposition, phase transformation, and grain growth in nanostructured materials caused by high-temperature [26]. Furthermore, the compressive stresses induced by cold spray deposition can significantly improve interlayer adhesion. By precisely controlling the deformation and bonding behavior of particles during deposition, it becomes feasible to regulate both coating porosity and interparticle interfacial bonding [27]. This offers a novel approach for fabricating HA composite coatings with diverse compositions and microstructures.

At present, numerous researchers are focusing primarily on experiments on the deposition behavior of cold-sprayed micronstructured HA/Ti (mHA/Ti) composite particles [34]. However, there is relatively limited research on the cold spray deposition behavior of nanostructured HA/Ti (nHA/Ti) composite particles, especially in the area of numerical simulations. Therefore, based on the characteristics of low-temperature deposition and particle impact compaction in cold spraying for enhancing interfacial bonding, as well as the features of nanocomposites that strengthen interfacial adhesion and improve coating interfacial strength through the addition of second-phase nanoparticles, the deposition behavior of nHA/Ti using both experimental and numerical simulation approaches was investigated. The main innovations and research objectives of this study are as follows: (1) dual-nano HA/Ti composite powder was successfully prepared using nano-HA and nano-Ti; (2) Numerical simulations of the deposition deformation of the dual-nano HA/Ti composite particles were conducted at different particle velocities to analyze the deformation mechanism and the stress evolution of both nano-HA and nano-Ti particles during deposition; (3) The bonding behavior during single- and multi-particle deposition of HA/Ti composites was examined using cold spray technology at different gas temperatures, thereby providing experimental validation for the numerical simulation results; (4) nanostructured HA/Ti composite coatings were successfully fabricated via cold spraying under varying gas temperatures, and the underlying coating bonding mechanism was elucidated.

2. Materials and Methods

Needle-shaped nano HA powder with an average particle size of 15 nm (Beijing Dk nano S&T Ltd., Beijing, China) and angular nano Ti powder with a mean particle size of 40 nm (Beijing Dk nano S&T Ltd., China) were selected as raw materials. Agglomeration-sintering method was used to prepare nHA/Ti composite powder with a HA: Ti mass ratio of 3:7. Figure 1 shows the surface morphology, particle size distribution, XRD pattern, and TEM morphology of nHA/Ti composite powder. The surface morphology shows that the nHA/Ti composite particles exhibits a spherical shape (Figure 1a) with a median particle size (D50) of 29.3 μm (Figure 1b). The XRD pattern (Figure 1c) confirms the exclusive presence of HA and Ti phases in the nHA/Ti composite powder, indicating no impurity formation during the preparation process. The TEM morphology of nHA/Ti composite particles (Figure 1d) reveals that the nano HA exhibits a needle-like morphology with sizes ranging between 7 and 22 nm, while the nano Ti displays an angular morphology with sizes of 30–50 nm. High-magnification TEM analysis of the nHA/Ti composite particles (Figure 1e) reveals well-defined lattice fringes, confirming the high crystallinity of both the nano HA and nano Ti starting powders. Furthermore, the absence of halo features in the corresponding selected-area electron diffraction (SAED) pattern (Figure 1f) provides conclusive evidence for the fully crystalline nature of these nanomaterials.

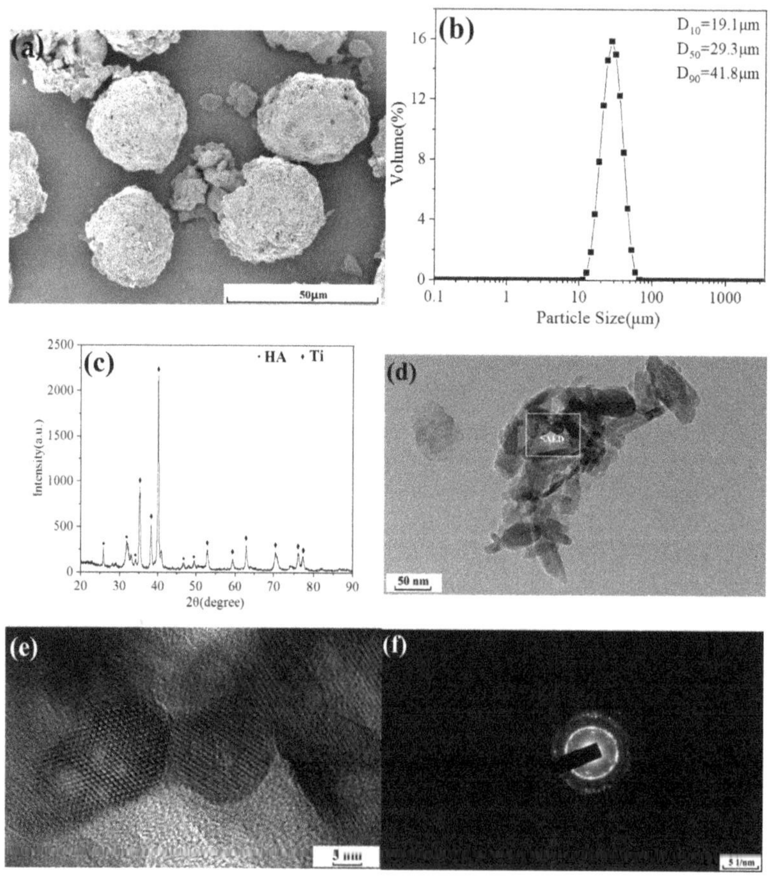

Figure 1. (**a**) Surface morphology; (**b**) particle size distribution; (**c**) XRD pattern; (**d–f**) TEM morphology of nHA/Ti composite powder.

The deposition of individual nHA/Ti splats and coating was performed using a cold spray system, which has been described in detail elsewhere [26,30]. Ti6Al4V was used as the substrate material. Prior to individual splat deposition, the substrate surface was polished. Before coating deposition, the substrate surface was sandblasted using 700 μm alumina particles, followed by deposition of a Ti buffer layer for subsequent nHA/Ti coating deposition. The detailed deposition parameters for both individual splats and coatings are listed in Table 1.

Table 1. The detailed deposition parameters for both individual splats and coatings.

Parameters	Unit	Value
Accelerating nitrogen gas pressure	MPa	2.0
Powder-feeding nitrogen gas pressure	MPa	2.2
Gas temperature for individual nHA/Ti splats	°C	300 ± 10, 500 ± 10, 700 ± 10
Gas temperature for nHA/Ti coating	°C	500 ± 30, 700 ± 30
Gas temperature for Ti buffer layer	°C	200 ± 30
Transverse speed for individual nHA/Ti splats	mm/s	500
Transverse speed for nHA/Ti coating	mm/s	30
Transverse speed for Ti buffer layer	mm/s	150
Spray distance	mm	20

The surface morphology of the nHA/Ti composite powder and individual splats and the surface and cross-sectional morphologies of the nHA/Ti composite coating were characterized by scanning electron microscopy (SEM, KYKY-EM8000, KYKY Technology Co., Ltd., Beijing, China). Transmission electron microscopy (Tokyo TEM, EM-2010, HR, JEOL Japan, Tokyo, Japan) was employed to analyze the microstructure and crystallographic properties of nHA/Ti composite powder. The phase composition of both the nHA/Ti composite powder and coatings was determined by X-ray diffraction (XRD, Bruker D8 Advance, Karlsruhe, Germany).

The particle size distribution of nHA/Ti composite powder was measured using a laser diffraction particle size analysis meter (GSL-1020, Liaoning Instrument Research Institute Co., Ltd., Dandong, China), and the in-flight velocity of composite particles under different temperature conditions during cold spraying was measured using a DPV eVOLUTION system (Tecnar Automation Ltd., Saint-Bruno-de-Montarville, QC, Canada).

In the ABAQUS/CAE module, the number of particle meshes is configured (constrained between 30,000 and 50,000) to generate a spherical particle model. The particle is then discretized using a hexahedral mesh, resulting in a fully meshed spherical particle (as shown in Figure 2a). As the substrate plate is a regular geometric model, hexahedral meshing was applied directly to the substrate, resulting in a fully meshed substrate model (as shown in Figure 2b). Through Python 3.9-based secondary development, the meshed spherical particle was randomly distributed according to a HA/Ti volume ratio of 1:1.59 (corresponding to a mass ratio of 30:70), categorizing the elements stochastically into HA and Ti phases, where yellow particles represent Ti and green particles denote HA (as shown in Figure 3). After importing the meshed composite particle model file into LS-PrePost 4.9-x64 software, a locally sectioned view of the spherical particle model could be obtained, where red particles represent Ti and blue particles denote HA (as shown in Figure 4). The established substrate model (Figure 2b) and randomly distributed spherical particle (Figure 3) were imported into LS-PrePost 4.9-x64 software to obtain an integrated composite particle–substrate model (as shown in Figure 5). For enhanced simulation accuracy, the parameters of nano-HA and nano-Ti particles in the nHA/Ti composite must be input into the LS-PrePost 4.9-x64 software, as both sets of nanoparticle parameters have been reported in the references [35].

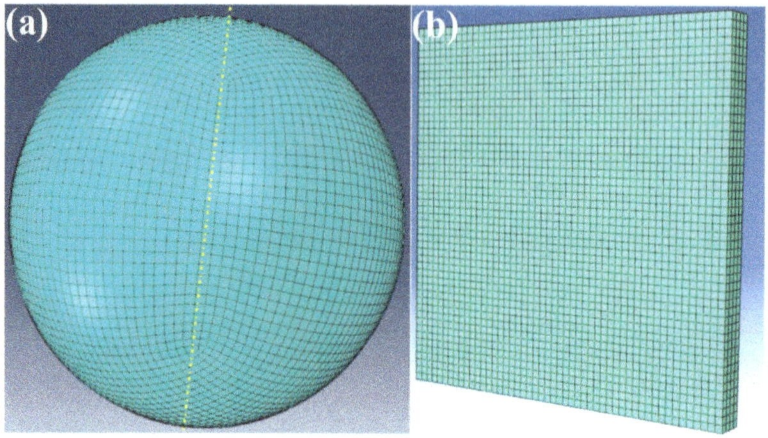

Figure 2. Meshing of (**a**) spherical particle; (**b**) substrate.

Figure 3. The surface morphology of a randomly distributed spherical particle.

Figure 4. A locally sectioned view of the spherical particle model.

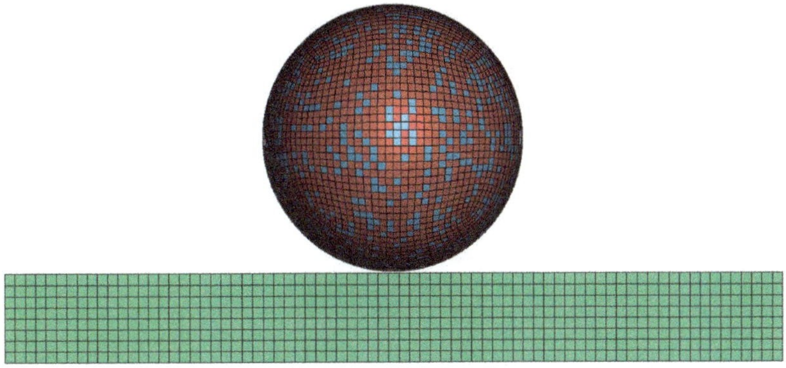

Figure 5. An integrated composite particle–substrate model.

Considering the simulation computation time, the powder feeding pressure for individual composite particle deposition simulation (with particle size set to 20 μm) was set at 2.2 MPa, with a spraying distance of 20 mm, nitrogen as the spraying gas, substrate initial temperature of 300 K, and particle initial temperature of 773 K. The deposition velocities of individual particles were set to 400 m/s, 500 m/s, 600 m/s, 700 m/s, and 800 m/s.

Particles deposited on the substrate surface are typically characterized using the particle compression ratio to quantify the degree of deformation. The particle compression ratio (η) is defined as shown in Equation (1):

$$\eta = (h_i - h_d)/h_i \quad (1)$$

where η is the particle compression ratio, h_i is the initial particle diameter, and h_d is the post-deformation height of the particle.

3. Results and Discussion

3.1. Numerical Simulation of Deposition Behavior for Individual Composite Particles at Varying Particle Velocities

Figure 6 illustrates the simulated instantaneous deformation morphologies of nHA/Ti composite particles after impacting on the substrate at velocities ranging from 400 to 800 m/s. The instantaneous simulation time was 3×10^{-8} s. The side views of the deformed particles are shown in Figure 6a,c,e,g,i, and the top views in Figure 6b,d,f,h,j. When the particles impacted the substrate at a relatively low velocity (400 m/s), the composite particles experienced limited deformation, largely maintaining their hemispherical morphology (Figure 6a). Only a small number of particles surrounding the deformation zone exhibited phenomena such as splattering (indicated by green arrows), rebounding-off (marked by blue arrows), or crack formation (denoted by yellow arrows). With increasing particle velocity, the deformed particles progressively evolved from hemispherical to flattened configurations (Figure 6c), while the splattering, rebounding-off and crack phenomena at the periphery of deformed particles became significantly more pronounced (Figure 6d). As the particle velocity further increased (Figure 6e,g,i), the phenomena of particle rebounding-off, splattering and cracking intensified significantly. Concurrently, the bonding between deformed particles and the substrate deteriorated progressively, leading to the formation of distinct gaps at the particle–substrate interface (indicated by purple arrows in Figure 6g,i), whose dimensions exhibited a positive correlation with impact velocity (Figure 6i). Ultimately, such interfacial failure may cause macroscopic spallation of deposited particles from the substrate surface. The compression ratios for deformed composite particles at different particle velocities (400 m/s to 800 m/s) were 37.0%, 47.0%, 59.9%, 64.3%, and 76.2%, respectively.

Figures 7 and 8 present the stress value curves of eight randomly selected Ti and HA particles from the composite particles deposited under varying particle velocity conditions, respectively. Figure 7 revealed that the stress value of the Ti particle unit rapidly peaked during the initial collision instant, demonstrating the occurrence of adiabatic shear instability (ASI). After sustaining a stabilized stress state for a duration, the stress curve of the Ti particle unit manifested a progressive decline. This phenomenon arose from thermomechanical coupling during impact: plastic work dissipates as thermal energy, inducing thermal softening in the Ti particles and consequently leading to stress value decrease. Figure 8 showed that the stress value in the HA particle unit also rose quickly to its peak value during the initial collision phase. Meanwhile, the peak stress in the HA particle unit increased with increasing particle velocity.

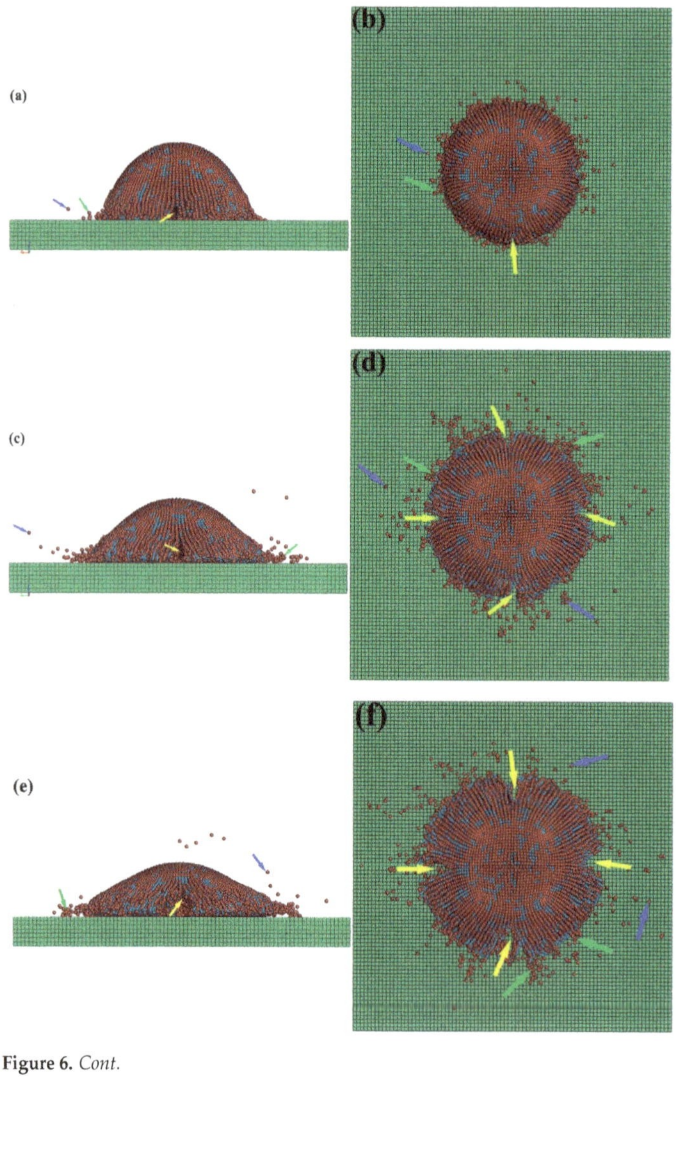

Figure 6. *Cont.*

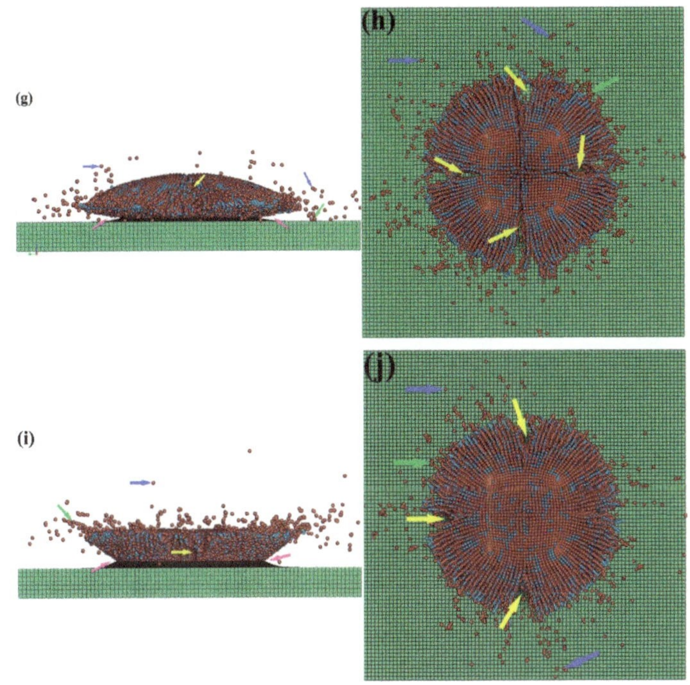

Figure 6. Simulated morphologies of HA-70wt.%Ti composite particles deposited under different particle velocity conditions: (**a**,**b**) 400 m/s; (**c**,**d**) 500 m/s; (**e**,**f**) 600 m/s; (**g**,**h**) 700 m/s; (**i**,**j**) 800 m/s.

3.2. Deposition Behavior Analysis of Cold-Sprayed Single nHA/Ti Composite Particle

Figure 9 shows the surface morphologies of cold-sprayed single nHA/Ti composite particle at different temperatures. When the gas temperature was relatively low (300 °C), the composite particle after deposition deformation exhibited a hump morphology at the center (as illustrated in Figure 9a), accompanied by peripheral spreading (as indicated by the black arrow), ejecta (as indicated by the blue arrow), and groove (as indicated by the yellow arrow) phenomena around the deposited particles. Compared to the deposition characteristics of composite particles at a gas temperature of 300 °C, elevating the gas temperature to 500 °C resulted in significantly enhanced deformation and more pronounced splattering phenomena upon particle deposition on the substrate surface (Figure 9b). Furthermore, the deformed particles exhibited a greater flattening degree, accompanied by the formation of more distinct grooves around the deposition zones. Compared with the deposition behavior at gas temperatures of 300 °C and 500 °C, as the gas temperature increased to 700 °C, the particles deposited on the substrate surface exhibited more pronounced deformation (Figure 9c). Upon impact with the substrate, individual particles underwent fragmentation (as indicated by the red arrow) and cracking (as indicated by the purple arrow), accompanied by a rebound-splattering phenomenon of the fractured particles (marked with a blue dot circle). This indicated that experimental results of composite particle deposition validated simulation predictions (as shown in Figure 6). Compared to previous studies [35], individual HA/70Ti composite particles exhibited relatively smaller deformation under the same cold spraying conditions. This indicates that the variation in the ceramic phase content of the composite powder affects particle deposition deformation. The increase in the hard and brittle HA ceramic phase makes it prone to cracking and rebound/splattering during the deposition process, which in turn leads to a greater extent of deformation. Figure 10 shows the bonding morphologies of nHA/Ti composite particles deposited on the substrate at different gas temperatures. It can be observed from the Figure that as the gas temperature increased from 300 °C to 500 °C, the density of composite particles bonded to the substrate surface gradually increased. However, when the gas temperature increased to 700 °C, the particle bonding density became slightly lower than that at 500 °C. Intense particle impaction at elevated temperature caused this phenomenon, inducing severe deformation, fragmentation, and cracking, which led to partial particle rebound and splashing from the substrate surface. Meanwhile, the measured average velocities of nHA/Ti composite particles at 300 °C, 500 °C, and 700 °C were 526 m/s, 685 m/s, and 758 m/s, respectively.

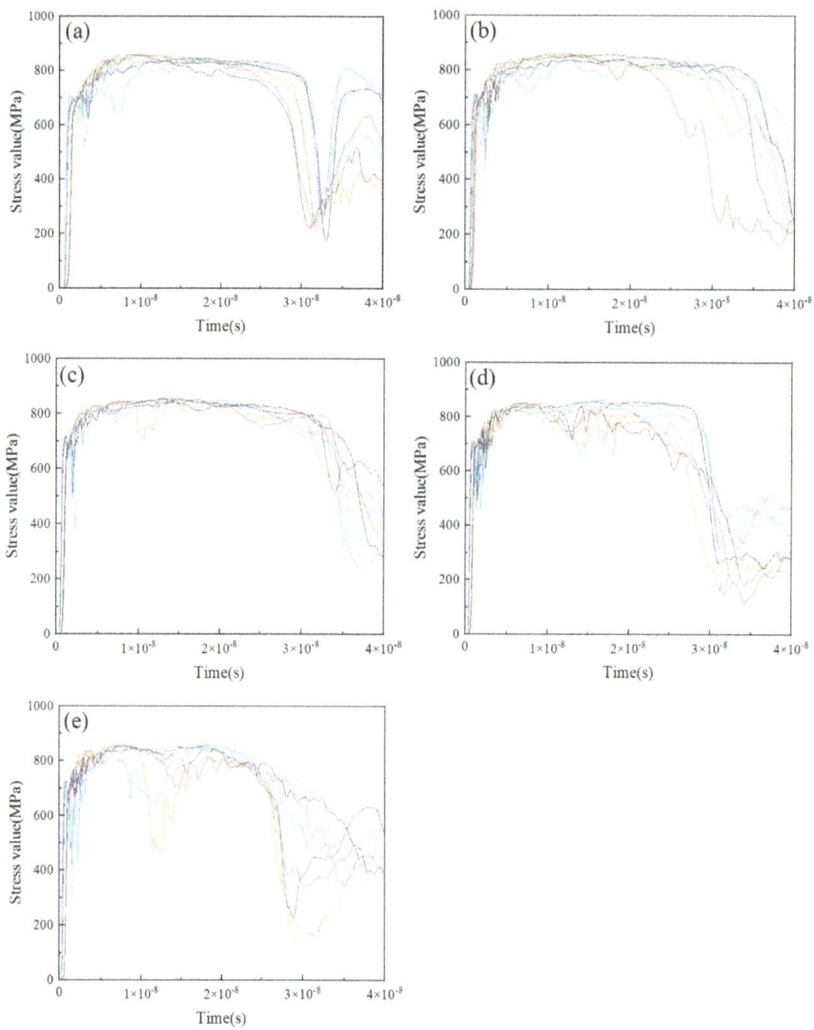

Figure 7. Stress curves of Ti particulate units in HA-70wt.%Ti composite particles after deposition under different particle velocity conditions: (**a**) 400 m/s; (**b**) 500 m/s; (**c**) 600 m/s; (**d**) 700 m/s; (**e**) 800 m/s.

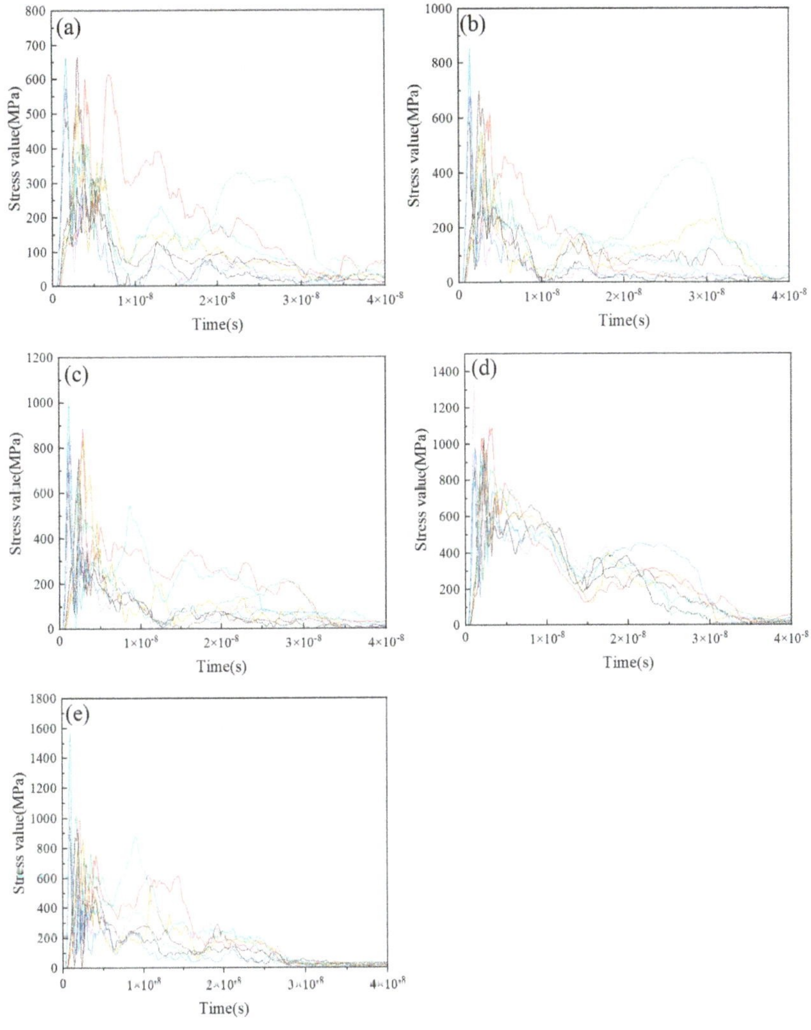

Figure 8. Stress curves of HA particulate units in HA-70wt.%Ti composite particles after deposition under different particle velocity conditions: (**a**) 400 m/s; (**b**) 500 m/s; (**c**) 600 m/s; (**d**) 700 m/s; (**e**) 800 m/s.

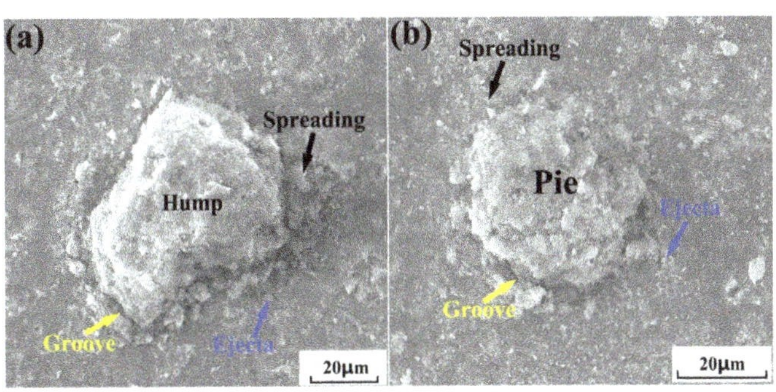

Figure 9. *Cont.*

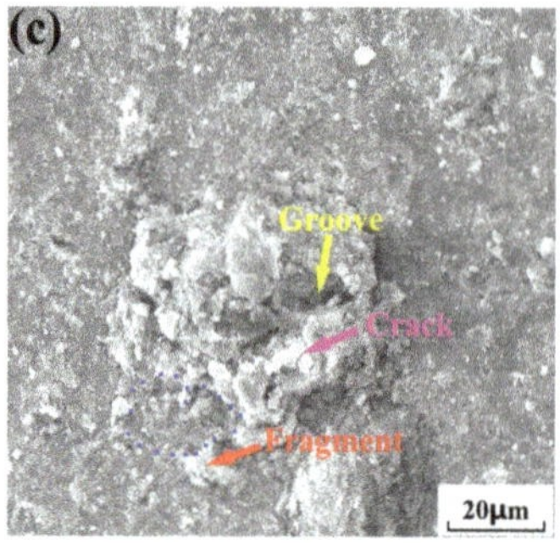

Figure 9. Surface morphologies of cold-sprayed single HA-70wt.%Ti composite particles at different gas temperatures: (**a**) 300 °C; (**b**) 500 °C; (**c**) 700 °C.

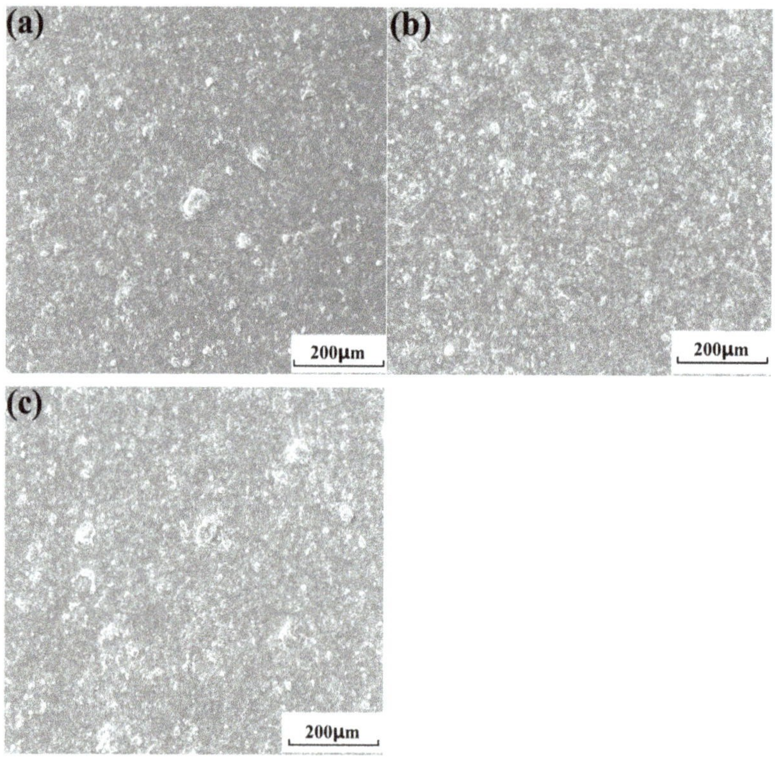

Figure 10. Bonding morphologies of cold-sprayed HA-70wt.%Ti composite particles at different gas temperatures: (**a**) 300 °C; (**b**) 500 °C; (**c**) 700 °C.

3.3. Deposition Behavior Analysis of Cold-Sprayed Multiple nHA/Ti Composite Particles

Figure 11 shows the surface morphologies of cold-sprayed multiple nHA/Ti composite particles at different temperatures. It revealed that at a relatively low gas temperature (300 °C), the deposited multiple particles exhibited limited deformation upon impact with the substrate. However, interparticle collisions resulted in noticeable splattering. As the gas temperature increased to 500 °C, the multiple particles' deformation and flattening became more pronounced, forming near-complete stacking between particles, accompanied by moderate splattering. When the gas temperature continued to rise to 700 °C, multiple particles underwent severe deformation after deposition, and the splashing phenomenon caused by the impact of deposited particles on the substrate and the mutual collision between particles also became more serious. This was consistent with the results of particle deposition simulations, further indicating that as the gas temperature increased, the degree of particle deformation increased accordingly, as did the intensity of mutual collisions between particles. Although the degree of multi-particle deposition deformation increased with the rise in gas temperature, the bonding effect of multi-particle deposition was better at a gas temperature of 500 °C. This also indicated that the bonding effect of particle deposition did not improve monotonically with increasing gas temperature. The main reason for this is that at a gas temperature of 700 °C, the poor deformability of both HA and Ti materials led to particularly severe particle splashing during deposition, which resulted in inferior bonding quality.

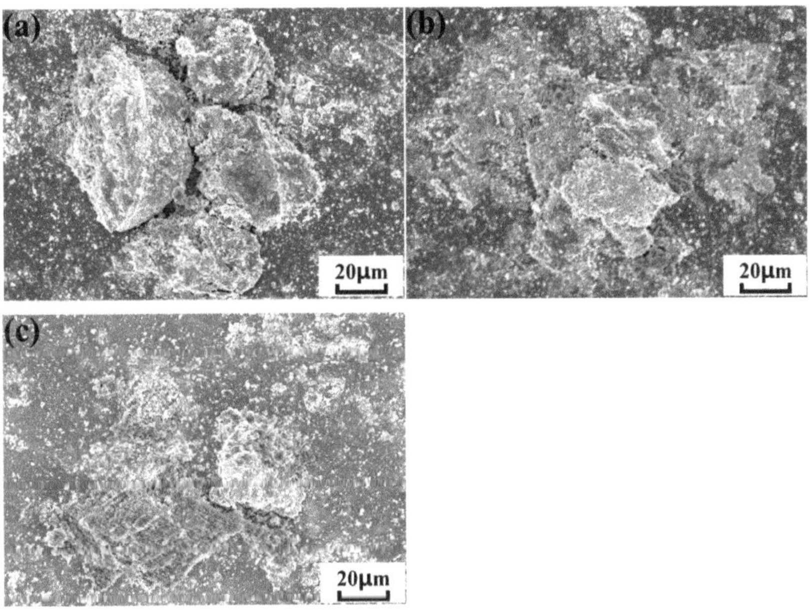

Figure 11. Surface morphologies of cold-sprayed multiple HA-70wt.%Ti composite particles at different gas temperatures: (**a**) 300 °C; (**b**) 500 °C; (**c**) 700 °C.

3.4. Microstructure of Cold-Sprayed nHA/Ti Composite Coatings

Based on the analysis of particle deposition deformation and bonding, it was found that both particle deformation and bonding were poor at a gas temperature of 300 °C. Therefore, the gas temperatures selected for coating preparation were 500 °C and 700 °C. Figure 12a,b show the surface morphologies of nHA/Ti coatings at gas temperatures of 500 °C and 700 °C, respectively. It can be observed that both coatings exhibit relatively dense and rough surface structures. However, the surface of the coating fabricated at a gas temperature of 500 °C exhibited a flatter morphology than that prepared at 700 °C. Figure 12c,d show the cross-sectional morphologies of nHA/Ti coatings at gas temperatures of 500 °C and 700 °C, respectively. It can be observed that at the gas temperature of 500 °C, the coating exhibited good bonding with the Ti buffer layer, and the internal density of the coating was relatively high. However, longitudinal cracks were generated in the coating during the deposition process. When the gas temperature was 700 °C, the bonding between the coating and the Ti buffer layer was poor, and the bonding between particles inside the coating was also weak. The main reason was that at a gas temperature of 700 °C, the splashing phenomenon caused during the particle deposition process led to poor bonding between internal particles and between the coating and the substrate. The average thicknesses of nHA/Ti coatings deposited at gas temperatures of 500 °C and 700 °C were 162.43 μm and 126.11 μm, respectively. The XRD analysis patterns of nHA/Ti coatings are shown in Figure 13. It can be seen that the phases of both coatings were consistent with those of the original powder, indicating that no phase transformation or new phase formation occurred in the nHA/Ti powder particle during the deposition process.

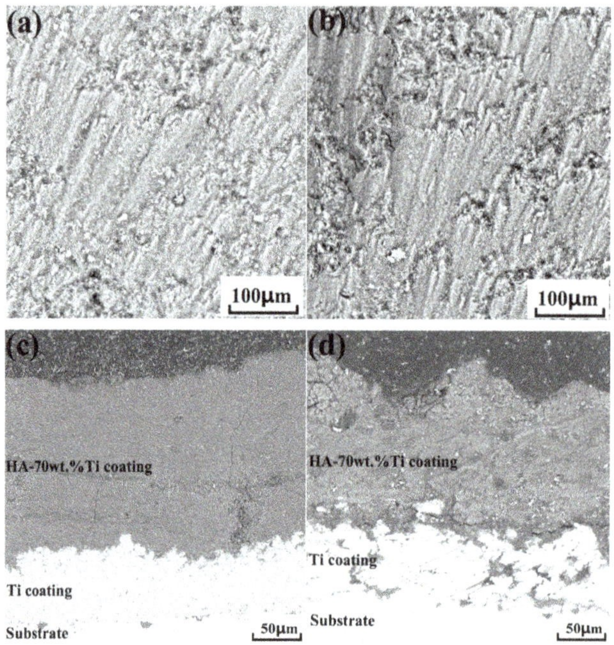

Figure 12. Surface and cross-sectional morphologies of cold-sprayed HA-70wt.%Ti composite coatings at different gas temperatures: (**a**,**c**) 500 °C; (**b**,**d**) 700 °C.

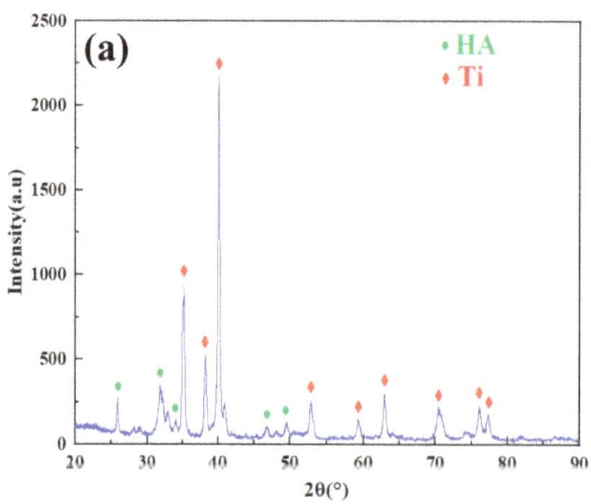

Figure 13. *Cont.*

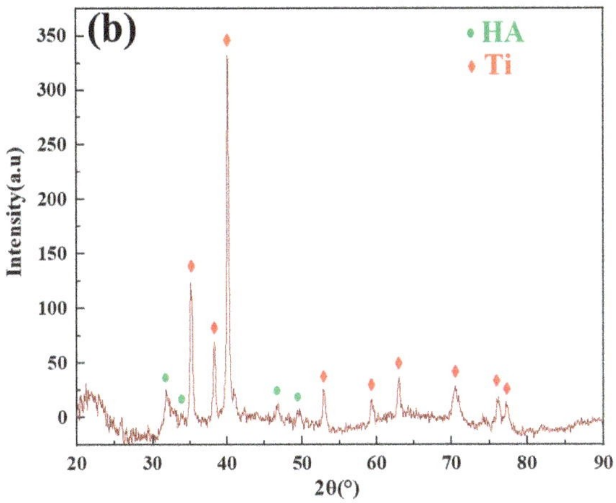

Figure 13. XRD patterns of HA-70wt.%Ti composite coatings at different gas temperatures: (**a**) 500 °C; (**b**) 700 °C.

4. Conclusions

In this study, the deposition behavior for cold sprayed nano-structured nHA/Ti composite coating using both experimental and numerical simulation approaches was investigated. The results of single particle deposition simulations indicated that as the particle velocity increased, the cracking phenomenon of the splat became more severe and the bond between the splat and the substrate deteriorated. As the gas temperature increased, the degree of particle deformation and the particle impact velocity both increased. The results from cold-sprayed multi-particle deposition experiments indicated that as the gas temperature increased, the extent of interparticle impact and deformation correspondingly increased. The coating fabricated at a gas temperature of 500 °C exhibited a flatter surface morphology and better bonding with the Ti buffer layer than that prepared at 700 °C. The phases present in the coatings were consistent with those of the original powder.

Author Contributions: Conceptualization, X.C.; methodology, X.C.; software, C.L.; validation, X.C., C.L. and S.Z.; investigation, X.C.; resources, C.L.; data curation, X.C. and Y.H.; writing—original draft preparation, X.C.; writing—review and editing, C.L. and P.A. All authors have read and agreed to the published version of the manuscript.

Funding: This research was funded by National Natural Science Foundation of China (no. 52161018).

Data Availability Statement: The data that support the findings of this study are available from the corresponding author upon reasonable request.

Conflicts of Interest: The authors declare no conflicts of interest.

References

1. Tian, Y.S.; Qian, X.L.; Chen, M.Q. Effect of saturated steam treatment on the crystallinity of plasma-sprayed hydroxyapatite coatings. *Surf. Coat. Technol.* **2015**, *266*, 38–41. [CrossRef]
2. Lin, Y.G.; Yang, Z.R.; Cheng, J. Preparation, characterization and antibacterial properties of cerium substituted hydroxyapatite nanoparticle. *J. Rare Earth* **2007**, *25*, 452–456. [CrossRef]
3. Yao, H.L.; Zou, Y.L.; Bai, X.B.; Wang, H.T.; Ji, G.C.; Chen, Q.Y. Microstructures, mechanical properties and electrochemical behaviors of nano-structured HA/Ti composite coatings deposited by high-velocity suspension flame spray (HVSFS). *Ceram. Int.* **2018**, *44*, 13024–13030. [CrossRef]

4. Yao, H.L.; Wang, H.T.; Bai, X.B.; Ji, G.C.; Chen, Q.Y. Improvement in mechanical properties of nano-structured HA/TiO_2 multilayer coatings deposited by high velocity suspension flame spraying (HVSFS). *Surf. Coat. Technol.* **2018**, *342*, 94–104. [CrossRef]
5. Harun, W.S.M.; Asri, R.I.M.; Alias, J.; Zulkifli, F.H.; Kadirgama, K.; Ghani, S.A.C.; Shariffuddin, J.H.M. A comprehensive review of hydroxyapatite-based coatings adhesion on metallic biomaterials. *Ceram. Int.* **2018**, *44*, 1250–1268. [CrossRef]
6. Ning, C.Y.; Wang, Y.J.; Lu, W.W.; Lam, R.W.M.; Chen, X.F.; Chiu, K.Y.; Ye, J.D.; Wu, G.; Wu, Z.H. Nano-structural bioactive gradient coating fabricated by computer controlled plasma-spraying technology. *J. Mater. Sci. Mater. Med.* **2006**, *17*, 875–884. [CrossRef]
7. Liu, B.; Xiao, G.Y.; Chen, C.Z.; Lu, Y.P. Nanostructured hydroxyapatite coating obtained from scholzite with improved corrosion resistance. *J. Mater. Res. Technol.* **2020**, *9*, 6928–6932. [CrossRef]
8. Jordi, G.M.; Núria, C.; Miquel, P.; Cano, I.G.; Gil, F.J.; Guilemany, J.M.; Dosta, S. Porous titanium-hydroxyapatite composite coating obtained on titanium by cold gas spray with high bond strength for biomedical applications. *Colloids Surf. B* **2019**, *180*, 245–253.
9. Ding, S.J. Properties and immersion behavior of magnetron-sputtered multi-layered hydroxyapatite/titanium composite coatings. *Biomaterials* **2003**, *24*, 4233–4238. [CrossRef]
10. Nguyen, T.H.N.D.; Nguyen, V.B.; Phuong, P.T.M.D.; Pham, S.H.; Tran, V.D.D.; Pham, A.T.; Pham, V.H.; Nguyen, T.T.H.; Nguyen, T.L.D.; Pham, V.H. Preparation and properties of Hydroxyapatite-Ti composite thin films by co-sputtering method for biocompatibility applications. *Mater. Res. Express* **2024**, *11*, 085404. [CrossRef]
11. Morteza, F. Electrophoretic deposition of fiber hydroxyapatite/titania nanocomposite coatings. *Ceram. Int.* **2018**, *44*, 622–630. [CrossRef]
12. Zeng, Y.X.; Pei, X.B.; Yang, S.Y.; Qin, H.; Cai, H.; Hu, S.S.; Sui, L.; Wan, Q.B.; Wang, J. Graphene oxide/hydroxyapatite composite coatings fabricated by electrochemical deposition. *Surf. Coat. Technol.* **2016**, *286*, 72–79. [CrossRef]
13. Chakraborty, R.; Raza, M.S.; Datta, S.; Saha, P. Synthesis and characterization of nickel free titanium–hydroxyapatite composite coating over Nitinol surface through in-situ laser cladding and alloying. *Surf. Coat. Technol.* **2019**, *358*, 539–550. [CrossRef]
14. Pei, X.B.; Wang, J.; Wan, Q.B.; Kang, L.; Xiao, M.; Hong, B. Functionally graded carbon nanotubes/hydroxyapatite composite coating by laser cladding. *Surf. Coat. Technol.* **2011**, *205*, 4380–4387. [CrossRef]
15. Behera, R.R.; Hasan, A.; Sankar, M.R.; Pandey, L.M. Laser cladding with HA and functionally graded TiO_2-HA precursors on Ti-6Al-4V alloy for enhancing bioactivity and cyto-compatibility. *Surf. Coat. Technol.* **2018**, *352*, 420–436. [CrossRef]
16. Singh, A.; Singh, G.; Chawla, V. Characterization and mechanical behaviour of reinforced hydroxyapatite coatings deposited by vacuum plasma spray on SS-316L alloy. *J. Mech. Behav. Biomed.* **2018**, *79*, 273–282. [CrossRef]
17. Li, H.; Khor, K.A.; Cheang, P. Effect of the powders'melting state on the properties of HVOF sprayed hydroxyapatite coatings. *Mater. Sci. Eng. A* **2000**, *293*, 71–80. [CrossRef]
18. Ji, G.C.; Zou, Y.L.; Chen, Q.Y.; Yao, H.L.; Bai, X.B.; Yang, C.; Wang, H.T.; Wang, F. Mechanical properties of warm sprayed HATi bio-ceramic composite coatings. *Ceram. Int.* **2020**, *46*, 27021–27030. [CrossRef]
19. Chen, X.Y.; Zhang, B.T.; Gong, Y.F.; Zhou, P.; Li, H. Mechanical properties of nanodiamond-reinforced hydroxyapatite composite coatings deposited by suspension plasma spraying. *Appl. Surf. Sci.* **2018**, *439*, 60–65. [CrossRef]
20. Li, W.Y.; Jiang, R.R.; Huang, C.J.; Zhang, Z.H.; Feng, Y. Effect of cold sprayed Al coating on mechanical property and corrosion behavior of friction stir welded AA2024-T351 joint. *Mater. Design* **2015**, *65*, 757–761. [CrossRef]
21. Suo, X.K.; Yu, M.; Li, W.Y.; Planche, M.P.; Liao, H.L. Effect of substrate preheating on bonding strength of cold-sprayed Mg coatings. *J. Therm. Spray Technol.* **2012**, *21*, 1091–1098. [CrossRef]
22. Chen, C.Y.; Xie, X.L.; Xie, Y.C.; Yan, X.C.; Huang, C.J.; Deng, S.H.; Ren, Z.M.; Liao, H.L. Metallization of polyether ether ketone (PEEK) by copper coating via cold spray. *Surf. Coat. Technol.* **2018**, *342*, 209–219. [CrossRef]
23. Li, C.J.; Li, W.Y. Deposition characteristics of titanium coating in cold spraying. *Surf. Coat. Technol.* **2003**, *167*, 278–283. [CrossRef]
24. Song, B.; Voisey, K.T.; Hussain, T. High temperature chlorine-induced corrosion of Ni50Cr coating: HVOLF, HVOGF, cold spray and laser cladding. *Surf. Coat. Technol.* **2018**, *337*, 357–369. [CrossRef]
25. Yang, G.J.; Li, C.J.; Liao, K.X.; He, X.L.; Li, S.; Fan, S.Q. Influence of gas flow during vacuum cold spraying of nano-porous TiO_2 film by using strengthened nanostructured powder on performance of dye-sensitized solar cell. *Thin Solid Films* **2011**, *519*, 4709–4713. [CrossRef]
26. Chen, X.; Wang, H.T.; Ji, G.C.; Bai, X.B.; Deng, Z.X. Deposition behavior of nanostructured WC-23Co particles in cold spraying process. *Mater. Manuf. Process.* **2016**, *31*, 1507–1513. [CrossRef]
27. Gao, P.H.; Li, C.J.; Yang, G.J.; Li, Y.G.; Li, C.X. Influence of substrate hardness on deposition behavior of single porous WC-12Co particle in cold spraying. *Surf. Coat. Technol.* **2008**, *203*, 384–390. [CrossRef]
28. Da Silva, F.S.; Cinca, N.; Dosta, S.; Cano, I.G.; Couto, M.; Guilemany, J.M.; Benedetti, A.V. Corrosion behavior of WC-Co coatings deposited by cold gas spray onto AA 7075-T6. *Corros. Sci.* **2018**, *136*, 231–243. [CrossRef]

29. Kim, H.J.; Lee, C.H.; Hwang, S.Y. Superhard nano WC-12%Co coating by cold spray deposition. *Mater. Sci. Eng. A* **2005**, *391*, 243–248. [CrossRef]
30. Chen, X.; Ji, G.C.; Bai, X.B.; Yao, H.L.; Chen, Q.Y.; Zou, Y.L. Microstructures and properties of cold spray nanostructured HA coatings. *J. Therm. Spray. Technol.* **2018**, *27*, 1344–1355. [CrossRef]
31. Chen, Q.Y.; Zou, Y.L.; Chen, X.; Bai, X.B.; Ji, G.C.; Yao, H.L.; Wang, H.T.; Wang, F. Morphological, structural and mechanical characterization of cold sprayed hydroxyapatite coating. *Surf. Coat. Technol.* **2019**, *357*, 910–923. [CrossRef]
32. Vilardell, A.M.; Cinca, N.; Garcia-Giralt, N.; Dosta, S.; Cano, I.G.; Nogués, X.; Guilemany, J.M. Functionalized coatings by cold spray: An in vitro study of micro- and nanocrystalline hydroxyapatite compared to porous titanium. *Mater. Sci. Eng. C* **2018**, *87*, 41–49. [CrossRef] [PubMed]
33. Ajdelsztajn, L.; Schoenung, J.M.; Jodoin, B.; Kim, G.E. Cold spray deposition of nanocrystalline aluminum alloys. *Met. Mat. Trans. A* **2005**, *36*, 657–666. [CrossRef]
34. Judd, K.G.; Sharma, M.M.; Eden, T.J. Multifunctional bioceramic composite coatings deposited by cold spray. *Key Eng. Mater.* **2019**, *813*, 228–233. [CrossRef]
35. Sun, M.; Chen, X.; Wu, Z.C.; Li, C.D.; Deng, X.F. Numerical simulation and experimental study of deposition behavior for cold sprayed dual nano HA/30 wt.% Ti composite particle. *Coatings* **2024**, *14*, 1330. [CrossRef]

nanomaterials

MDPI

Article

Nanostructured Silver Found in Ancient Dacian Bracelets from Cehei Hoard—Salaj, Romania

Ioan Petean [1,*], Emanoil Pripon [2], Horea Pop [2], Codruta Sarosi [3,*], Gertrud Alexandra Paltinean [3], Simona Elena Avram [4], Nicoleta Ignat [1], Lucian Barbu Tudoran [5,6] and Gheorghe Borodi [6]

1 Faculty of Chemistry and Chemical Engineering, Babes-Bolyai University, 11 Arany Janos Street, 400028 Cluj-Napoca, Romania; nicoleta.cotolan@ubbcluj.ro
2 Zalau County Museum of History and Art, 9 Unirii Street, 450042 Zalau, Romania; emanoilpripon@gmail.com (E.P.); horeapopd@yahoo.com (H.P.)
3 Department of Polymer Composites, Institute of Chemistry Raluca Ripan, Babes Bolyai University, 30 Fantanele Street, 400294 Cluj-Napoca, Romania; gertrud.paltinean@ubbcluj.ro
4 Faculty of Materials and Environmental Engineering, Technical University of Cluj-Napoca, 103–105 Muncii Boulevard, 400641 Cluj-Napoca, Romania; simona.avram@imadd.utcluj.ro
5 Faculty of Biology and Geology, Babes-Bolyai University, 44 Gheorghe Bilaşcu Street, 400015 Cluj-Napoca, Romania; lucian.barbu@ubbcluj.ro
6 National Institute for Research and Development of Isotopic and Molecular Technologies, 65–103 Donath Street, 400293 Cluj-Napoca, Romania; borodi@itim-cj.ro
* Correspondence: ioan.petean@ubbcluj.ro (I.P.); liana.sarosi@ubbcluj.ro (C.S.)

Abstract: Nanomaterials are usually associated with modern technologies and advanced processing methods. Three silver Dacian bracelets within Cehei hoard (Salaj County, Romania) are tougher than they should be according to the apparently higher silver content. The microstructural investigation reveals that all three bracelets have silver content of about 90 wt.%. The metallographic inspection of a bracelet sample reveals a very refined microstructure of α grain while fewer eutectic grains are almost undetectable, indicating intensive plastic deformation. XRD patterns of the bracelets reveal relevant peaks for silver (without copper) having a much-broadened aspect indicating nanostructural level. The nano-grains were evidenced at high magnification of SEM imaging: 55 nm for bracelet 1, 95 nm for bracelet 2 and 75 nm for bracelet 3. Elemental maps reveal that the nanograins are basically formed by α phase; the finest eutectic traces are situated and uniformly dispersed within α phase, appearing as small red spots. Vickers μHV10 micro indentation was calibrated on a pure silver 999.9 ‰ in annealed state, resulting in 37 HV10. The nanostructured bracelets have about 56 μHV10 for bracelet 1; 50 μHV10 for bracelet 2 and 52 μHV10 for bracelet 3. Dyrrachium drachmas have Vickers microhardness of about 37 μHV10. The obtained results confirm the historian's supposition that Dyrrachium drachmas could be the source for silver but also clearly indicate that the final steps of bracelets manufacturing were effectuated by cold deformation with intensive cold hardening. It results that cold deformation of the bracelets rods induces a nanostructural state that significantly increases their microhardness instead of their higher silver title.

Keywords: nanostructured silver; Dacian bracelets; Vickers microhardness

1. Introduction

Archeological science needs help from the other scientific fields for a proper understanding of the artifacts: how were they made and from which materials? Metallic artifacts' shape and microstructure strongly depend on the manufacturing technological procedures and require specific investigations such as X-ray diffraction (XRD), Scanning Electron

Microscopy (SEM) coupled with Energy Dispersive Spectroscopy (EDS), X-ray fluorescence (XRF) and metallographic microscopy when it is suitable [1,2].

It is well known that the intensive cold working of the metals and alloys induces the cold hardening effect which makes the alloys tenacious and steely because of the intensive refining of the grain structure such as in, for example, the cold rolling [3] or cryomilling process [4]. It has certain benefits in tailoring the alloys microstructure, but the excessive cold hardening causes the inter-granular failure and breaking of the worked metallic parts [5,6]. An increased tenacity of the metallic parts requires proper annealing between the cold working sessions in a similar manner to the Japanese samurai swords [7] to allow internal restructuring of the grains [8,9].

The search within international journals databases did not retrieve any results regarding the metallic artifacts' nanostructure resuming the investigated aspects to their microstructure, a fact understandable if considering the relatively poor metallurgical processes in antiquity compared with the key parameters control nowadays. But the search returns many new results regarding the nanostructured system used for the restoring of ancient artifacts like nanostructured metal oxides used for conservation of cultural heritage [10] and the effect of organo-silicon on the nanostructure of waterlogged archeological oak [11]. Some of the ancient artifact production techniques might cause the grains size after refinement to drop below 99 nm, making them nanostructural by chance, not by intended purpose. For example, Dacian civilization has less written archeological sources but certain references to their culture and practice are found in Roman and Greek sources. There are also a lot of metallic archeological findings (including gold and silver beside bronze) including metallurgical workshops [12,13]. It is very interesting to look for a Dacian hoard discovery.

Discovered on 25 April 1986, in Cehei (the town of Șimleu Silvaniei), the hoard consists of a brooch, three bracelets and a chain, all made of silver, as well as 552 Dyrrhachium drachmas, of which 445 are original and contain no less than 26 associations of names of the issuing magistrates and 7 are imitations [14]. The hoard was discovered in the area of the Dacian settlement of Șimleu Silvaniei-Observator, located on an important trade route: the salt road, which entered Sălaj county from the southeast, passed through the Meseș Gates and headed towards the Crasna Valley through Sărmășag or along the Barcăului Valley towards Marca, reaching Șimleu Silvaniei along the way [15].

The Cehei hoard is part of a series of numerous monetary hoards buried in the Second Iron Age, La Tène D1, sometime between the end of the 2nd century BC and the first two decades of the 1st century BC [15]. Since its publication, it has been hypothesized that the good quality silver in the five objects of adornment (jewelry), valued at a fineness of 800/1000, came from the coins that make up the hoard. At that time, the idea was that a good part of the Dacian ornaments came from the coins that entered Dacia after the middle of the 2nd century BC, namely tetradrachms of Macedonia I, drachms of Dyrrhachium and Apollonia, tetradrachms of Thasos and Roman Republican denarii [14].

The authors assumed that to make the five ornaments, the coins were first melted and then transformed into silver bars during "a quasi-mandatory intermediate stage" for the transformation of coins into ornaments. Taking into account a 10% loss as a result of melting the coins and transforming them into bars, the authors concluded that the raw material for making the five ornaments in the hoard was the equivalent of 120 drachmas of Dyrrhachium [14].

The actual evaluation of the hoard components reveals an increased tenacity of the bracelets, compared to the other jewelry from the hoard (e.g., necklace and fibulae) having a strong steely behavior unspecific to silver. Thus, simply melting the Dyrrhachium drachma and subsequently casting into bars is an insufficient assumption to fully characterize the bracelets. Such toughness would require alloying with hardening elements or, as previously mentioned, the cold working induces an advanced refining of the microstructure, improving its tenacity. Therefore, assuming the typical protocol for investigation of nanostructured materials, the current investigation aims to reveal the precise composition of the bracelets and to figure out the ancient manufacturing technology. The intensive cold working of the bracelets might turn their material into a nanostructured one, a very possible fact which could be revealed by advanced investigations like XRD associated with SEM and EDS.

2. Materials and Methods

2.1. Samples Description and Preparation

The investigated bracelets belong to the Cehei Hoard, Figure 1, in custody of Zalau Museum of History and Art, Romania. Their shape is typical for the Dacian culture, the terminations being decorated with specific motifs resembling snake heads (bracelet 1) and fir branches in bracelets 2 and 3, Figure 1c,d. These are simple bracelets without spirals, allowing their easy investigation through the nondestructive methods. All hoard pieces are perfectly cleaned and preserved according to the standard of practice for being able to be presented in exhibitions; therefore, they are suitable for direct investigations without any other preparations.

Figure 1. The investigated pieces from Cehei Hoard: (**a**) Dyrrhachium drachma, (**b**) bracelet 1, (**c**) bracelet 2 and (**d**) bracelet 3.

A very small splinter about 1 mm length was gently sampled from the inside part of each bracelet for metallographic investigation. The sampling procedure was effectuated with care to not affect the artifact state. The collected splinters were embedded in Bakelite and subsequently grinded with different granulation sandpapers: 1000, 2000, 4000 and 8000. The final polishing step was performed on a felt disc impregnated with colloidal alumina. The metallographic samples were etched using a swabbing method with a reagent based on ferric chloride and sodium thiosulfate [16].

Since the hypothesis made by the authors who published the hoard discovery mentioned that Dyrrhachium drachmas were used as raw material for bracelets, we selected a representative coin which was investigated for comparison. Their metallographic aspect was revealed on its side through a minimally invasive procedure by vertical embedding in Bakelite, and care was taken to ensure minimal loss during grinding and polishing procedures.

2.2. Investigation Methods

The alloy phases and oxide traces from the artifact natural patina were evidenced by the X-ray diffraction (XRD) which was performed on a Bruker D8 Advance (Bruker Company, Karlsruhe, Germany) using Cu kα radiation (1.540562 Å). The patterns were registered at a speed of 1 degree/minute. Peaks assignment was performed with Match 1.0 software equipped with a PDF 2.0 database (Crystal impact Company, Bonn, Germany).

Metallographic microscopy was effectuated on a specific optical microscope IOR MC8 (IOR company, Bucharest, Romania) equipped with a digital capture of the images Sony 14MPX (Sony company, Minato, Japan).

Scanning Electron Microscopy (SEM) coupled with Energy Dispersive Spectroscopy (EDS) was used for the investigation of the morphological aspects correlated with the elemental distribution. Therefore, the morphological aspects were observed on the Secondary electron images (SEI), and the elemental distribution map was displayed on the Backscattered Electron Image (BSE) taken on the same area as SEI image. All these operations were performed with a Hitachi SU8230 microscope (Hitachi Company, Tokyo, Japan) operated in high vacuum mode at an acceleration voltage of 30 kV. Elemental analysis was effectuated with EDS detection module X-Mas 1160 EDX (Oxford Instruments, Oxford, UK) installed on the SEM device.

Transmission Electron Microscopy (TEM) was performed with Hitachi TEM series HD-2700 (Hitachi Company, Tokyo, Japan) on the collodion replicas taken from the samples surface. The images were further analyzed with specific software Image J version 1.53t (National Institute of Health, Bethesda, Rockville, MD, USA).

Microhardness was effectuated with a Duramin40 AC3 device (Struers Company, Ballerup, Denmark) operated with Vickers indenter µHV10 type at a dwell time of 10 s and a load of 10 g. There were at least three indentations taken on each sample (in the less visible side of the bracelets), and the mean value was calculated along with the standard deviation. The mean values were statistically analyzed using Anova method followed by Tukey post hoc test at a relevance level of 0.05. The graphical plot and statistical analysis were effectuated with Origin 9b software (Microcal Co., Amherst, MA, USA).

3. Results

3.1. Alloy Phases and Their Microstructural Constituents

The alloys phases (e.g., pure metals or intermetallic compounds) can be detected by XRD with high accuracy, giving advanced qualitative characterization. Thus, Dyrrhachium drachma has a diffraction pattern featuring very well-developed peaks with strong intensities and narrow shape, Figure 2. It corresponds to a well-annealed metallic sample. The drachma composition is dominated by Ag while the smaller amount of copper appears to be oxidized as CuO (crystallized as Tenorite), a fact in good agreement with the antique look of the coin. It makes it difficult to establish a semi-quantitative of the Ag/Cu ratio within.

On the other hand, the XRD patterns resulting from bracelets have a completely changed shape with less intense peaks and broadened width. This aspect clearly indicates advanced refinement of the microstructure most likely induced by the cold working, as shown in Figure 2.

XRD peaks of all three bracelets are dominated by the Ag peaks followed by significantly smaller Cu peaks. The oxide amount on their surface is below diffractometer detection limit and therefore the specific peaks are not observed. It allows applying the Relative Intensity Ratio (RIR) method for establishing the silver amount [17,18]. Briefly, there are two phases which generated diffraction peaks, and they are co-related as follows:

$$\frac{Ia}{Ib} = \left(\frac{xa}{xb}\right) \cdot \left(\frac{fa}{fb}\right). \tag{1}$$

where Ia is the most representative intensity of phase a, and Ib is the same for phase b; xa and xb are the mass percentage concentrations for the two phases and fa and fb are their Corundum factors. Finally, it results that bracelets have the silver amount of about 90–95% regarding copper.

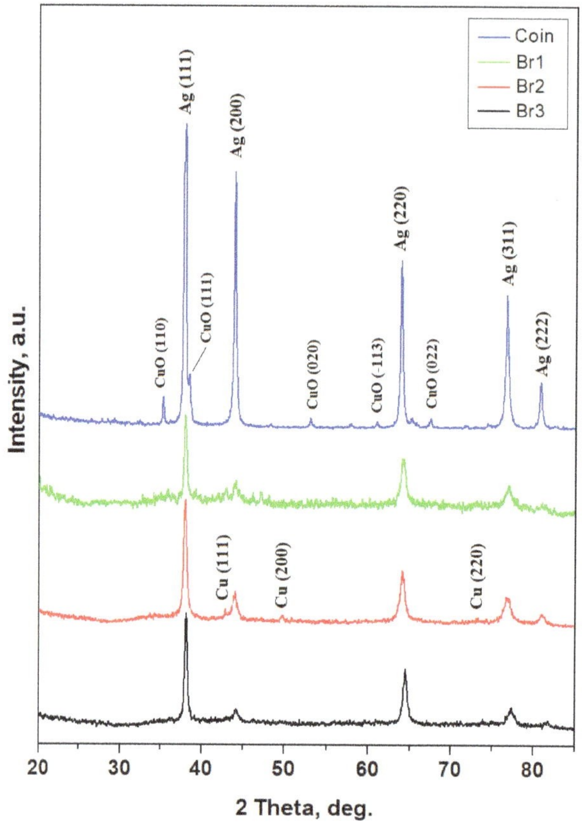

Figure 2. XRD patterns obtained for the investigated pieces from Cehei Hoard. The following cards from the PDF database were used for the peaks assignment: Ag PDF 65-2871; Cu PDF 89-2838 and CuO PDF 89-2529.

The full width at half height (FWHM) of the diffraction peaks allows calculation of the crystallite size. Scherrer formula is generally used for the mineral samples and for crystalline matter which was not subjected to the intense mechanical stress [19,20]; it is described by Equation (2):

$$D = \frac{k\lambda}{B\cos\theta}. \tag{2}$$

where D represents the crystallite size, k is the Scherrer constant typical 0.9 for metals and alloys, B is FWHM and θ is the Bragg angle.

When the alloy amorphization is induced by mechanical stress like in the mechanical alloying, the remnant strains accumulated into the crystal lattice superimpose a widening effect of the diffraction peaks beside the crystallite refinement. In consequence, the strain component must be subtracted from the crystallite size calculation. Williamson–Hall is one of the effective methods of fulfilling this requirement [21,22]. This method assumes that the FWHM is induced by crystallites size refinement B_{size} as well as due to the remnant strains B_{strain} at the crystal lattice level described as

$$B_{total} = B_{size} + B_{strain}, \tag{3}$$

which further becomes

$$B_{total} = \frac{k\lambda}{D\cos\theta} + C_{\varepsilon}\tan\theta. \tag{4}$$

The remnant strain component is represented by C_{ε}. Equation (4) has a straight-line form; the plot intersection to the origin gives the crystallite size D, and the slope of the line gives the strain component C_{ε}.

Both methods, Scherrer and Williamson–Hall, were applied to the bracelets XRD pattern, and the obtained values are centralized in Table 1.

Table 1. The data calculated on the XRD patterns.

Sample	Ag, wt.%	Crystallite Size, nm	
		Scherrer	Williamson–Hall
Bracelet 1	91	5.22	50
Bracelet 2	90	17.46	80
Bracelet 3	95	8.49	70

Scherrer equation results in very small nano-crystallites of only a few nanometers which are not likely to be found in the alloy structure and represent a misquantification of the strains induced at the crystal phase's level. Therefore, the Williamson Hall method provides more reliable values, as shown in Table 1, having a great chance to be observed by advanced microstructural investigation. However, these strains induced by the intense cold working cannot be relaxed in a sustainable manner unless the annealing treatment is applied.

The metallographic investigation, shown in Figure 3, provides more insight into the phase's distribution within the investigated artifacts. Dyrrhachium drachma has a typical cast microstructure with a large amount of α phase organized in polyhedral grains of about 8 μm in diameter with borders less visible due to the silver matrix coherence and to the reduced etching of the used reagent. The eutectic grains corresponding to the equilibrium α + β phase (e.g., organized in a very refined lamellar structure alternating Ag with Cu sheets [23,24]) have rounded shape and are uniformly distributed in α matrix having brown color due to the ferric chloride etching copper components of the microstructure. Their diameter is situated preponderantly on 8–10 μm, but some smaller eutectic grains are observed around 2.5–3 μm. Small inclusions are observed having grey–black nuance under chemical etch and most likely belong to the sand grains from the casting mold.

The dies struck applied on the cast round pellet has a minor influence on the microstructure, causing only a small elongation of some eutectic grains due to the suddenly compressive effort induced by the hammering.

All bracelet samples have similar microstructures revealed by the metallographic investigation in Figure 3b–d. White spots correspond to α grains; eutectic grains appear in orange hue due to the refined intercalation of α with β sheets. The grain structure is much kneaded with the eutectic grains, severely destructured and scrambled into the α matrix. Therefore, α grains are considerably divided into small fractions difficult to be quantified by their diameter because of the irregular shape but still allow observing that the sizes are situated below 1 µm. Several dark spots appear associated with a grey hue, indicating the significant presence of nonmetallic inclusions.

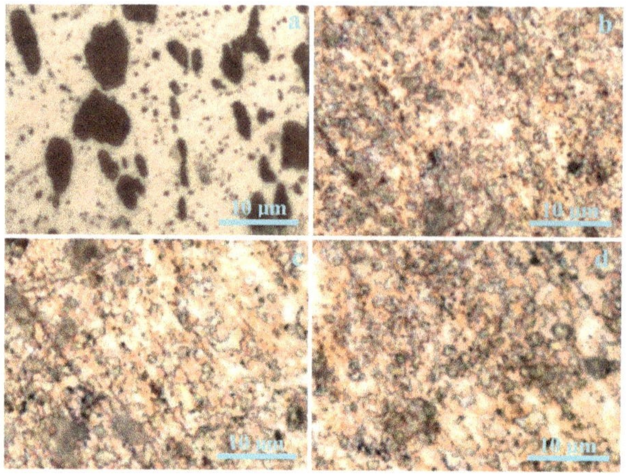

Figure 3. Metallographic images of the investigated pieces volume: (**a**) Dyrrhachium drachma, (**b**) bracelet 1, (**c**) bracelet 2 and (**d**) bracelet 3.

The dark aspect indicates the presence of amorphous carbon or traces of iron from the metallurgic process like the literature data concerning ancient artifacts [25]. Such an appearance would have been found mainly on the bracelet surface, not in the polished metallographic samples. Moreover, their dissolution within adjacent grains like grey hue indicates an intensive intermixing of the microstructural constituents. Generally, the cold working of alloy slabs induces the grains texturing under the drawing of rolling direction [26,27]. The nonmetallic inclusions and the lack of specific texturing within the bracelets indicates that they were included into the alloy bulk during working procedures, a fact which requires detailed discussion regarding a close correlation with the morphological aspects.

3.2. Sample's Morphology and Nanostructure

Dyrrhachium drachma surface has a worn surface that affects the grain structure, shown in Figure 4a, with partly scratched α grains maintaining their cohesion. The worn effect erodes the eutectic grains which appear slightly eroded, forming local topographic depressions.

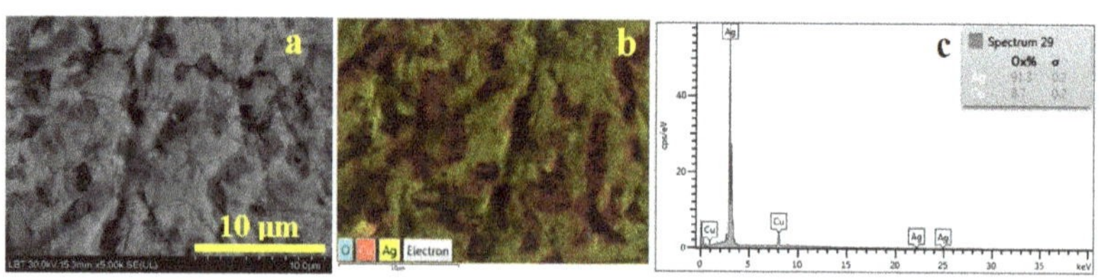

Figure 4. SEM investigation of the Dyrrhachium drachma surface: (**a**) SEI image revealing the morphological aspects, (**b**) BSE image with the elemental distribution map and (**c**) EDS spectrum.

The elemental map in Figure 4b clearly reveals the α matrix (yellow nuance) containing rounded eutectic grains (red coloration) in good agreement with the metallographic microscopy. The turquoise label belongs to oxygen which is less observed on the surface, bringing greenish hues to the deeper scratches having mild oxidation (below the XRD detection limit). Therefore, the elemental composition of the alloy was determined by subtracting the oxygen component within the spectrum in Figure 4c. It results in a silver content of 91.3 wt.% Ag and 8.7 wt.% Cu. It certainly corresponds to the hypoeutectic region of the Ag–Cu phase diagram [28]. Overall, the cast characteristics of the drachma clearly appear on its surface microstructure, having significant influence on the worn characteristics. Also, it indicates that the coin was in circulation for a relatively long time.

The surface of bracelet 1 has a totally changed morphology, as shown in Figure 5a. It has worn marks represented by several scratches randomly disposed on the image's observation field. There are also some small depressions with a darkened aspect which correlates with the nonmetallic inclusions observed at the metallographic observation. These nonmetallic inclusions keep their equiaxed aspect but with irregular sides sustaining their dispersion through intensive cold working of the metal piece. Alloy grains cannot be distinguished at this magnification. However, the alloy composition was measured by the EDS spectrum in Figure 5b in a region free of nonmetallic inclusions. The bracelet's surface has a mild oxidation below the XRD detection limit but sufficient to be detected by SEM–EDX, and thus oxygen was subtracted from the counting.

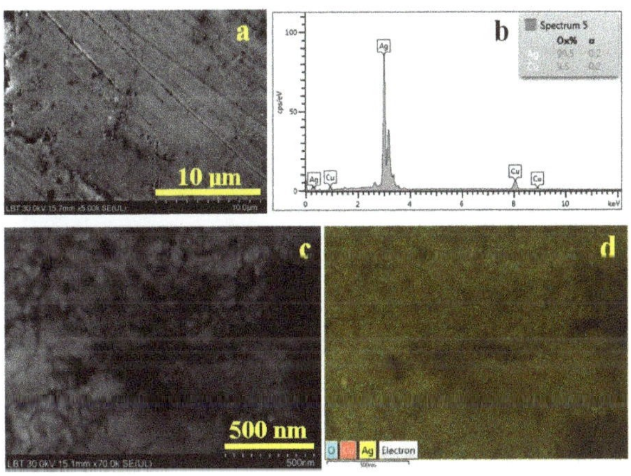

Figure 5. SEM image of the bracelet 1 surface: (**a**) SEI image revealing the morphological aspects, (**b**) EDS spectrum, (**c**) nanostructural detail and (**d**) BSE image with the elemental distribution map for the nanostructural detail.

The bracelet 1 surface was observed at the same magnification as Dyrrhachium drachma. Therefore, we proceed with inspection of the bracelet's surface in high-magnification operation mode at ×70,000 (a scale bar of 500 nm), as shown in Figure 5c. The nanostructural detail was taken in a smooth area free of scratches and nonmetallic inclusions. It reveals a dense structure of nanograins having irregular shapes caused by the advanced destructuring of the initial cast α phase and eutectic grains. Overall, assuming an equiaxed shape of the nano grain, their size results around 55 nm. The elemental map in Figure 5 indicates an advanced mixing of the eutectic grains cannot be detected from the α phase; in fact, copper spots appear uniformly distributed within silver mass. It confirms the metallographic observations.

Bracelet 2 has more randomly dispersed worn scratches on its surface but less non-metallic inclusions, as shown in Figure 6a. It makes it easy to choose the proper area for the EDS spectrum in Figure 6b by revealing the alloy compound. The surface also has minor oxidation which must be subtracted from the elemental analysis counting.

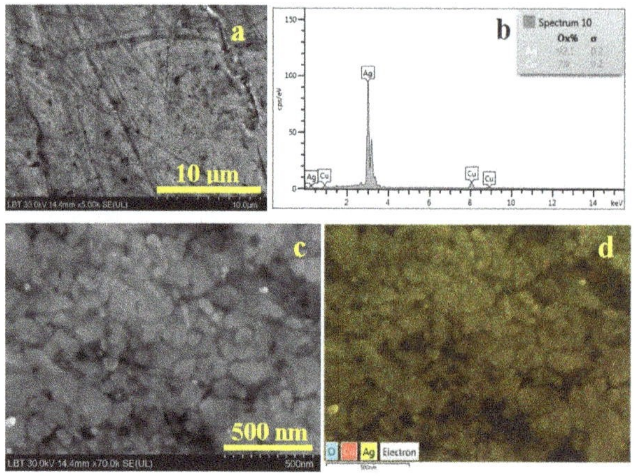

Figure 6. SEM image of the bracelet 2 surface: (**a**) SEI image revealing the morphological aspects, (**b**) EDS spectrum, (**c**) nanostructural detail and (**d**) BSE image with the elemental distribution map for the nanostructural detail.

The nanostructural details in Figure 6c reveal a more heterogeneous arrangement of the nanograins, with most of them having boulder-like shape and sizes preponderantly about 95 nm, but very often they are surrounded by some finer ones indicating advanced intra-granular fragmenting. It would indicate that the eutectic grains were more intensively fragmented than α grains. This assumption is infirmed by the elemental map in Figure 6d revealing a uniform distribution of copper in the silver mass.

The surface of bracelet 3 reveals parallel flow marks indicating the alloy texturing during the cold working. The edges of the flow marks have worn marks making them brighter in SEI image. The nonmetallic inclusions are scarcer; few of them can be observed in the upper left corner in Figure 7a.

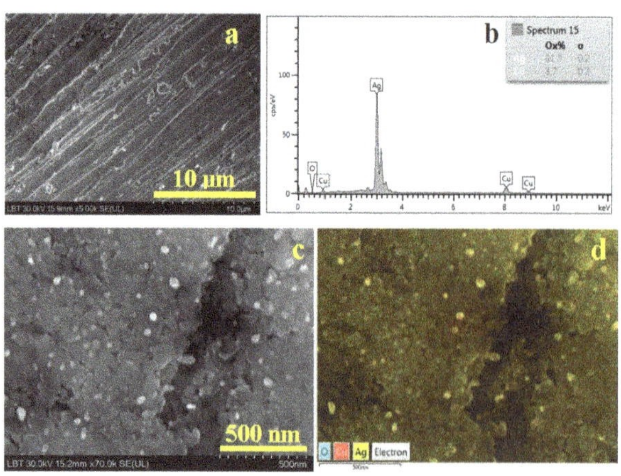

Figure 7. SEM image of the bracelet 3 surface: (**a**) SEI image revealing the morphological aspects, (**b**) EDS spectrum, (**c**) nanostructural detail and (**d**) BSE image with the elemental distribution map for the nanostructural detail.

EDS spectrum in Figure 7b reveals the alloy composition taking into consideration the mild oxidation of the surface. The nanostructural detail in Figure 7c reveals a cold deformation flow mark looking like a depression positioned vertically in the left side of the image. It looks like two layers of the same nanostructured material were pressed one to another until they became the same bulk. The nanograins also have fragmented boulder-like shapes and sizes of about 75 nm. It is interesting that the nanostructure has embedded some white rounded nanograins into the material bulk. The elemental map in Figure 7d allows us to observe that these rounded grains are in fact pure α phase; the yellow nuance clearly shows that they are free of mild oxidation. The rest of the bulk mass has a uniform distribution of copper into silver mass.

SEM observations are centralized in Table 2, revealing that the overall composition of the alloy from the coin is found in all bracelets with minor variations. The grain size decreases about 100 times from the coin's microstructure to the bracelet's nanostructure. It is a severe structural modification induced by severe deformation.

Table 2. The data resulted from SEM investigation.

Sample	Ag, wt.%	Cu, wt.%	Grain Size, nm	
			SEM	TEM
Drachma	91.3	8.7	8000	-
Bracelet 1	90.5	9.5	55	55 ± 15
Bracelet 2	92.1	7.9	95	95 ± 25
Bracelet 3	91.3	8.7	75	75 ± 20

TEM investigation of solid bulk materials is impossible because the accelerated electron beam cannot penetrate the samples depth. Therefore, a thin film of collodion was applied on the bracelet samples in the area observed by SEM. Collodion film solidification ensure a thin replica of the surface microstructure, allowing its observation under TEM conditions obtaining the images in Figure 8.

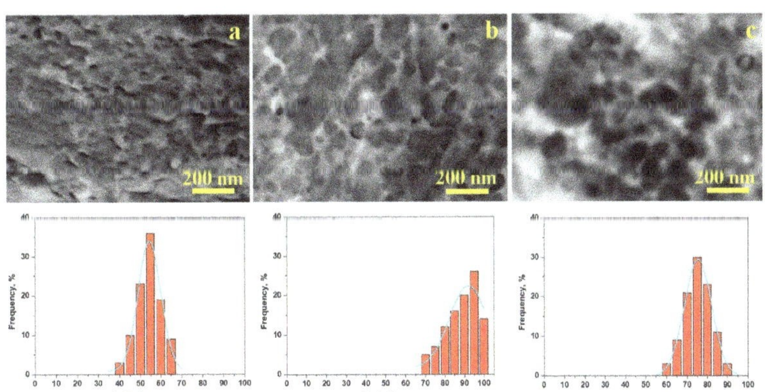

Figure 8. TEM image of the bracelet samples: (**a**) bracelet 1, (**b**) bracelet 2 and (**c**) bracelet 3. The grain size distribution plots are presented below each TEM image. The blue line in the grain size distribution histograms represent the Gaussian fitting curve.

TEM images in Figure 8 allow better observing of the grain's nanostructure insight. Ferric chloride reagent used for metallographic investigation etches only copper, bringing dark aspects to the eutectic grains and keeping α phase grains un-etched. Thus, the eutectic distribution within α phase grains borders is evidenced by bright contrast as a consequence of the chemical etching process, while the silver grains appear darker. The uniform distribution of eutectic grains regarding α phase grains gives a valuable clue indicating that it was partly melted during the silver bulk consolidation, suggesting that the coins were transformed into bracelet bars by pressure welding. It most likely happened closely to the eutectic temperature of 779 °C. On the other hand, α phase grains have irregular margins as a consequence of the intensive cold working associated with fine parallel lines inside the grain, indicating their random orientation. Unfortunately, the collodion replica does not allow their observation at high resolution. SEM–EBSD investigation is a powerful tool able to elucidate the crystallographic orientation of grains, but unfortunately it is unavailable at the current stage of investigation. Thus, HR-TEM associated with SEM-EBSD would be the next step of advancement of these precious artifacts for Romanian Cultural Heritage.

Nanostructural differences were also evidenced in TEM images, indicating an advanced grain refinement with a narrow distribution of the grain sizes in bracelet 1, as shown in Figure 8a, and the Gaussian fitting confirmed the SEM observation. The precise mean value with the corresponding standard deviation is presented in Table 2. Bracelet 2, shown in Figure 8b, has a broader grain size distribution, but the average value is given by the dominant size of 95 nm in good observation with SEM measurements. The histogram distribution in Figure 8c also has a narrow aspect with dominant size of about 75 nm, the mean value and corresponding standard deviation being presented in Table 2.

3.3. Mechanical Properties

Unfortunately, the bracelets cannot be subjected to tensile testing for investigating their mechanical properties, but microhardness testing allows a proper quantitative observation of their mechanical behavior, as shown in Figure 9.

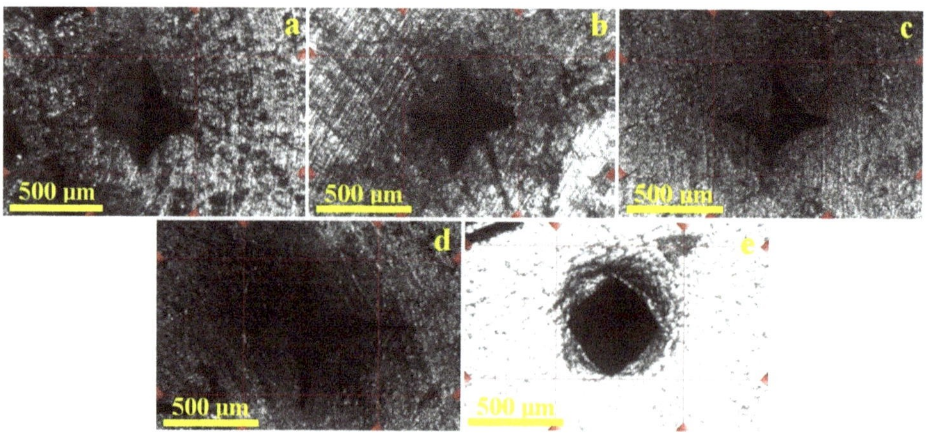

Figure 9. Reflected light microscopy images of the Vickers microhardness indentations: (**a**) bracelet 1, (**b**) bracelet 2, (**c**) bracelet 3, (**d**) Dyrrhachium drachma and (**e**) Annealed silver ingot 999.9‰.

Such cold-hardened nanostructure definitely would resist better to the penetration of the Vickers indenter compared to the cast microstructure of the Dyrrhachium drachma. Still, a comparative standard must be investigated. Therefore, a pure 999.9 silver ingot in annealed state was used for this purpose. The aspect of indentation marks reveals slim cross marks for all bracelets, as shown in Figure 9a–c, while the coin's indentation has a more deeply rhombic shape and a bigger size, as shown in Figure 9d. The annealed ingot has a greater plasticity than the indentation cross and is less visible, while the rhombic shape of the edges becomes more visible, as shown in Figure 9e.

The mean values of microhardness resulted for the investigated bracelets ranges from 50 to 56 μHV10 in the samples, as shown in Figure 10. Bracelet 1 is the tougher one because of its smaller nanostructure, while bracelet 2 is the softer one because of the bigger nanograins. However, the microhardness resulted for all bracelets forms a statistically relevant group ($p > 0.05$). On the other hand, the Dyrrachium drachma has a lower microhardness comparable with the annealed pure silver ingot, being situated at about 37 HV10 and forming the second statistically relevant group. The statistical analysis also reveals a significant difference between the two identified groups ($p < 0.05$). Thus, the microhardness investigation confirms the assumption regarding the cold hardening and its subsequent nanostructure development.

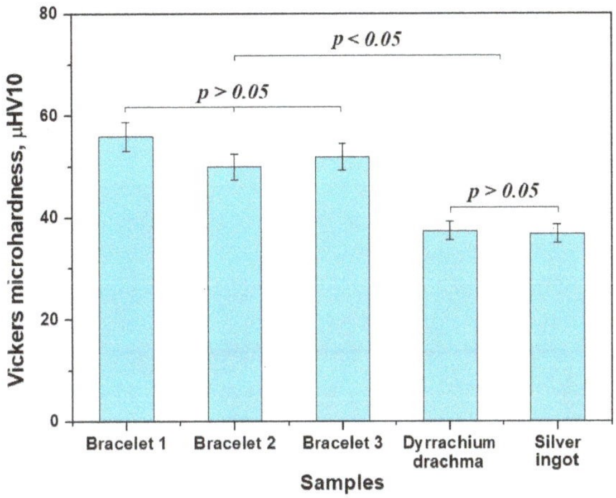

Figure 10. Mean values of Vickers microhardness variation.

4. Discussion

Dyrrhachium drachmas were very popular coins within antiquity because of their stable weight of 2.98 g (diameter about 18 mm having side approximate round) and a silver title about 90–95 wt.% depending on the literature source and their archeological deposits [13,29]. These drachmas were issued during the Roman protectorate on the ancient Dyrrhachium during 229–30 BC, and their quality was controlled and ensured by two magistrates, their names being inscribed on the obverse and reverse, in our case Damenos on the obverse and Meniskos on the reverse. The magistrate names combination and particular style of the coins allow a proper date of their issue during the circulation period [30]. The good quality of Dyrrhachium drachmas make them want for commercial exchange into Dacian territories being related to the salt and slave commerce as well as stipendium for the mercenary activities [31].

As observed, the historical context explains the massive presence of such drachmas in the Cehei hoard. The higher amount of good silver coins sustains the archeologist's arguments that they were used as raw material for jewelry production. The initial assumption made by the archeologists in 1986 indicates their re-melting and casting in ingots which were further processed by hammering. It is a reasonable assumption which is partly unsustained by the current results. The coins re-melting would collect the nonmetallic impurities and oxides into the slag, conducting to a cleaner cast microstructure [32,33]. It would result in a better microstructural distribution than the one observed in drachma because of advanced removal of mineral inclusions embedded from the original casting mold. The successive hammering (there is no archeological evidence for metal rolling in Dacian workshops) would have textured the grain shapes under the metal flow direction [34], a fact not sustained by our observations. The intermediary re-crystallization annealing would restore the equiaxed shape of the grains, inducing a regulated microstructure regarding α and eutectic grains in a similar manner with the data reported in the literature [35].

The bracelets' kneaded microstructure does not resemble any conventional technological pathways available nowadays. There is a forgotten welding technique by hammering the welded parts when they are heated at high temperature but below the melting point, ensuring the alloy diffusion from one part into another into a continuous microstructure. This technique was used by blacksmiths for welding the chain links known as pressure welding [36,37]. Thus, we proposed another technological method for the conversion of Dyrrhachium drachmas into bracelets bars without re-melting. The coin pile is heated at high temperature below the melting point in a range beginning closely below eutectic temperature of 779 °C and closing slightly above this temperature. It allows partial melting of the eutectic grains while α grains remain in solid state but have an increased plasticity. Subsequently, hammering of the heated coin pile causes their welding into a single metal piece which is continuously modeled into a bar. This procedure ensures the kneading of α grains while partly melted eutectic ensures their cohesion, a fact sustained by the TEM images in Figure 8. Heating in coal-auctioned furnaces and hammering on the iron anvil induces significantly the observed nonmetallic inclusions. The proposed technological pathway is fully sustained by the metallographic analysis results.

The further cold hammering of the bracelets activates the cold hardening mechanism, causing an extra refinement of the microstructure, turning its grains into nanostructured ones. The nano-crystallites induce diffraction peaks broadening due to their size reduction and because of the remnant strains at the crystal lattice level, and thus the Williamson–Hall method allows a proper quantification of the nanograins size [21,22], values confirmed by the high magnification SEM images. Thus, the nanograins' size is about 55 nm for bracelet 1, 95 nm for bracelet 2 and 75 nm for bracelet 3. They have preponderantly boulder-like shapes with irregular margins well embedded within the eutectic grains looking diffused within the α matrix. The elemental maps clearly indicate uniform distribution of Cu within silver mass without contouring precise eutectic grains. This nanostructural refinement indicates that the silver alloy bar was produced by pressure welding instead of re-melting followed by casting.

It is well known that silver melting has significant loss through evaporation if a proper artificial slag layer is induced during crucible heating [38,39]; thus, the ancient re-melting would lead to a significant loss in the silver amount. Applying the pressure welding method has the advantage of keeping the silver amount about the same, ensuring a maximization of the coins' conversion to jewelry. Drachmas processing through melting as Chirila and Matei stated in 1986 [14] would have caused a sensitive decrease in the silver title because of its peculiar evaporation in a melted state at high temperatures. Our SEM observation reveals that the drachma's silver title is similar with the silver title of the bracelets. Thus, our results sustain that the drachmas were processed into raw bars for bracelets through pressure welding techniques which avoid silver evaporation through the melted bath.

Nanostructuring through cold hardening of the silver alloy induces a significant increase in the Vickers microhardness of the bracelets in comparison with the coin having similar microhardness with the annealed pure silver ingot, confirming once more that the proposed technology was applied to obtain the bracelets. The bracelets' increased values of the Vickers microhardness clearly sustain their nano-structuring compared to Dyrrhachium drachma and annealed silver ingot. The literature reveals that the coin's surface is very affected by the strong indentations like the demonetization marks imprinted in the modern issues [40,41]. Thus, smaller and thin indentations on the bracelets contrast with the bigger and wider indentations of the coin's surface.

The current study's limitations reside in nondestructive methods which are absolutely necessary when valuable archeological artifacts are involved. Tensile strength testing and subsequent SEM fractography would have brought additional information of the production technology, but they are not available for now. The future archeological finds maybe will bring out some similar bracelets fragments which perhaps allow tensile testing. A further nondestructive investigation should be conducted using Electron Backscatter Diffraction (EBSD) for evidencing the crystallographic orientation of the grains within the investigated alloy samples [42,43] and to find similar archeological artifacts to verify if the proposed technology is confirmed in other similar artifacts. The validation of the proposed manufacturing technology should be proved in further research by replication of the process with materials as close as possible to the original ones and their comparison with the other metallurgical archeological finds in Salaj County.

5. Conclusions

The high tenacity of almost pure silver bracelets from the Cehei Dacian hoard is caused by the advanced refinement of the microstructure induced by the cold hardening process occurring during cold working of the final steps. XRD and SEM observation reveal that all three bracelets are in fact nanostructured with nanograins of 55 nm for bracelet 1, 95 nm for bracelet 2 and 75 nm for bracelet 3. The bracelet alloy nano-structuring causes its microhardness to increase to 56 μHV10 for bracelet 1; 50 μHV10 for bracelet 2 and 52 μHV10 for bracelet 3, while the Dyrrhachium drachma and annealed pure silver ingot hardness is situated around 37 HV10. Finally, it results that the coins were used as raw material for the bracelets, but it was processed by pressure welding instead of casting preserving the metal amount. The advanced cold working during the final step of bracelets manufacturing ensures the grains nanostructuring.

Author Contributions: Conceptualization, I.P. and E.P.; methodology, I.P., C.S., and S.E.A.; software, G.B.; validation, I.P., C.S., and E.P.; formal analysis, G.A.P., N.I., and H.P.; investigation, L.B.T., G.B., C.S., and S.E.A.; resources, E.P. and H.P.; data curation, N.I. and G.A.P.; writing—original draft preparation, I.P. and E.P.; writing—review and editing, I.P. and N.I.; visualization, L.B.T., and I.P.; supervision, C.S., S.E.A., and I.P.; project administration, E.P. and I.P. All authors have read and agreed to the published version of the manuscript.

Funding: This research received no external funding.

Data Availability Statement: The original contributions presented in the study are included in the article.

Acknowledgments: The authors gratefully acknowledge that the XRD diffractometer maintenance which was supported by the Ministry of Research, Innovation, and Digitization through Program 1-Development of the National Research and Development System, Subprogram 1.2-Institutional Performance-Funding Projects for Excellence in RDI, Contract No. 37PFE/30.12.2021.

Conflicts of Interest: The authors declare no conflicts of interest.

Abbreviations

The following abbreviations are used in this manuscript:

BSE	Backscatter Electron Image
EBSD	Electron Backscatter Diffraction
EDS	Energy Dispersive Spectroscopy
RIR	Relative Intensity Ratio
SEM	Scanning Electron Microscopy
SEI	Secondary Electron Image
TEM	Transmission Electron Microscopy
XRD	X-ray Diffraction

References

1. Dwivedi, D.; Mata, J.P. Archaeometallurgical investigation of ancient artefacts' degradation phenomenon. *NPJ Mater. Degrad.* **2019**, *3*, 35. [CrossRef]
2. Cosano, D.; Esquivel, D.; Tena, C.; Arrebola, J.R.R. Chemical analysis of archaeological metals found in a Roman burial site in Castro del Río (Córdoba, Spain). *Next Mater.* **2025**, *8*, 100721. [CrossRef]
3. Zhang, P.; Liu, D.; Zhang, L.; Liu, K.; Zhang, J.; Si, Y.; Chen, G.; Zhu, Q. Microstructural Evolution and Tensile Deformation Behavior of FeCoNiCrTi0.2 High-Entropy Alloys Regulated by Cold Rolling and Annealing. *Metals* **2025**, *15*, 1037. [CrossRef]
4. da Silva, V.R.; Fjellvåg, Ø.S.; Švec, P.; Švec, P., Jr.; Hauback, B.C.; Deledda, S. The Effect of Mo and Al Substitution in Cryomilled and Cold-Rolled FeNi Alloys. *Metals* **2025**, *15*, 996. [CrossRef]
5. Xie, Z.; Bi, Y.; Fan, Y.; Gao, C.; Zhang, X.; Feng, Y.; Zhou, D.; Dong, L. Finite Element Investigation of a Novel Cold-Formed Steel Shear Wall. *Buildings* **2024**, *14*, 1691. [CrossRef]
6. Hang, H.; Gao, T.; Chen, J.; Li, X.; Song, H.; Huang, G. Investigation on Strain Hardening and Failure in Notched Tension Specimens of Cold Rolled Ti6Al4V Titanium Alloy. *Materials* **2022**, *15*, 3429. [CrossRef]
7. Wadsworth, J. Archeometallurgy related to swords. *Mater. Charact.* **2015**, *99*, 1–7. [CrossRef]
8. Yang, P.-H.; Chiu, M.-C.; Huang, H.-C.; Liang, C.-L. The behaviors and mechanisms of current-enhanced recrystallization and grain growth in as-rolled pure nickel during electrical annealing. *Mater. Sci. Eng. A* **2025**, *946*, 149173. [CrossRef]
9. Liang, P.; Zhang, J.; Kong, N.; Li, H.; Zhu, H. The Microstructure Characteristics Evolution of Bulk High-Purity Silver for High Relief Application. *Metals* **2023**, *13*, 463. [CrossRef]
10. Mandal, S.; Kumar, P.; Satpathy, B.; Das, K.; Das, S. Nanostructured metal oxide based coating for the protection and conservation of cultural heritage: A comprehensive review. *J. Cult. Herit.* **2024**, *69*, 94–112. [CrossRef]
11. Plaza, N.Z.; Pingali, S.V.; Broda, M. The effect of organosilicon compounds on the nanostructure of waterlogged archeological oak studied by neutron and X-ray scattering. *J. Cult. Herit.* **2025**, *71*, 203–210. [CrossRef]
12. Petean, I.; Pețan, A.; Pop, H.; Avram, S.E.; Tudoran, L.B.; Borodi, G. Nondestructive investigations of tin-rich bronze mirrors discovered in Dacian sites from Sălaj County—Romania. *J. Archaeol. Sci. Rep.* **2025**, *67*, 105424. [CrossRef]
13. Constantinescu, B.; Bugoi, R.; Cojocaru, V.; Simon, R.; Grambole, D.; Munnik, F.; Oberländer-Târnoveanu, E. Elemental analysis through X-ray techniques applied in archeological gold authentication—The case of Transylvanian gold and of the Dacian bracelets. *Spectrochim. Acta Part B At. Spectrosc.* **2009**, *64*, 1198–1203. [CrossRef]
14. Chirilă, E.; Matei, V.A. Tezaurul dacic de la Cehei. *Acta Mvsei Porolissensis* **1986**, *10*, 95–117.
15. Pop, H. Tezaurul Cehei-Deluț. In *Șimleu Silvaniei, jud. Sălaj în Aurul și Argintul Antic al României, Catalog de Expoziție*; Marghitu, R.A., Ed.; Muzeul Național de Istorie a României: București, Romania, 2013; pp. 323–325.
16. Liu, S.; Gong, Z.; Lu, H.; Zhang, W.; Ma, Y.; Yang, X.; Xie, Z.; Hu, G.; Hu, D. Optimizing Metallographic Etchants for Ancient Gold and Silver Materials. *Metals* **2022**, *12*, 1229. [CrossRef]
17. Xiao, J.; Song, Y.; Li, Y. Comparison of Quantitative X-ray Diffraction Mineral Analysis Methods. *Minerals* **2023**, *13*, 566. [CrossRef]
18. Farcas, I.A.; Dippong, T.; Petean, I.; Moldovan, M.; Filip, M.R.; Ciotlaus, I.; Tudoran, L.B.; Borodi, G.; Paltinean, G.A.; Pripon, E.; et al. Material Evidence of Sediments Recovered from Ancient Amphorae Found at the Potaissa Roman Fortress. *Materials* **2023**, *16*, 2628. [CrossRef] [PubMed]
19. Avram, S.E.; Tudoran, L.B.; Borodi, G.; Petean, I. Microstructural Characterization of the Mn Lepidolite Distribution in Dark Red Clay Soils. *Appl. Sci.* **2025**, *15*, 6445. [CrossRef]
20. Dippong, T.; Toloman, D.; Lazar, M.D.; Petean, I. Effects of Lanthanum Substitution and Annealing on Structural, Morphologic, and Photocatalytic Properties of Nickel Ferrite. *Nanomaterials* **2023**, *13*, 3096. [CrossRef]
21. Hassanzadeh-Tabrizi, S.A. Precise calculation of crystallite size of nanomaterials: A review. *J. Alloys Compd.* **2023**, *968*, 171914. [CrossRef]
22. Kibasomba, P.M.; Dhlamini, S.; Maaza, M.; Liu, C.-P.; Rashad, M.M.; Rayan, D.A.; Mwakikunga, B.W. Strain and grain size of TiO_2 nanoparticles from TEM, Raman spectroscopy and XRD: The revisiting of the Williamson-Hall plot method. *Results Phys.* **2018**, *9*, 628–635. [CrossRef]
23. Northover, S.M.; Northover, J.P. Microstructures of ancient and modern cast silver–copper alloys. *Mater. Charact.* **2014**, *90*, 173–184. [CrossRef]
24. Liu, S.; Shi, M.; Zhou, Y.; Li, R.; Xie, Z.; Hu, D.; Zhang, M.; Hu, G. Scanning electrochemical cell microscopy: A powerful method to study the intergranular corrosions of archaeological silver artifacts. *J. Cult. Herit.* **2020**, *46*, 176–183. [CrossRef]
25. Ortega-Avilés, M.; Tenorio-Castilleros, D.; Segura-Venzor, I.L.; Miranda-Hernández, J.G.; Velasco, M.A. An archaeometallurgical study of Mixtec silver gold alloy foils from Tomb No. 7, Monte Alban, Oaxaca, México. *Mater. Chem. Phys.* **2020**, *242*, 122475. [CrossRef]
26. Zhang, L.; Liu, W.; Xia, E.; Chen, W.; He, X.; Tang, D. Effect of Solution and Aging Treatment on the Microstructural Evolution and Mechanical Properties of Cold-Rolled 2024 Aluminum Alloy Sheets. *Coatings* **2025**, *15*, 1139. [CrossRef]

27. Liu, H.; Yang, Z.; Yang, Z.; Wu, Y.; Ju, J. Effect of Cold Rolling on Microstructure Evolution and Mechanical Properties of Zn-3Cu-1Mg-0.3Nd Alloy. *Crystals* **2025**, *15*, 769. [CrossRef]
28. Kogtenkova, O.; Straumal, B.; Korneva, A.; Czeppe, T.; Wierzbicka-Miernik, A.; Faryna, M.; Zięba, P. Grain Boundary Complexions and Phase Transformations in Al- and Cu-Based Alloys. *Metals* **2019**, *9*, 10. [CrossRef]
29. Uzonyi, I.; Bugoi, R.; Sasianu, A.; Kiss, A.Z.; Constantinescu, B.; Torbágyi, M. Characterization of Dyrrhachium silver coins by micro-PIXE method. *Nucl. Instrum. Methods Phys. Res. Sect. B Beam Interact. Mater. At.* **2000**, *161–163*, 748–752. [CrossRef]
30. Arghir, G.; Petean, I.; Pripon, E.; Pascuta, P. On the mathematical algorithm of some ancient drachma minting time. *Acta Tech. Napoc. Appl. Math. Mech. Eng.* **2014**, *57*, 141–148.
31. Marzano, A. Marine salt production in the Roman world: The salinae and their ownership. *Quat. Sci. Rev.* **2024**, *336*, 108776. [CrossRef]
32. Chiarantini, L.; Benvenuti, M.; Bianchi, G.; Dallai, L.; Volpi, V.; Manca, R. Medieval Pb (Cu-Ag) Smelting in the Colline Metallifere District (Tuscany, Italy): Slag Heterogeneity as a Tracer of Ore Provenance and Technological Process. *Minerals* **2021**, *11*, 97. [CrossRef]
33. Westner, K.J.; Klein, S.; Sergeev, D.; Müller, M. Temperature estimates of historical Pb-Ag smelting slags: A multi-methodological approach. *J. Archaeol. Sci. Rep.* **2022**, *46*, 103654. [CrossRef]
34. Ning, Z.; Wang, Q.; Zhao, D.; Pei, W.; Wen, M. Structural Evolution of Bulk Silver during Cold Rolling and Annealing. *Metals* **2022**, *12*, 1525. [CrossRef]
35. Angella, G.; Esfandiar Jahromi, B.; Vedani, M. A comparison between equal channel angular pressing and asymmetric rolling of silver in the severe plastic deformation regime. *Mater. Sci. Eng. A* **2013**, *559*, 742–750. [CrossRef]
36. Weman, K. Chapter 11—Pressure welding methods. In *Woodhead Publishing Series in Welding and Other Joining Technologies, Welding Processes Handbook*, 2nd ed.; Weman, K., Ed.; Woodhead Publishing: Sawston, UK, 2012; pp. 119–132. [CrossRef]
37. Yongbing, L. Pressure Welding. In *ECPH Encyclopedia of Mining and Metallurgy*; Xu, K., Ed.; Springer: Singapore, 2024. [CrossRef]
38. Zhang, Y.; Zhang, L.; Zhang, D.; Guo, X.; Du, J.; Wang, E. Effect of Sn Addition on Microstructural Evolution and Properties of Cu–Ag Alloys. *JOM* **2025**. [CrossRef]
39. Fischer-Bühner, J.; Bertoncello, R.; Basso, A.; Poliero, M. *Silver Casting Revisited: The Alloy Perspective*; The Bell Group, Inc.: Glasgow, UK, 2011.
40. Gavrilă, C.C.; Lateş, M.T.; Grebenişan, G. Sustainable Approach to Metal Coin Canceling Methods, Using 3D Modeling and Finite Element Method Analysis. *Sustainability* **2024**, *16*, 2322. [CrossRef]
41. Gavrilă, C.C.; Lateş, M.T. Metal coin error study using 3D modelling and FEM Analysis, Annals of the Oradea University. *Fascicle Manag. Technol. Eng.* **2019**, *29*, 3394. [CrossRef]
42. Victoria-Hernández, J.; Jeong, Y.; Letzig, D. Role of recovery in the microstructure development and mechanical behavior of a ductile Mg Zn Nd Y Zr alloy: An analysis using EBSD data and crystal plasticity simulations. *Int. J. Plast.* **2025**, *191*, 104380. [CrossRef]
43. Ji, Y.; Suo, H.; Ma, Z.; Huang, B.; Liu, J.; Zhang, Z.; Wang, Q. In-situ tensile analysis of microstructure evolution in Ni, Ni5W, and Ni9W using EBSD. *J. Mater. Res. Technol.* **2025**, *36*, 1475–1490. [CrossRef]

nanomaterials

MDPI

Article

Low-Friction and Corrosion-Resistant Orthodontic Stainless Steel Archwires with Functional Carbon Films

Pengfei Wang *, Minghui Hao and Shiqi Cheng

Institute of Nanosurface Science and Engineering (INSE), State Key Laboratory of Radio Frequency Heterogeneous Integration, College of Mechatronics and Control Engineering, Shenzhen University, Shenzhen 518060, China; 17625797242@163.com (M.H.); chengshiqi0212@163.com (S.C.)
* Correspondence: wangpf@szu.edu.cn; Tel.: +86-755-22673825

Abstract: To mitigate the adverse effects of immersion in fluoride-containing solutions on the surface corrosion of orthodontic stainless steel archwires, carbon films were fabricated on these archwires under various deposition times and substrate bias voltages using a self-designed plasma sputtering system. Structural analysis revealed that the carbon films deposited at lower substrate bias voltages were classified as amorphous carbon films, whereas those fabricated at higher substrate bias voltages were identified as graphene nanocrystalline carbon films. Particularly, immersion tests and electrochemical experiments demonstrated that carbon film prepared at a substrate bias voltage of +50 V for 80 min exhibited exceptional corrosion resistance. Furthermore, a low friction coefficient and low wear rate were obtained even after soaking in a fluoride toothpaste mixed solution. The mechanisms underlying the corrosion resistance and friction properties of these superior carbon films were thoroughly investigated. This study provides valuable insights into the application of carbon film for reducing friction and wear while enhancing corrosion resistance, thus promoting their practical clinical applications in coated orthodontic stainless steel archwires.

Keywords: orthodontic archwire and bracket; carbon film; graphene sheets; fluoride toothpaste; corrosion; peeling; friction and wear

1. Introduction

Malocclusion is one the three most common oral diseases in clinical application (global prevalence of 56% [1–5]), which not only significantly affects the facial aesthetics of patients, but also affects their mental health and functionality of the oral and maxillofacial systems. Nowadays, clinical orthodontic treatment serves as a prevalent method for correcting malocclusion. A metallic fixed appliance, which consists of brackets bonded to the teeth and archwires that connect the brackets, is designed to align teeth and restore the occlusal relationship by precisely controlling the three-dimensional tooth movement [6–12].

The oral cavity is a highly complex biological environment. Temperature fluctuations, pH changes, high humidity, mechanical stress, and the presence of microorganisms in the oral environment all make it highly conducive to the promotion of the corrosion of metallic orthodontic appliances, with electrochemical corrosion being the main type [13–16]. Fluorine, as an active chemical element, demonstrates a pronounced anti-caries effect, on the one hand, while on the other hand, it could induce severe corrosion damage in metallic materials, thus accelerating material degradation and metal ion release [17–21].

The corrosion resistance of the archwires and brackets has been significantly enhanced through the application of surface coatings, such as metallic layers, metal oxides, polymers, and diamond-like carbon (DLC) [22–24]. Specifically, Ryu et al. [25] incorporated platinum (Pt) into silver (Ag) films by using the physical vapor deposition technique to enhance the chemical stability. Potentiodynamic polarization tests revealed that the corrosion resistance of the AgPt-film-coated stainless steel archwires was significantly improved with the Pt content reaching up to 7%. Wu et al. [26] deposited Al–SiO_2 composite coatings onto the surfaces of the NiTi and stainless steel archwires using a magnetron sputtering technique. The surface smoothness and overall corrosion resistance of the materials were strongly enhanced, while exhibiting no significant cytotoxicity. Espinar et al. [27] achieved a nickel-free surface through a novel oxidation treatment method. The generated titanium oxide layer markedly improved the corrosion resistance of the NiTi archwires and suppressed the release of nickel ions. Zhang et al. [28] deposited DLC coatings onto the surface of stainless steel archwires using plasma-enhanced chemical vapor deposition (PECVD), and applied salt bath nitrocarburizing to achieve surface hardening of the archwires. It was found that both DLC coating and nitrocarburizing treatment significantly enhanced surface hardness, reduced the friction coefficient, and exhibited favorable biocompatibility. A reduction in corrosion current density by two orders of magnitude was also observed for the DLC-coated archwire. Sui et al. [29] fabricated DLC coatings onto the surface of the NiTi archwires using an ion implantation technique and clarified that the DLC coating not only significantly improved the corrosion resistance of the NiTi archwires but also effectively inhibited the release of nickel ions. Kobayashi et al. [30] immersed both DLC-coated and uncoated NiTi archwires in physiological saline at 37 °C for a duration of up to six months and assessed the cytotoxic effects of nickel ion release using squamous carcinoma cells. It was revealed that the DLC coating reduced the nickel ion concentration in the solution by one-sixth, indicating its role as a diffusion barrier in corrosive environments and confirming its favorable biocompatibility. Fróis et al. [31] deposited the a-C:H coatings on the SS316L substrate with a Cr-based interlayer. The common corrosion, delamination, and coating detachment from the substrates could not be observed after a 30-day immersion test in artificial saliva solution. However, it is still a great challenge for fabricating an exceptional functional coating that could satisfy various functions, such as low friction, wear resistance, corrosion resistance, and even biocompatibility.

In our previous work, graphene nanocrystalline films were deposited onto stainless steel archwires to reduce the friction coefficients of the archwire–bracket sliding contacts in artificial saliva environment [32,33]. Therefore, the main objective of this work was to achieve low friction, low wear rate, and corrosion resistance of stainless steel archwires with outstanding carbon film for potential clinical applications. Versatile carbon films were fabricated on the working surfaces of the stainless steel archwires to mitigate the adverse effects of fluoride-containing solution immersion on the surface corrosion of the stainless steel archwires. Subsequently, friction and wear experiments as well as electrochemical analyses were conducted. Finally, the stainless steel archwires were characterized, and the corrosion prevention mechanism as well as low friction mechanism were systematically discussed.

2. Materials and Methods

A commercial AISI 304 stainless steel archwire (06Cr19Ni10) with rectangular cross-sectional dimensions of 0.017 inch × 0.025 inch (0.42 mm × 0.64 mm) (HXH-0023, Xihubiom, Hangzhou, China) was applied. The orthodontic archwire was engaged in an AISI 304 stainless steel bracket (0Cr18Ni9) with a slot size of 0.022 inches (0.56 mm) (ALS-0008, Alice Dental, Hangzhou, China). Prior to being loaded into the deposition chamber, the archwire and bracket were sequentially cleaned in an ultrasonic bath (M2800-C, BRANSON,

Shanghai, China) using acetone, ethanol, and deionized water, each for a duration of 20 min.

A photograph of the customized mirror confinement electron cyclotron resonance (MCECR) plasma sputtering apparatus is shown in Figure S1. Detailed information on the ECR sputtering system has been reported in our previous studies [34,35]. The main working principle involves the formation of electron cyclotron oscillation acceleration under the control of a mirror confined magnetic field (875 G). When a positive bias voltage is applied to the substrate, the oscillating electrons are attracted to irradiate the substrate surface, thus enabling the deposition of the carbon films under the low energy electron irradiation working mode. The detailed parameters for preparing the carbon films on the surface of stainless steel archwires are presented in Table S1.

To evaluate the corrosion resistance of carbon-film-coated archwires, the specimens were soaked in a fluoride toothpaste mixed solution (i.e., fluoride toothpaste of 10 g and deionized water of 30 mL) for 16 h [36]. The working surface of the archwire was carefully observed using an optical microscope (LV-150N, NIKON, Tokyo, Japan) and a scanning electron microscope (SEM, Scios, FEI, Hillsboro, OR, USA) before and after the immersion test.

A custom-designed tribo-system was developed and constructed to assess the friction and wear behavior of orthodontic archwire–bracket contact combinations under an artificial saliva environment, as schematically illustrated in Figure 1. Comprehensive details regarding the structure of the tribo-system have been reported in our previous studies [32,33]. The system primarily comprised an electrically controlled displacement platform, an archwire fixation and pre-tensioning module, a bracket fixation unit, a three-axis stage, a data acquisition system, a thermostatic water bath, and a peristaltic pump. During the experimental procedure, the bracket was secured to a cantilever beam connected to the three-axis stage, while the archwire was actuated by a stepper motor-driven displacement platform, which induced reciprocating motion in the archwire fixation module. Concurrently, strain signals produced during the sliding process were captured by the strain gauge affixed to the cantilever beam and transmitted in real time via the data acquisition system to the computer for continuous recording. The peristaltic pump delivered commercial artificial saliva (ISO.10271.2011, pH = 6.9, dripping speed of 3.6 mL/min), preheated to a controlled temperature, onto the contact interface between the archwire and the bracket. The temperature of the droplet artificial saliva at the contact interface was maintained at approximately 37 °C, closely simulating the physiological temperature of the human oral cavity. The tribological experiment was conducted at a tensile force of 10 N, a normal load of 1 N, a reciprocating displacement of 0.15 mm, a sliding speed of 0.19 mm/s, and a sliding period of 5000 cycles.

The nanostructures of the deposited carbon films were examined using a double spherical aberration corrected transmission electron microscope (TEM, Titan3 Cubed Themis G2 300, THERMO FISHER, Hillsboro, OR, USA) operated at an accelerating voltage of 80 kV. The bonding structure configuration of the fabricated carbon film was obtained using a Raman spectroscope (LabRAM HR Evolution, HORIBA, Kyoto, Japan).

A 3D laser confocal microscope (VK-X250K, KEYENCE, Osaka, Japan) was applied to measure the surface roughness of the carbon-film-coated archwires (scanning area of 600 mm × 300 mm) and observe the wear scars on the coated archwires (scanning area of 200 mm × 150 mm) with a depth resolution of 0.2 mm. The observed raw data were evaluated with a standard F-filter filtering process under a cut-off value of 80 mm. The worn surfaces of the archwires were analyzed using an optical microscope and an SEM equipped with an energy dispersive X-ray spectroscope (EDS, THERMO FISHER, Hillsboro, OR, USA).

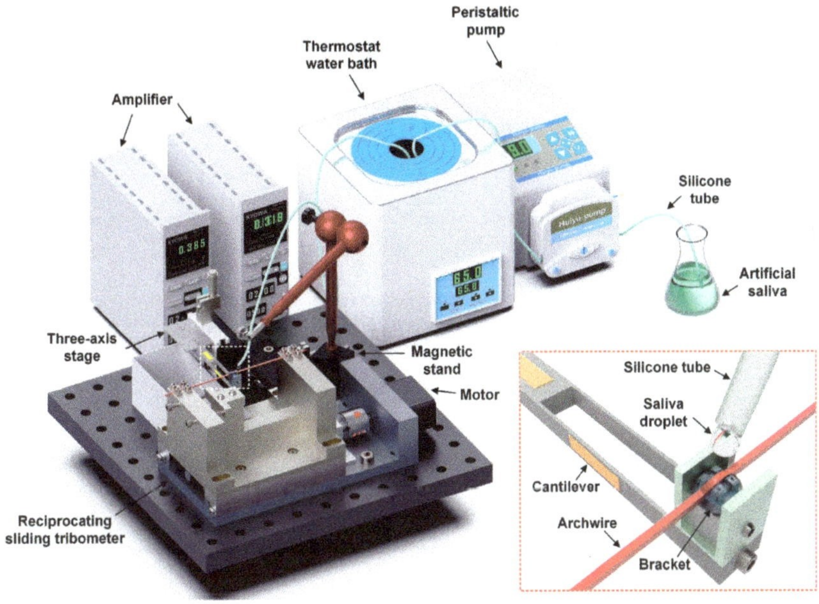

Figure 1. Schematic illustration of the orthodontic stainless steel archwire–bracket reciprocating sliding tribometer. Inset shows an enlarged diagram of the sliding contact of the archwire and bracket.

The electrochemical experiments were performed on an electrochemical workstation (Reference 600+, Gamry, PA, USA). A standard three-electrode cell configuration was established, comprising a platinum wire as the counter electrode (CE), a saturated Ag/AgCl electrode serving as the reference electrode (RE), and the surface of carbon-film-coated archwire as the working electrode (WE). The corrosion solution was a specially prepared fluoride toothpaste mixed solution (the mass ratio of fluoride toothpaste to deionized water is 1:3). In the potentiodynamic polarization test, the measured data were imported into Gamry Echem Analyst software (Version 7.05) to obtain the Tafel curve. A suitable linear region of the curve was selected for Tafel extrapolation to determine the corrosion potential and corrosion current density of the archwire.

3. Results

3.1. Characterization of the Carbon-Film-Coated Archwires

The surface characterization of the as-deposited carbon films prepared under various substrate bias voltages is summarized in Figure 2. The surface morphology and roughness profiles of the carbon-film-coated stainless steel archwires under different substrate bias voltages are shown in Figure 2a–e. Clearly, the surface roughness (Ra) of the stainless steel archwire increased gradually from 0.686 μm to 0.843 μm with an increase in substrate bias voltage from +5 V to +50 V. The increase in surface roughness could be attributed to the increased electron irradiation intensity associated with higher substrate bias voltage. Stronger electron irradiation promoted the growth of nanoscale graphene sheets, leading to a larger grain size, as described in previous studies [35].

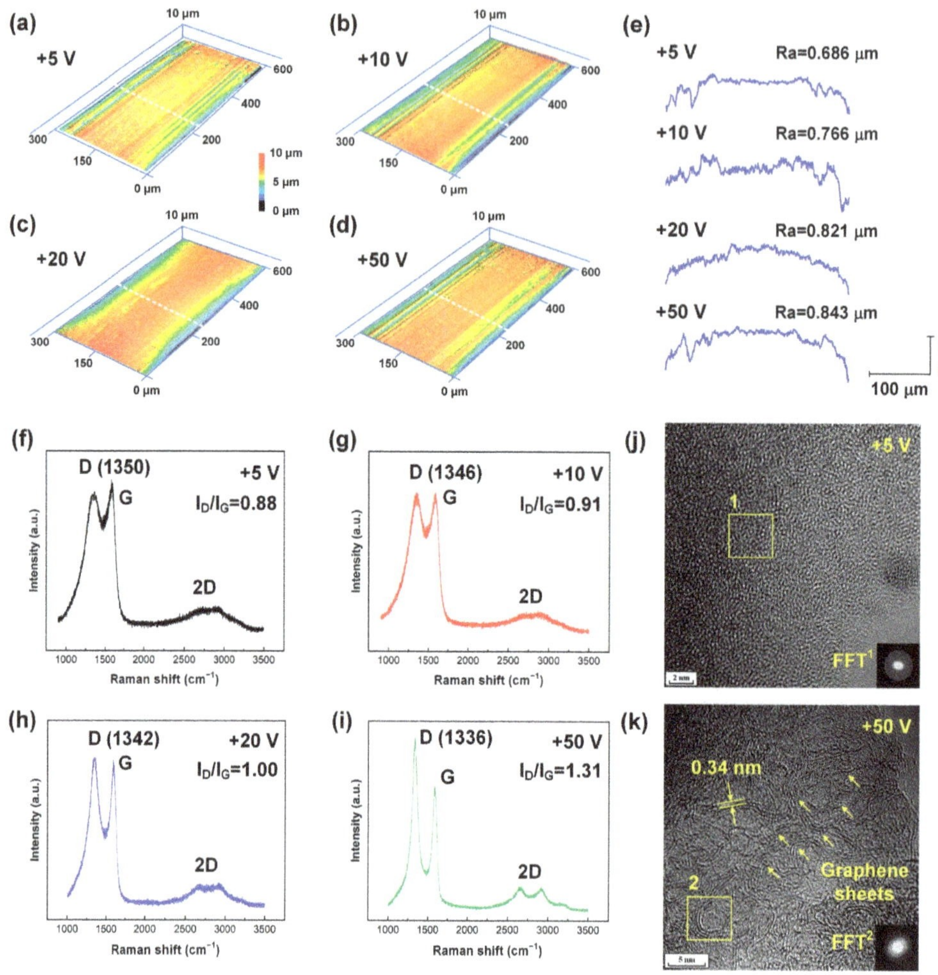

Figure 2. Characterization of the as-deposited carbon films prepared under various substrate bias voltages. Three-dimensional laser microscopy image of the carbon film fabricated at a substrate bias voltage of (**a**) +5 V, (**b**) +10 V, (**c**) +20 V, and (**d**) +50 V. (**e**) Two-dimensional cross-sectional profiles of the fabricated carbon films from the selected positions marked with white dotted lines in the corresponding optical images. Raman spectrum of the carbon film fabricated at a substrate bias voltage of (**f**) +5 V, (**g**) +10 V, (**h**) +20 V, and (**i**) +50 V. Plan-view HRTEM images of the carbon film fabricated at a substrate bias voltage of (**j**) +5 V and (**k**) +50 V. Insets in (**j**) and (**k**) show FFT images of the marked regions in the corresponding TEM images.

Structural information on the carbon film was further characterized using Raman spectroscopy, as shown in Figure 2f–i. Specifically, the D peak corresponded to the degree of disorder in sp^2 hybridized carbon bonding, and the G peak represented the vibrational modes associated with the sp^2-bonded carbon network. The appearance of the 2D peak distinctly indicated the existence of graphitic or graphene-like structures [37]. Particularly, higher substrate bias voltage led to sharpening of both the D and G peaks, suggesting an enhancement in the crystallinity of the carbon film. Moreover, the progressive intensification of the 2D peak implies a structural evolution from a disordered to a more ordered arrangement of carbon atoms [38]. The I_D/I_G ratio increased straightly from 0.88 to 1.31 with the increase in substrate bias voltage from +5 V to +50 V. Contrastingly the D peak position decreased continuously from 1350 cm^{-1} to 1336 cm^{-1} with the increase in the substrate bias voltage from +5 V to +50 V. Thus, the carbon films deposited at +5 V and +10 V were clarified as amorphous carbon films, whereas those deposited at +20 V and +50 V were identified as graphene nanocrystalline carbon films. Furthermore, high-magnification TEM imaging revealed that the carbon film prepared under a substrate bias voltage of +5 V exhibited a highly disordered state, lacking any distinct structural features, as evidenced by the dispersive ring pattern observed in the Fourier transformed filtering (FFT) image, as shown in Figure 2j. In contrast, the carbon film fabricated under a substrate bias voltage of +50 V displayed a well-defined layered crystalline structure, characterized by two distinct points in the FFT image. The interplanar spacing was measured to be approximately 0.340 nm, as illustrated in Figure 2k.

3.2. Immersion Test of the Carbon-Film-Coated Archwires

Firstly, the stainless steel archwires were coated with carbon films fabricated under a substrate bias voltage of +5 V and deposition time varied from 25 min to 80 min (thereafter denoted as +5 V & 25 min, +5 V & 35 min, +5 V & 50 min, and +5 V & 80 min), and subsequently soaked in a fluoride toothpaste mixed solution for 16 h. After the immersion test, the surface morphology of the archwires was observed using an optical microscope and an SEM–EDS. Figure 3 presents the optical image and corresponding SEM–EDS analysis results of the coated stainless steel archwires. Severe peeling of the carbon film from the stainless stee archwire occurred after soaking in mixed solution for 16 h, and the peeling phenomenon of the carbon films mainly occurred at the edges of the stainless steel archwire, as typically shown in Figure 3f,k,p,u. The detachment of the carbon film from the archwire surface resulted in the depletion of the carbon (C) element (e.g., Figure 3h), and the enrichment of the oxygen (O, e.g., Figure 3i) and fluorine (F, e.g., Figure 3j) elements. It was also found that the extent of the peeling from the archwire surface gradually decreased with the increase in deposition time from 25 min to 80 min. The great change in element distribution was mainly attributed to the direct exposure of the stainless steel archwire to the fluoride toothpaste solution after the peeling of the carbon film, thus triggering the corrosion of fluoride ions on the stainless steel [23,39]. Therefore, more attention was paid to further optimize the deposition parameters to improve the corrosion resistance and long-term stability of the carbon film in a fluoride-containing environment.

Figure 4 presents the optical images and corresponding SEM–EDS analysis results of the coated stainless steel archwires with carbon film prepared under the deposition time of 80 min and various substrate bias voltages (+5 V to +50 V, and thereafter denoted as +5 V & 80 min, +10 V & 80 min, +20 V & 80 min, +50 V & 80 min) and subsequently soaked in a fluoride toothpaste mixed solution for 16 h. The degree of the carbon film peeling from the stainless steel archwire after the long-term immersion test was relatively lower than that shown in Figure 3. Moreover, the extent of carbon film peeling after soaking in the fluoride toothpaste mixed solution gradually diminished with the increase in the substrate bias voltage from +5 V to +50 V. At a substrate bias voltage of +20 V, only minor peeling occurred at the edges of the carbon film, as shown in Figure 4p, whereas at a substrate bias voltage of +50 V, the carbon film remained largely intact on the stainless steel archwire, as shown in Figure 4u, exhibiting excellent protective performance. With the mitigation of the peeling degree of the carbon film on the archwire, the distribution of carbon element greatly increased (e.g., Figure 4w) followed by a strong decrease in fluoride element (e.g., Figure 4y) on the archwire. Therefore, it could be concluded that the presence of the GSEC film significantly minimized the corrosive effects of the fluoride-containing solution on the stainless steel archwire.

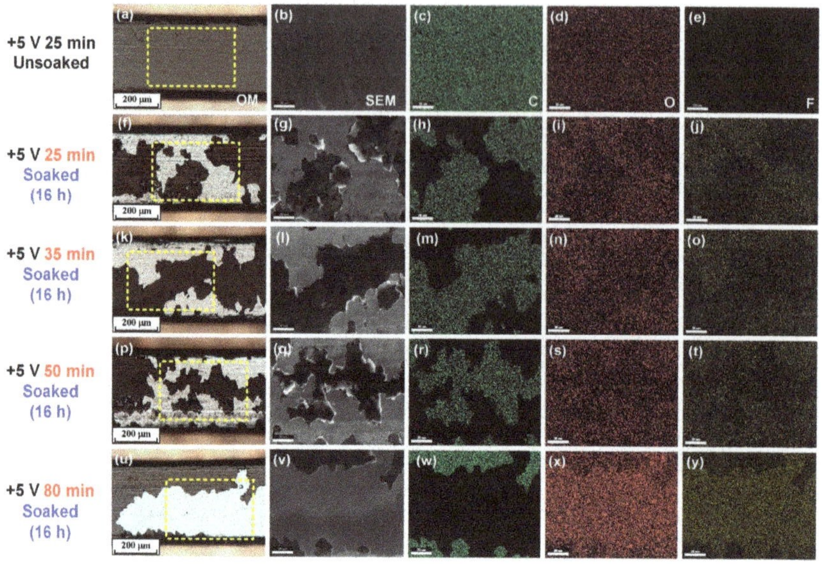

Figure 3. Optical images (**a**,**f**,**k**,**p**,**u**) and corresponding SEM–EDS images ((**b**–**e**,**g**–**j**,**l**–**o**,**q**–**t**,**v**–**y**), marked with yellow rectangular dotted line) of the carbon-film-coated stainless steel archwires after soaking in a fluoride toothpaste mixed solution for 16 h. (**f**–**j**) +5 V & 25 min, (**k**–**o**) +5 V & 35 min, (**p**–**t**) +5 V & 50 min, and (**u**–**y**) +5 V & 80 min. The unsoaked results of the carbon-film-coated stainless steel archwire (+5 V & 25 min) are also shown in (**a**–**e**) for reference. The scale bar in the SEM–EDS image is 20 mm.

Figure 4. Optical images (**a**,**f**,**k**,**p**,**u**) and corresponding SEM–EDS images ((**b**–**e**,**g**–**j**,**l**–**o**,**q**–**t**,**v**–**y**), marked with yellow rectangular dotted line) of the carbon-film-coated stainless steel archwires after soaking in fluoride toothpaste mixed solution for 16 h. (**f**–**j**) +5 V & 80 min, (**k**–**o**) +10 V & 80 min, (**p**–**t**) +20 V & 80 min, and (**u**–**y**) +50 V & 80 min. The unsoaked results of the carbon-film-coated stainless steel archwire (+5 V & 80 min) are also shown in (**a**–**e**) for reference. The scale bar in SEM–EDS image is 20 mm.

3.3. Friction Behavior of the Carbon-Film-Coated Archwires

The friction curves and corresponding friction coefficients of the uncoated and unsoaked archwire (thereafter denoted as uncoated/unsoaked), coated and unsoaked archwire (thereafter denoted as uncoated/unsoaked), as well as coated and soaked archwire (thereafter denoted as uncoated/unsoaked) sliding against stainless steel bracket under artificial saliva environment are shown in Figure 5. The uncoated/unsoaked archwire exhibited a relatively high and stable friction coefficient throughout the entire sliding friction test, as shown in Figure 5a. On the contrary, the coated/unsoaked archwire exhibited a fast running-in process and then stabilized after approximately 25 cycles, as shown in Figure 5b. Similarly, the coated/soaked archwire maintained a stable stage after 75 cycles of the running-in process, as shown in Figure 5c. The above results definitely showed that lower friction coefficients were observed with the carbon-film-coated archwire, regardless of the soaking process. Specifically, at the early stage, the friction curve of the uncoated archwire remained at a relatively high level compared to the coated archwires, although the difference was minor, as shown in Figure 5d. However, at the final stage, the friction coefficient of the uncoated archwire became significantly higher than that of the coated archwire, which maintained a consistently stable friction coefficient for the whole test, as shown in Figure 5e.

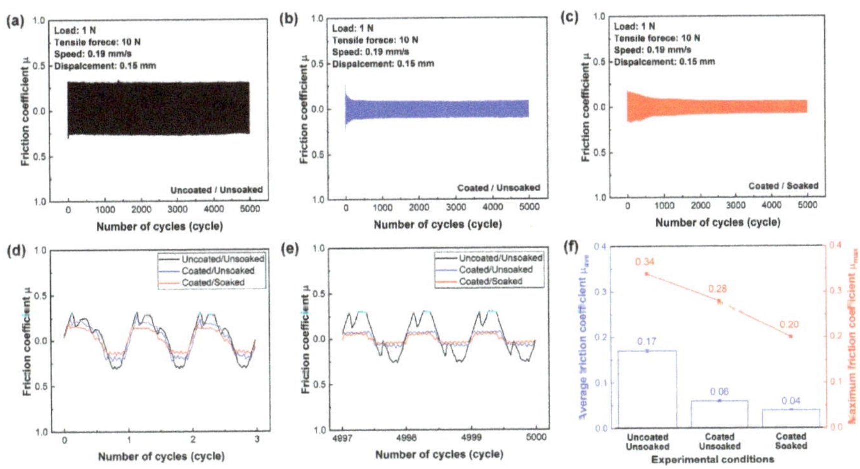

Figure 5. Friction behaviors of the reciprocating sliding contacts of three types of stainless steel archwires and brackets in artificial saliva environment. The carbon film was deposited onto the stainless steel archwire surface at a substrate bias voltage of +50 V and a deposition time of 80 min. The term "soaked" means the carbon-film-coated stainless steel archwire was soaked in a fluoride toothpaste mixed solution for 16 h. (**a**) Typical friction curve of the uncoated and unsoaked stainless steel archwire. (**b**) Typical friction curve of the carbon-film-coated and unsoaked stainless steel archwire. (**c**) Typical friction curve of the carbon-film-coated and soaked stainless steel archwire. (**d**) Typical friction curve of the initial three cycles (i.e., 1–3). (**e**) Typical friction curve of the final three cycles (i.e., 4997–4999). (**f**) Maximum static friction coefficients and stable average friction coefficients.

The maximum static friction coefficient and average friction coefficient of three types of sliding contact combination are summarized in Figure 5f. The maximum static friction coefficient and average friction coefficient of 0.34 and 0.17 were observed with the uncoated/unsoaked archwire, respectively. Interestingly, the lowest maximum static friction coefficient and average friction coefficient of 0.20 and 0.04 were obtained with the coated/soaked archwire, respectively. There was no degradation of the frictional performance of the carbon-film-coated archwire after the soaking treatment; nevertheless, the friction coefficient was slightly decreased from 0.06 to 0.04. Hence, the carbon-film-coated archwire fabricated under a substrate bias voltage of +50 V and a deposition time of 80 min not only exhibited excellent corrosion resistance, but also retained friction-reducing property even after the long-term exposure to a fluoride-containing environment.

Optical images of the wear scars on three types of stainless steel archwires after sliding against stainless steel brackets in an artificial saliva environment are shown in Figure 6a–f. It was clearly observed that carbon film still covered the whole wear scar on the coated stainless steel archwire, regardless of soaking or not, after running against a stainless steel bracket in an artificial saliva environment, as shown in Figure 6d,f. The peeling off or detachment of the carbon film from the archwires could not be found for the soaked carbon film. Moreover, the SEM–EDS images (i.e., Figure 6g–n) of the wear scar on the carbon-film-coated stainless steel archwires further confirmed that the carbon film did not wear out after the reciprocating sliding friction test in an artificial saliva environment. Mild wear of the carbon film was achieved in an artificial saliva environment whether soaked or not in a fluorine-containing environment. With respect to the wear scar dimensions, the width of the wear scars remained consistent for both soaked and unsoaked carbon-film-coated archwires; however, the length of the wear scar on the unsoaked archwire (i.e., Figure 6d) was approximately twice that of the soaked one (i.e., Figure 6f). Moreover, mild wear was clearly observed on the wear tracks of the brackets, as shown in Figure S2.

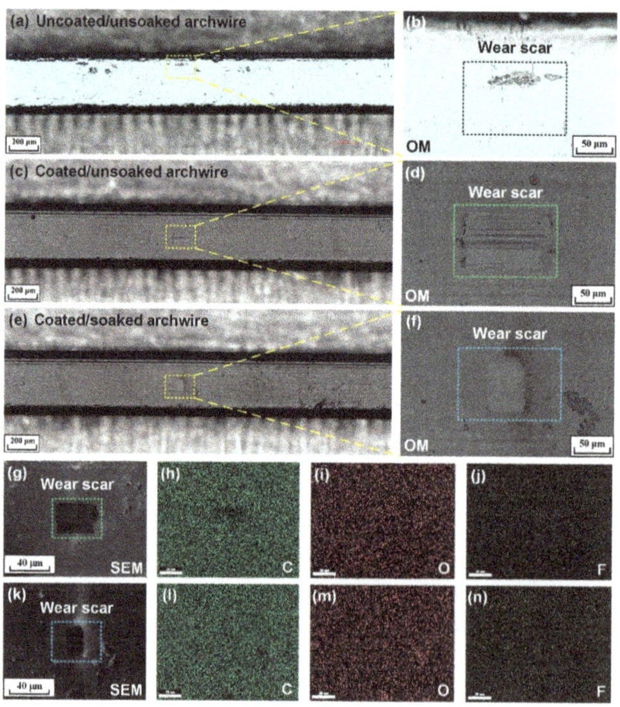

Figure 6. Optical images and SEM–EDS images of the worn surfaces on the stainless steel archwires after running against stainless steel brackets in an artificial saliva environment. (**a**,**b**) Wear scar on the uncoated/unsoaked archwire. (**c**,**d**,**g**–**j**) Wear scar on the coated/unsoaked archwire. (**e**,**f**,**k**–**n**) Wear scar on the coated/soaked archwire. (**a**–**f**) Optical images and (**g**,**k**) SEM images of the wear scars on the archwires. (**h**,**l**) Carbon element, (**i**,**m**) oxygen element, and (**j**,**n**) fluorine element distributions of the wear scars on the carbon-film-coated stainless steel archwires. The scale bar in EDS image is 20 mm.

4. Discussion

4.1. Electrochemical Corrosion of the Carbon-Film-Coated Archwires

Figure 7 presents the polarization curves of the carbon-film-coated stainless steel archwires prepared under different substrate bias voltages, which exhibit distinct corrosion behaviors in a fluoride toothpaste mixed solution. Detailed electrochemical corrosion parameters are summarized in Table 1. It was clarified that the carbon film deposited at the substate bias voltage of +50 V exhibited the highest corrosion potential (−135.1 mV), whereas the carbon film deposited at the substrate bias voltage of +5 V showed the lowest value (−226.3 mV), both significantly higher than that of the uncoated stainless steel archwire (−303.1 mV). Regarding corrosion current density, the highest value (0.229 μA/cm^2) was observed for the +5 V specimen, while the lowest value (0.072 μA/cm^2) was recorded for the +50 V specimen, all of which were lower than the uncoated stainless steel archwire (0.336 μA/cm^2). A higher corrosion potential together with a lower corrosion current density was obtained with the increase in the substrate bias voltage from +5 V to +50 V. These findings indicated that increasing the substrate bias voltage led to a gradual increase in corrosion potential and a corresponding decrease in corrosion current density, thus demonstrating the enhanced corrosion resistance of the carbon-film-coated stainless steel archwires. Moreover, in the previous studies of corrosion resistance of stainless steel archwire, the potential and current density of AISI 304 stainless steel (18% Cr and 8% Ni) were evaluated to be −447 mV and 0.08 μA/cm^2, respectively, in an artificial saliva environment [40]. The potential and current density of AISI 316L stainless steel (18% Cr and 12% Ni) were measured to be −483 mV and 0.02 μA/cm^2, respectively, in an artificial saliva environment [41]. Therefore, a relative increase in the current density and corrosion rate in a fluoride-containing environment is reasonable in the present work. Hence, it was confirmed that the carbon film deposited for 80 min at a substrate bias voltage of +50 V exhibited the best corrosion-resistant performance.

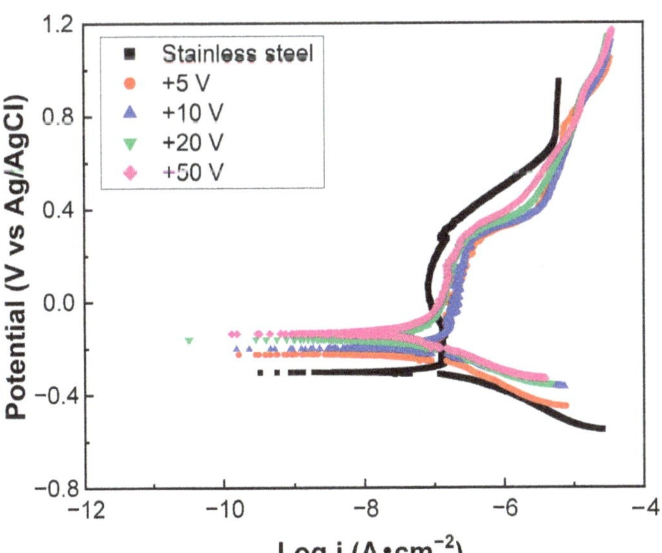

Figure 7. Polarization curves of the uncoated and carbon-film-coated stainless steel archwires in a fluoride toothpaste mixed solution.

Table 1. Electrochemical corrosion parameters of uncoated and carbon-film-coated stainless steel archwires in a fluoride toothpaste mixed solution.

Substrate Bias Voltage (V)	Potential (mV/SCE)	Current Density (μA/cm^2)	Corrosion Rate (mpy)
None	−303.1	0.336	62.81×10^{-3}
+5	−226.3	0.229	34.40×10^{-3}
+10	−200.8	0.253	37.97×10^{-3}
+20	−159.5	0.084	12.63×10^{-3}
+50	−135.1	0.072	10.74×10^{-3}

The experimental results revealed that when the substrate bias voltage was set at +5 V, the carbon films detached after soaking in the fluorine-containing environment for 16 h, regardless of the deposition time. Raman spectroscopy and TEM characterizations confirmed that the carbon films deposited under +5 V exhibited an amorphous carbon structure, which demonstrated poor stability in the fluorine-containing environment and failed to provide effective protection for the underlying archwire substrate. In contrast, when the deposition time was fixed at 80 min, the degree of film detachment decreased progressively with increasing substrate bias voltage, and no detachment was observed at +50 V. Furthermore, Raman and TEM analyses indicated that the carbon films deposited at +50 V exhibited a nanocrystalline graphene structure, suggesting enhanced corrosion resistance and the ability to effectively resist the aggressive effects of active fluorine species.

A hypothesis for the corrosion mechanisms of amorphous carbon-film-coated and GSEC-film-coated stainless steel archwires in fluoride-containing solutions is proposed, as schematically shown in Figure 8a,b, respectively. Specifically, Figure 8(a1,a2) illustrates that when soaking in a fluoride-containing environment, the amorphous carbon film (e.g., +5 V & 25 min) undergoes partial delamination due to the aggressive attack of active fluoride ions. This delamination re-exposed the underlying stainless steel archwire to the corrosive medium. Surface composition analysis revealed enrichment of F and O elements in the exposed regions, whereas their concentrations remained relatively low in the regions still covered by the intact carbon film. It was clarified that the amorphous carbon film provided a certain degree of protection as long as it remained intact. However, once delamination occurred, the protective capability was markedly reduced, allowing fluoride ions to further corrode surface of the stainless steel archwire. It was suggested that amorphous carbon film was inherently unstable in the fluoride-containing environment and was unable to provide sustained protection. On the contrary, Figure 8(b1,b2) demonstrates that the GSEC film exhibits a vertically aligned and densely packed microstructure on the stainless steel archwire surface, indicating superior structural integrity. Surface composition characterization confirmed that the levels of F and O elements on the GSEC film were minimal, suggesting strong adhesion even under fluoride exposure. While active fluoride ions may slightly affect the amorphous carbon regions at the edges, the nanocrystalline structure itself remained largely unaffected. It was revealed that the graphene nanocrystalline carbon film effectively acted as a barrier against fluoride ion penetration, therefore preventing severe corrosion of the stainless steel archwire in a fluoride-containing environment.

4.2. Wear Rate of the Carbon-Film-Coated Archwires

To further investigate the wear resistance of carbon-film-coated archwires, 3D laser confocal microscopy was employed to systematically characterize the surface morphology of the worn surfaces on the stainless steel archwires. Figure 9a–d illustrates the three-dimensional images and corresponding two-dimensional cross-sectional profiles at five distinct points across the wear scar, measured along the white dashed lines. The wear depth and width of the carbon-film-coated archwires remained relatively consistent at approximately 0.65 μm (less than the thickness of the film, as shown in Figure S3) and 70 μm, respectively. However, a notable difference was observed in the wear scar length, where the length of the unsoaked coated archwire (100 mm) was approximately twice that of the soaked one (50 mm). As shown in Figure 9e, the wear volume of the unsoaked/coated archwire was 5.30×10^{-6} mm^3, whereas that of the soaked/coated archwire was 2.30×10^{-6} mm^3, with the former being around twice that of the latter. In terms of wear rate, the values were 4.82×10^{-6} mm^3/Nm and 4.43×10^{-6} mm^3/Nm, respectively, indicating a relatively small difference between the two experimental conditions. It was confirmed that soaking in a fluoride toothpaste mixed solution for 16 h did not significantly degrade the performance of the GSEC-film-coated stainless steel archwire, which retained its excellent wear resistance.

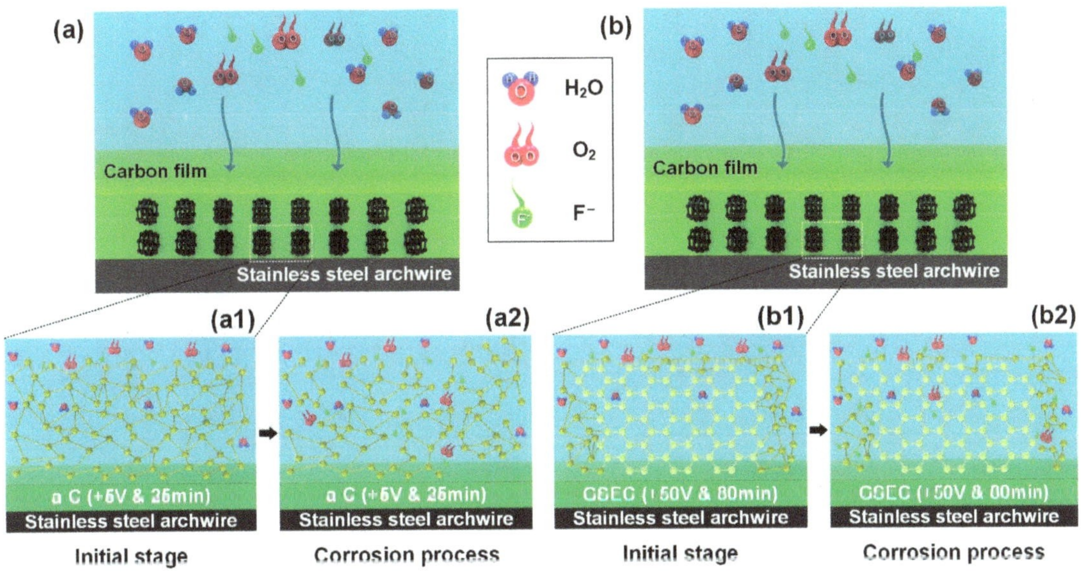

Figure 8. Schematic illustrations of the corrosion mechanism of the carbon-film-coated stainless steel archwire in a fluoride-containing environment. (**a**) Amorphous carbon film (+5 V & 25 min)-coated stainless steel archwire. (**a1**) Initial stage and (**a2**) corrosion process for amorphous carbon film. (**b**) Graphene-sheet-embedded-carbon film (+50 V & 80 min)-coated stainless steel archwire. (**b1**) Initial stage and (**b2**) corrosion process for GSEC film.

4.3. Low Friction Mechanism of the Carbon-Film-Coated Archwires

Carbon films with varying deposition times and substrate bias voltages were deposited on the surface of orthodontic archwires, followed by the immersion test in a fluorine-containing environment. The results revealed that the carbon film prepared under a deposition time of 80 min and a substrate bias voltage of +50 V exhibited superior corrosion resistance. Following 16 h of soaking, a friction and wear test was performed between the archwire and bracket. The experimental data indicated no significant change in the friction coefficient or wear rate before and after soaking. Specifically, the friction coefficient was measured at 0.06 prior to soaking and decreased to 0.04 after soaking, while the corresponding wear rates were 4.82×10^{-6} mm^3/Nm and 4.43×10^{-6} mm^3/Nm, respectively. Similarly, no statistically significant difference in static friction force was observed among the artificial saliva corrosion, mouthwash corrosion, and non-corrosion groups upon introduction of the DLC-coated metal bracket [42]. It was demonstrated that the graphene nanocrystalline carbon film retained excellent tribological performance even after exposure to a fluorine-containing environment, highlighting its dual advantages of corrosion resistance as well as anti-friction and anti-wear properties. Notably, it could effectively reduce the archwire–bracket friction coefficient from 0.17 to 0.06. Moreover, surface composition analysis of the carbon film-coated archwires (80-min deposition time and +50 V substrate bias) before and after immersion revealed no appreciable alteration in elemental composition, further confirming the structural stability and chemical inertness of the carbon film during the friction and wear process.

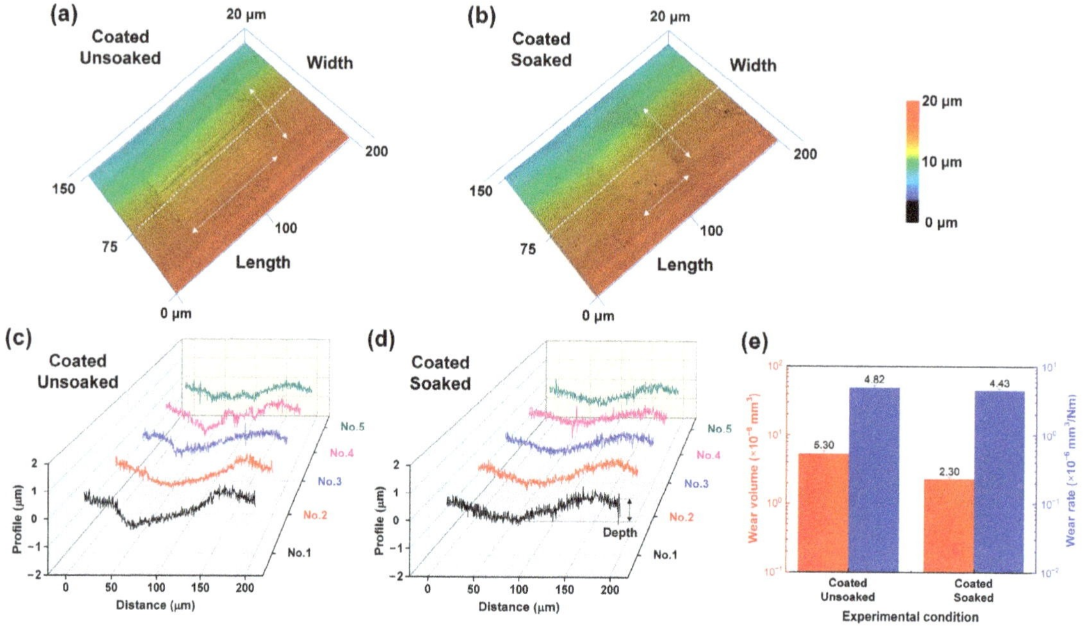

Figure 9. Calculation of the wear volume and specific wear rate of the unsoaked and soaked carbon-film-coated (i.e., +50 V & 80 min) stainless steel archwires after sliding against stainless steel brackets in an artificial saliva environment. (**a**) Three-dimensional optical image and corresponding (**b**) two-dimensional cross-sectional profile of the wear scar on the unsoaked/coated stainless steel archwire. (**c**) Three-dimensional optical image and corresponding (**d**) two-dimensional cross-sectional profile of the wear scar on the soaked/coated stainless steel archwire. (**e**) Calculated wear volume and specific wear rate.

A hypothesis for the low friction mechanism involving GSEC-film-coated stainless steel archwire and a stainless steel bracket is proposed, as schematically shown in Figure 10. The presence and/or formation of graphene nanocrystallites at the contact interfaces of amorphous carbon films, GSEC films, and graphene nanocrystalline carbon nitride (GNCN) films has been demonstrated to contribute significantly to achieving friction coefficients below 0.10 [32,33,35]. TEM analysis revealed that the graphene nanocrystalline carbon film exhibited a vertically oriented growth morphology on the surface of stainless steel archwire. Its densely packed microstructure contributed to excellent adhesion stability, enabling the film to maintain superior tribological performance even after soaking in a fluoride-containing environment. Figure 10a illustrates the contact interface between the archwire and bracket, where artificial saliva is continuously supplied and distributed across the contact zone. Under applied loading, intimate contact was established between the two components, with the archwire surface protected by a vertically aligned graphene nanocrystalline carbon film.

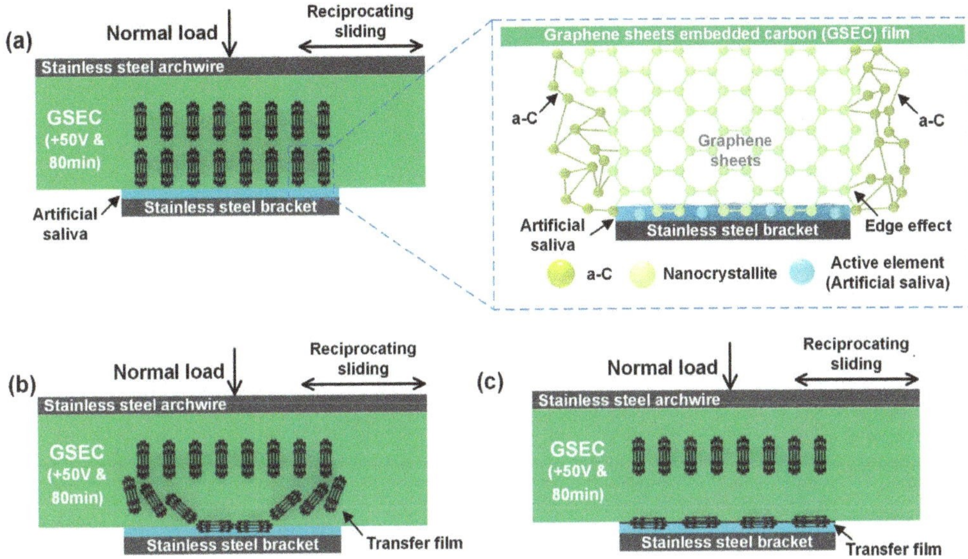

Figure 10. Schematic diagram of the low friction mechanism of the GSEC film (i.e., +50 V & 80 min) coated stainless steel archwire sliding against stainless steel bracket in artificial saliva environment. (**a**) Initial stage of the sliding contact combination of the stainless steel archwire–bracket. Inset illustration is the detailed atomic structure of the GSEC film. (**b**) Formation of the transfer film at the running-in stage. (**c**) Steady-state low friction coefficient of the GSEC-film-coated archwire with the covering of the transfer film rich in graphene sheets and adsorbed salivary layer.

Figure 10b presents a schematic representation of the carbon film's microstructure, highlighting how the edge effects of the densely packed nanocrystals promote the formation of an irregular amorphous carbon layer. During the reciprocating sliding between the archwire and bracket, a transfer film rapidly formed on the contact interface, leading to a significant reduction in the friction coefficient. Concurrently, the continuous supply of artificial saliva above the contact region facilitated the formation of an adsorbed lubricating film on the bracket surface [43,44], further enhancing the lubrication conditions, as illustrated in Figure 10c,d. The synergetic lubricating effects of the transfer film and the saliva-derived lubricating film resulted in a rapid decrease in friction coefficient, as shown in Figure 5b,c, which was subsequently maintained at a consistently low level throughout the entire sliding process. These findings not only elucidated the influence of substrate bias voltage on the stability of the carbon film but also highlighted its potential as a protective coating for metallic materials in corrosion-prone environments.

In this study, the GSEC film prepared under a substrate bias voltage of +50 V and a deposition duration of 80 min was found to exhibit excellent friction, wear, and corrosion resistance properties in a customized laboratory artificial saliva environment. These outstanding performances are favorable for the potential design of low-friction, light-force orthodontic appliances aimed at enhancing treatment efficiency and reducing clinical risks [6]. However, it is crucial to emphasize that the experimental results discussed above were obtained under controlled laboratory conditions (i.e., in vitro), which still differ significantly from the complex intraoral environment (i.e., in vivo). Furthermore, in practical clinical applications of carbon-based films, critical factors such as long-term stability, durability, and biocompatibility must be rigorously evaluated. These aspects require systematic investigation in future research.

5. Conclusions

In this study, carbon films were successfully fabricated on the working surfaces (width of 0.017 inch) of the stainless steel archwires under various deposition times (10 min to 80 min) as well as substrate bias voltages (+5 V to +50 V) using a self-designed MCECR plasma sputtering system with low energy electron irradiation. Structural analysis revealed that the carbon films deposited at the substrate bias voltages of +5 V and +10 V were amorphous carbon, whereas those fabricated at the substrate bias voltages of +20 V and +50 V were identified as graphene nanocrystalline carbon (i.e., GSEC) films. It was found that the area of carbon film detachment from the stainless steel archwire gradually decreased with increasing substrate bias voltage after soaking in a mixed solution of fluoridated toothpaste for 16 h. Notably, the GSEC film prepared at the substrate bias voltage of +50 V for 80 min exhibited no detachment at all, indicating excellent corrosion resistance. Electrochemical experiments further confirmed that the corrosion resistance of the fabricated carbon films improved significantly with higher substrate bias voltage. Moreover, the friction and wear performances of the GSEC films remained stable even after soaking in a fluorine-containing environment for 16 h. The friction coefficients before and after soaking were measured as 0.06 and 0.04, respectively, while the corresponding wear rates were 4.82×10^{-6} mm^3/Nm and 4.43×10^{-6} mm^3/Nm. Based on the characterization results, the corrosion resistance and friction mechanisms of the outstanding GSEC films were investigated. The nanocrystalline carbon film grew vertically on the archwire surface, and its dense microstructure contributed to its stability and resistance to attack by the active fluoride ions. Under the combined effects of applied load and shear stress, a transfer film formed rapidly on the sliding contact interface. The continuous supply of artificial saliva enhanced the lubrication properties of the adsorbed saliva film on the bracket. These two mechanisms synergistically led to a rapid reduction in the friction coefficient and its subsequent maintenance at a stable low level. Therefore, this work provides new insights into the application of GSEC films in friction and wear reduction as well as corrosion resistance, promoting the practical clinical application for utilizing the coated orthodontic stainless steel archwires.

Supplementary Materials: The following supporting information can be downloaded at https://www.mdpi.com/article/10.3390/nano15211615/s1: Figure S1: Photograph of the customized electron cyclotron resonance plasma sputtering apparatus. Specifically, 1 is left coil, 2 is middle coil, 3 is right coil, 4 is main vacuum chamber, 5 is pre-vacuum chamber, 6 is program control panel, 7 is substrate holder, 8 is voltage and current control panel, and 9 is the microwave generator; Table S1: Experimental conditions of the carbon films deposition; Figure S2: Optical images of the wear tracks on the stainless steel brackets after sliding against different archwires in an artificial saliva environment. (a) Uncoated/unsoaked archwire. (b) Coated/unsoaked archwire. (c) Coated/soaked archwire. The wear tracks, which are primarily located at the edges of the slot surfaces, are indicated by red dotted rectangles; Figure S3: Cross-sectional profile of the GSEC film fabricated under a substrate bias voltage of +50 V and a deposition time of 80 min. The calculated film thickness was determined to be 0.66 mm.

Author Contributions: Conceptualization, M.H., S.C. and P.W.; methodology, M.H., S.C. and P.W.; validation, M.H. and S.C.; investigation, M.H. and S.C.; resources, P.W.; data curation, M.H. and S.C.; writing—original draft preparation, M.H. and P.W.; writing—review and editing, P.W.; visualization, M.H. and P.W.; supervision, P.W.; project administration, P.W.; funding acquisition, P.W. All authors have read and agreed to the published version of the manuscript.

Funding: This research was funded by the Shenzhen Natural Science Found Project (grant number: 20231121101714001).

Data Availability Statement: The original contributions presented in this study are included in the article/Supplementary Material. Further inquiries can be directed to the corresponding author.

Acknowledgments: The Electron Microscopy Center (EMC) of Shenzhen University is kindly acknowledged for technical support in TEM and SEM analysis.

Conflicts of Interest: The authors declare no conflicts of interest.

References

1. Niu, Q.; Chen, S.; Bai, R.; Lu, Y.; Peng, L.; Han, B.; Yu, T. Dynamics of the oral microbiome during orthodontic treatment and antimicrobial advances for orthodontic appliances. *Iscience* **2024**, *27*, 111458. [CrossRef]
2. Cenzato, N.; Nobili, A.; Maspero, C. Prevalence of dental malocclusions in different geographical areas: Scoping review. *Dent. J.* **2021**, *9*, 117. [CrossRef]
3. Lombardo, G.; Vena, F.; Negri, P.; Pagano, S.; Barilotti, C.; Paglia, L.; Colombo, S.; Orso, M.; Cianetti, S. Worldwide prevalence of malocclusion in the different stages of dentition: A systematic review and meta-analysis. *Eur. J. Paediatr. Dent.* **2020**, *21*, 115–122. [CrossRef]
4. Zou, J.; Meng, M.; Law, C.S.; Rao, Y.; Zhou, X. Common dental diseases in children and malocclusion. *Int. J. Oral Sci.* **2018**, *10*, 7. [CrossRef]
5. Guo, L.; Feng, Y.; Guo, H.G.; Liu, B.W.; Zhang, Y. Consequences of orthodontic treatment in malocclusion patients: Clinical and microbial effects in adults and children. *BMC Oral Health* **2016**, *16*, 112. [CrossRef]
6. Burrow, S.J. Friction and resistance to sliding in orthodontics: A critical review. *Am. J. Orthod. Dentofac. Orthop.* **2009**, *135*, 442–447. [CrossRef]
7. Arici, N.; Akdeniz, B.S.; Oz, A.A.; Gencer, Y.; Tarakci, M.; Arici, S. Effectiveness of medical coating materials in decreasing friction between orthodontic brackets and archwires. *Korean J. Orthod.* **2021**, *51*, 270–281. [CrossRef]
8. Espinoza-Montero, P.J.; Montero-Jiménez, M.; Fernández, L.; Paz, J.L.; Piñeiros, J.L.; Ceballos, S.M. In vitro wearing away of orthodontic brackets and wires in different conditions: A review. *Heliyon* **2022**, *8*, e10560. [CrossRef]
9. Xie, Q.; Li, P.; Zhou, Z.; Bai, G.; Sun, K.; Li, X.; Tao, H.; Yang, H.; Zou, D.; Yang, C. Fabrication of three-dimensional orthodontic force detecting brackets and preliminary clinical test for tooth movement simulation. *Heliyon* **2023**, *9*, e19852. [CrossRef]
10. He, L.; Zhang, W.; Liu, J.; Pan, Y.; Li, S.; Xie, Y. Applications of nanotechnology in orthodontics: A comprehensive review of tooth movement, antibacterial properties, friction reduction, and corrosion resistance. *BioMed. Eng. OnLine* **2024**, *23*, 72. [CrossRef]
11. Ardila, C.M.; Arrubla-Escobar, D.E.; Vivares-Builes, A.M. Efficacy of microchips and 3D sensors for orthodontic force measurement: A systematic review of in vitro studies. *Orthod. Craniofac. Res.* **2024**, *27*, 88–102. [CrossRef]
12. Ardila, C.M.; Jiménez-Arbeláez, G.A.; Vivares-Builes, A.M. Efficacy of wireless sensors in assessing occlusal and bite forces: A systematic review. *J. Oral Rehabil.* **2024**, *51*, 1337–1347. [CrossRef]
13. Khoury, E.S.; Abboud, M.; Bassil-Nassif, N.; Bouserhal, J. Effect of a two-year fluoride decay protection protocol on titanium brackets. *Int. Orthod.* **2011**, *9*, 432–451. [CrossRef]
14. Khoury, E.S.; Abboud, M.; Bassil-Nassif, N.; Bouserhal, J. Effect of eliminating the residual fluoride gel on titanium bracket corrosion. *Int. Orthod.* **2011**, *9*, 298–315. [CrossRef]
15. Huang, H.H.; Lee, T.H.; Huang, T.K.; Lin, S.Y.; Chen, L.K.; Chou, M.Y. Corrosion resistance of different nickel-titanium archwires in acidic fluoride-containing artificial saliva. *Angle Orthod.* **2010**, *80*, 547–553. [CrossRef]
16. Huang, H.H. Variation in surface topography of different NiTi orthodontic archwires in various commercial fluoride-containing environments. *Dent. Mater.* **2007**, *23*, 24–33. [CrossRef]
17. Trolic, I.M.; Serdarevic, N.L.; Todoric, Z.; Budimir, A.; Spalj, S.; Curkovic, H.O. Corrosion of orthodontic archwires in artificial saliva in the presence of Lactobacillus reuteri. *Surf. Coat. Technol.* **2019**, *370*, 44–52. [CrossRef]
18. Ique, M.M.A.; Ferreira, M.F.; Delbem, A.C.B.; de Mendonça, M.R. Corrosion-induced changes in surface properties and roughness of orthodontic wires. *Am. J. Orthod. Dentofacial Orthop.* **2024**, *165*, 565–575. [CrossRef]
19. Mikulewicz, M.; Suski, P.; Tokarczuk, O.; Warzyńska-Maciejewska, M.; Pohl, P.; Tokarczuk, B. Metal ion release from orthodontic archwires: A comparative study of biocompatibility and corrosion resistance. *Molecules* **2024**, *29*, 5685. [CrossRef]
20. Wijesinghe, D.M.L.; Gunathilake, W.S.S.; Weerasekera, W.B.M.C.R.D.; Jayasinghe, U.J.M.A.L. Corrosion analysis of orthodontic brackets and arch wires–An in vitro study. *Ceylon J. Sci.* **2024**, *53*, 75–85. [CrossRef]
21. Stoyanova-Ivanova, A.; Georgiev, V.; Martins, J.N.R. Nickel ion release in nickel-containing orthodontics archwires: A narrative review of in vitro and in vivo studies. *Dent. J.* **2025**, *13*, 206. [CrossRef] [PubMed]

22. Bącela, J.; Łabowska, M.B.; Detyna, J.; Zięty, A.; Michalak, I. Functional coatings for orthodontic archwires—A review. *Materials* **2020**, *13*, 3257. [CrossRef] [PubMed]
23. Fróis, A.; Santos, A.C.; Louro, C.S. Corrosion of fixed orthodontic appliances: Causes, concerns, and mitigation strategies. *Metals* **2023**, *13*, 1955. [CrossRef]
24. Zhang, R.; Han, B.; Liu, X. Functional surface coatings on orthodontic appliances: Reviews of friction reduction, antibacterial properties, and corrosion resistance. *Int. J. Mol. Sci.* **2023**, *24*, 6919. [CrossRef]
25. Ryu, H.; Bae, I.; Lee, K.; Hwang, H.; Lee, K.; Koh, J.; Cho, J. Antibacterial effect of silver-platinum coating for orthodontic appliances. *Angle Orthod.* **2012**, *82*, 151–157. [CrossRef]
26. Wu, H.; Yang, J.; Yan, Y.; Zheng, B.; Algahefi, A.L.; Ma, S.; Liu, Y. Study of Al–SiO_2 aesthetic composite coating on orthodontic metal archwire. *Coatings* **2022**, *12*, 746. [CrossRef]
27. Espinar, E.; Llamas, J.M.; Michiardi, A.; Ginebra, M.P.; Gil, F.J. Reduction of Ni release and improvement of the friction behaviour of NiTi orthodontic archwires by oxidation treatments. *J. Mater. Sci Mater. Med.* **2011**, *22*, 1119–1125. [CrossRef]
28. Zhang, H.; Guo, S.; Wang, D.; Zhou, T.; Wang, L.; Ma, J. Effects of nanostructured, diamondlike, carbon coating and nitrocarburizing on the frictional properties and biocompatibility of orthodontic stainless steel wires. *Angle Orthod.* **2016**, *86*, 782–788. [CrossRef]
29. Sui, J.H.; Cai, W. Formation of diamond-like carbon (DLC) film on the NiTi alloys via plasma immersion ion implantation and deposition (PIIID) for improving corrosion resistance. *Appl. Surf. Sci.* **2006**, *253*, 2050–2055. [CrossRef]
30. Kobayashi, S.; Ohgoe, Y.; Ozeki, K.; Hirakuri, K.; Aoki, H. Dissolution effect and cytotoxicity of diamond-like carbon coatings on orthodontic archwires. *J. Mater. Sci. Mater. Med.* **2007**, *18*, 2263–2268. [CrossRef]
31. Fróis, A.; Aleixo, A.S.; Evaristo, M.; Santos, A.C.; Louro, C.S. Can a-C:H-sputtered coatings be extended to orthodontics? *Coatings* **2021**, *11*, 832. [CrossRef]
32. Pan, Z.; Zhou, Q.; Wang, P.; Diao, D. Robust low friction performance of graphene sheets embedded carbon films coated orthodontic stainless steel archwires. *Friction* **2022**, *10*, 142–158. [CrossRef]
33. Wang, P.; Luo, X.; Qin, J.; Pan, Z.; Zhou, K. Effect of graphene sheets embedded carbon films on the fretting wear behaviors of orthodontic archwire–bracket contacts. *Nanomaterials* **2022**, *12*, 3430. [CrossRef] [PubMed]
34. Zhang, W.Q.; Diao, D.F.; Sun, K.; Fan, X.; Wang, P. Study on friction-electrification coupling in sliding-mode triboelectric nanogenerator. *Nano Energy* **2018**, *48*, 456–463. [CrossRef]
35. Wang, P.; Xue, P.; Chen, C.; Diao, D. Structural and tribological behaviors of graphene nanocrystallited carbon nitride films. *Appl. Surf. Sci.* **2019**, *495*, 143591. [CrossRef]
36. Liu, C.; Chu, P.K.; Lin, G.; Yang, D. Effects of Ti/TiN multilayer on corrosion resistance of nickel-titanium orthodontic brackets in artificial saliva. *Corrosion Sci.* **2007**, *49*, 3783–3796. [CrossRef]
37. Wang, C.; Zhang, X.; Diao, D. Nanosized graphene crystallite induced strong magnetism in pure carbon films. *Nanoscale* **2015**, *7*, 4475–4481. [CrossRef]
38. Saleemi, A.S.; Singh, R.; Sun, W.; Luo, Z.; Zhang, X. Large magnetoresistance of amorphous carbon films. *Carbon* **2017**, *122*, 122–127. [CrossRef]
39. Mirjalili, M.; Momeni, M.; Ebrahimi, N.; Moayed, M.H. Comparative study on corrosion behaviour of Nitinol and stainless steel orthodontic wires in simulated saliva solution in presence of fluoride ions. *Mater. Sci. Eng. C* **2013**, *33*, 2084–2093. [CrossRef]
40. Begum, A.A.; Vahith, R.M.; Mohamed, M.K.; Kotra, V.; Shaik, B.; Al-Kahtani, A. Corrosion mitigation on orthodontic wire made of SS 18/8 alloy using esomeprazole tablet (Esiloc-40 mg) in artificial saliva. *J. Saudi Chem. Soc.* **2023**, *27*, 101681. [CrossRef]
41. Rajendran, S.; Raja, U.T.; Majellah, L. Corrosion resistance of orthodontic wire made of SS 316L alloy in artificial saliva in the presence of amlodipine tablet by electrochemical studies. *Zast. Mater.* **2020**, *61*, 251–258. [CrossRef]
42. Huang, T.H.; Guo, J.U.; Kao, C.T. A comparison of the friction associated with diamond-like carbon (DLC) or titanium nitride (TiN) plating metal brackets. *Surf. Coat. Technol.* **2010**, *205*, 1917–1921. [CrossRef]
43. Dridi, A.; Bensalah, W.; Mezlini, S.; Tobji, S.; Zidi, M. Influence of bio-lubricants on the orthodontic friction. *J. Mech. Behav. Biomed. Mater.* **2016**, *60*, 1–7. [CrossRef]
44. Yakubov, G.E.; Macakova, L.; Wilson, S.; Windust, J.H.; Stokes, J.R. Aqueous lubrication by fractionated salivary proteins: Synergistic interaction of mucin polymer brush with low molecular weight macromolecules. *Tribol. Int.* **2015**, *89*, 34–45. [CrossRef]

nanomaterials

Article

Influences of Initial Stresses on Formation of Shear Bands and Mechanical Properties in Binodal Decomposed Metallic Glass Composites

Yongwei Wang [1,*], Guangping Zheng [2] and Mo Li [3,*]

[1] Collaborative Innovation Center of Steel Technology, University of Science and Technology Beijing, Beijing 100083, China
[2] Department of Mechanical Engineering, Hong Kong Polytechnic University, Hung Hom, Kowloon, Hong Kong, China; mmzheng@polyu.edu.hk
[3] School of Materials Science and Engineering, Georgia Institute of Technology, Atlanta, GA 30332, USA
* Correspondence: yw_wang@ustb.edu.cn (Y.W.); mo.li@gatech.edu (M.L.)

Abstract: Structural heterogeneity is useful for improving the plasticity of metallic glasses (MGs) by blocking the propagation of shear bands (SBs). The introduction of a heterogeneous structure often introduces residual stresses, which significantly influences the deformation behaviors of MGs; however, the quantitative impact of residual/initial stresses on shear banding remains unclear. In this work, through finite element models, we demonstrate that residual/initial stresses can promote the initiation of SBs at the interfaces between droplet or particle reinforcements and the matrix in Binodal decomposed metallic glass composites (BDMGCs). These reinforcements do not effectively block the SBs when the fraction of particle reinforcement is very low. We demonstrate that a heterogeneous distribution of initial tensile stresses reduces the strength of BDMGCs, particularly in those containing a homogenous matrix. This profound understanding of the synergistic effects arising from a heterogeneous microstructure and initial stresses could effectively promote the design and optimization of MGs and their composites.

Keywords: initial stress; Binodal decomposed metallic glass composites; free volume theory; shear banding; strength

1. Introduction

Metallic glasses (MGs), or amorphous alloys, have many excellent mechanical and physical properties in comparison to their coarse-grained crystalline counterparts, which are derived from a disordered atomic structure [1]. However, one major drawback is their limited macroscopic plastic deformability at room temperature due to the formation of highly localized shear bands (SBs) [2]. Enormous efforts have been devoted in recent years to improving the plasticity of MGs. The concept of heterogeneity, including composites, is widely accepted to be useful for improving the plasticity of MGs by blocking SBs [3–26]. For instance, the introduction of a second phase or inclusions, surface treatments, or mechanical deformations such as shot peening or grinding in the surface region can induce structural, residual stress or chemical heterogeneity that can contribute to enhanced ductility [25–43].

Residual stress is usually accompanied by the generated heterogenous structure. For example, varying cooling rates during quick quenching not only create the gradient in free volumes (FVs) but also induce residual stress [28]. In an as-cast MG cylinder, the profile of residual stress exhibits a parabolic form, and the maximum tensile stress at the center is roughly half the maximum compressive stress at the surface [36–38]. Shot peening can also

 https://doi.org/10.3390/nano15221725

introduce surface compressive stress and promote the heterogenous distribution of many SBs [26]. In metallic glass composites (MGCs), particularly in ex situ MGCs, residual stress arises from differences in thermal conductivity and the elastic modulus between the matrix and reinforced phase [36].

In particle-reinforced MGCs, including in situ and ex situ ones, the secondary reinforcement is anticipated to block the propagation of SBs in the matrix and subsequently enhance ductility or toughness [24]. In ex situ MGCs, interfacial reactions take place at particle–matrix interfaces, leading to low wettability and making it challenging to enhance plasticity [6–9]. In contrast, in in situ MGCs, there are strong atomic-level bonding interfaces between the secondary reinforcement and glassy matrix. When alloys contain a large positive enthalpy of mixing and possess high glass forming ability, metallic glass-forming alloys can separate into two different amorphous phases during solidification [15–22]. An interconnected-type structure (Spinodal decomposition) or a droplet-type (Binodal decomposition) structure can be obtained [20,21]. Heterogeneous microstructures can be at various scales ranging from micro-scales to nano-scales. For the differences in coefficient of thermal expansion and mechanical property, the glassy reinforcement and glassy matrix will both exsit the compressive and tensile re-sidual stresses.

We note that certain Binodal decomposed metallic glass composites (BDMGCs), which have a large fraction of reinforcement, show essentially zero ductility in tension and limited plasticity in compression [22]. Most experimental and theoretical analyses focus on the effects of size and fraction and the morphology of the particles. However, the impact of the spatial residual stresses induced by the mismatch of thermophysical properties on the mechanical properties and shear banding process is rarely quantitatively investigated. This work will quantitatively assess the impact of initial spatial stresses (both compressive and tensile stresses) on the initiation and propagation of SBs in BDMGCs. Additionally, we will explore and investigate the synergistic effects arising from a heterogeneous microstructure and initial stresses. It is essential to develop new theories that effectively incorporate the influences of residual/initial stress in the structural design of MGs or MGCs.

2. Materials and Methods

Recently, the effects of compressive residual stress on the hardness of MGs have been extensively studied [25,26,29–45]. For shot-peened MGs, the following argument is proposed: softening via the introduction of SBs and hardening via compressive residual stress [26]. This phenomenon has also been observed in silicate glass. For instance, the exceptional strength of Prince Rupert's drops is attributed to substantial surface compressive stresses [31]. Moreover, the magnitude of the surface compression developed in tempered glass is significantly higher than that in float glass. However, as the tensile zone of silicate glass is subjected to load, it becomes more brittle. As we know, as the stress equilibrium exist in a stable or metastable system, surface pre-compression should be counterbalanced by an associated subsurface tensile stress field. Consequently, both tensile and compressive residual stresses must be incorporated into models of MGs or MGCs.

In this work, we systematically explore how residual/initial stresses (both tensile and compressive) modulate SB dynamics and the mechanical performance of BDMGCs through finite element analysis. BDMGCs have particle/drop-like reinforcements. By assigning the FV and stress values to each element, we establish the initial microstructure (Figure 1a) and spatial stress distribution (Figure 1b). To understand the synergistic effects of residual stress and heterogeneous microstructure on mechanical performance and shear banding, all systems are designed to have the same particle diameter, maintaining an identical volume fraction of the particles. Through finite element models (FEMs) that emphasize the FV parameter as a characteristic quantity, the variation in and evolution of the structure of BDMGCs can be gauged.

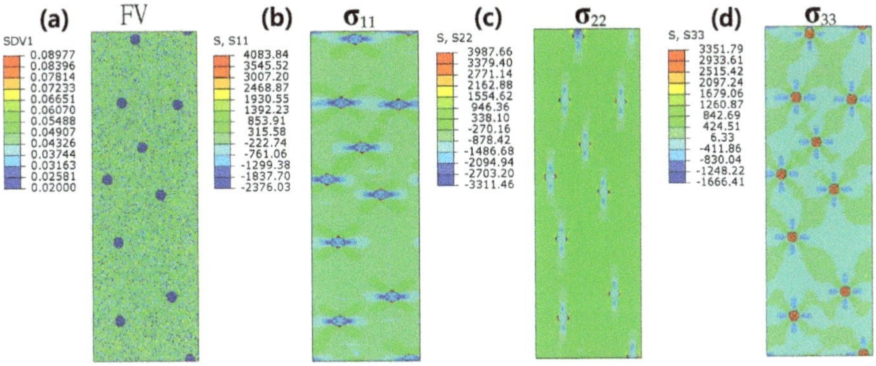

Figure 1. The initial FV state (**a**), residual stress σ_{11} (**b**), residual stress σ_{22} (**c**), and residual stress σ_{33} (**d**) of the MA system with soft particles subjected to hydrostatic pressure.

To obtain the synergistic effects of initial stress and heterogeneous microstructure on the mechanical properties of BDMGCs with various particle properties on relatively large scales, we employ a continuum approach to model the mechanical behaviors of the BDMGCs that contain submicron-sized particles. The constitutive models that explicitly incorporate FV variations in MGs [25,44–52] are used in the FEMs. We assign FV values at the mesh element according to certain spatial distributions, such as those following the drop-like structure (Figure 1). In the FEMs, all elements are square, and the side length is designated as d = 0.1 μm. By distributing different FV densities ρ_M and ρ_p in the glassy matrix and particles, we can make a BDMGC heterogeneous structure with spatial patterns. The FV density in a matrix can be determined from statistical distributions, which are obtained via the transformations of the beta distribution, 0.04 × *Beta*(50,50) + 0.03, 0.04 × *Beta*(1,1) + 0.03, or 0.04 × *Beta*(0.1,0.1) + 0.03, which is labeled as matrix A, B, or C, respectively. The FV distribution in matrix A resembles a truncated Gaussian, B a random distribution, and C a bimodal distribution, all of which have a mean FV value of 0.05 [49]. The range of FVs in these matrices is limited from 0.03 to 0.07. The BDMGC samples, which consist of matrices A, B, and C, are labeled as systems MA, MB, and MC, accordingly. The FV values ρ_p for the particle range from 0.02 to 0.10, so we can see how the varying mechanical properties of reinforcements affect the overall mechanical properties of BDMGCs with an FV change from a small ρ_p (=0.02), or "hard" particles, to a larger ρ_p (=0.10), or "soft" particles. Particle diameter D can be measured using the number of elements across the particle, so D is expressed as a multiple of d. Here D is taken as 8, and the total volume fraction of 10 particles is about 2.3%.

Figure 1a shows the MA BDMGC system containing soft particles. An isolated particle with diameter D is compressed by hydrostatic pressure P (in GPa). The tensile stress distribution of the matrix in the spherical coordinate system is as follows:

$$\sigma_r = P\frac{D^3}{8r^3} \tag{1}$$

$$\sigma_\theta = \sigma_\varphi = P\frac{D^3}{16r^3} \tag{2}$$

where D is the particle diameter, and r is the distance from the center of the particle [53]. Based on Saint-Venant's principle, the range or diameter of residual stress distribution is

5 times the particle diameter. In the plane strain model, the compressive stresses of inner isolated particles are as follows:

$$\sigma_r = \sigma_\theta = P \quad (3)$$

$$\sigma_z = 2\mu P \quad (4)$$

while the tensile stresses in the matrix are as follows:

$$\sigma_r = (1 - \frac{25D^2}{4r^2})\frac{P}{24} \quad (5)$$

$$\sigma_\theta = (1 + \frac{25D^2}{4r^2})\frac{P}{24} \quad (6)$$

$$\sigma_z = 2\mu\frac{P}{24} \quad (7)$$

The principal stresses I, II, and III are illustrated in Figure 1b–d under a Cartesian coordinate system, respectively.

As shown in previous work, the choice of these parameters allows us to carry out a systematic and quantitative investigation of the initial stress effect on shear banding and the synergistic effects of initial stress and heterogeneous microstructure on mechanical properties. To obtain the mechanical properties of all samples, we use an elastoplastic constitutive model that incorporates FV as an internal state variable. The deformation strain in this model includes an elastic and a plastic part:

$$\varepsilon_{ij} = \varepsilon_{ij}^{el} + \varepsilon_{ij}^{pl} \quad (8)$$

where ε_{ij}^{el} is the elastic strain. We obtain the plastic strain from the plastic flow equation:

$$d\varepsilon_{ij}^{pl} = d\lambda\frac{\partial g}{\partial \sigma_{ij}} \quad (9)$$

where g is the plastic potential function.

$$g(\sigma_{ij}) = b'I_1 + \sqrt{3J_2} - K \quad (10)$$

where λ is the plastic deformation parameter related to FV change, σ_{ij} is the Cauchy stress, I_1 is the first invariant of the stress tensor σ_{ij}, and J_2 is the second invariant of the deviatoric stress. The effective stress and increase in equivalent plastic strain are obtained through the following relations:

$$\sigma_{DP} = a'I_1 + \sqrt{3J_2} \quad (11)$$

$$d\varepsilon_{eff}^{pl} = \sqrt{(2/3)d\varepsilon_{ij}^{pl}d\varepsilon_{ij}^{pl}} \quad (12)$$

where a', b', and K are constant, and $a' = b'$ for the associated flow rule. The plastic strain is a function of FV production via the following relation:

$$\dot{\varepsilon}_{eff}^{pl} = 2f\exp(-\alpha v^*/\overline{v}_f)\exp\left(-\frac{\Delta G^m}{k_B T}\right)\sinh\left(\frac{\tau\Omega}{2k_B T}\right) \quad (13)$$

$$\frac{\partial \overline{v}_f}{\partial t} = v^* f\exp(-\alpha v^*/\overline{v}_f)\exp\left(-\frac{\Delta G^m}{k_B T}\right)\left[\frac{2\alpha k_B T}{\overline{v}_f S}\left(\cosh\left(\frac{\tau\Omega}{2\alpha k_B T}\right) - 1\right) - \frac{1}{n_D} + \kappa\nabla^2\overline{v}_f\right] \quad (14)$$

where $\overline{v}_f$ is the mean FV; α a geometrical factor close to 1; v^* the hard-sphere volume of the atom; k_B the Boltzmann constant; Ω the atomic volume; τ the equivalent shear stress; ΔG^m

the activation energy; f the frequency of atomic vibration; T temperature; n_D the number of atomic jumps needed to annihilate a free volume equal to v^* which ranges between 3 and 10; $S = \frac{E}{3(1-\mu)}$, where E is Young's modulus and μ Poisson's ratio; and κ a free volume gradient coefficient.

Equations (8)–(14) prescribe the constitutive relationship for the MG. Whether the deformation behavior is homogeneous or inhomogeneous is not explicitly distinguished. The FV and the stress field will be computed using a finite element method. Given the history of applied load, we aim to calculate the displacements, strains, and stresses satisfying the following governing equations:

$$\frac{\partial \sigma_{\mathrm{ji}}}{\partial x_{\mathrm{i}}} + \frac{\partial \rho v_i}{\partial \mathrm{t}} + \rho f_i = 0 \tag{15}$$

$$\varepsilon_{\mathrm{ij}} = \frac{1}{2}(u_{i,j} + u_{j,i}) \tag{16}$$

where the boundary condition is as follows: $\sigma^*{}_{ij}\mathrm{n}_j = t_i$. Here we assume that the body force is zero, i.e., $f_i = 0$, and ignore the density change rate, i.e., $\frac{\partial \rho}{\partial \mathrm{t}} = 0$. Based on the variation principle, we transfer the partial differential equations of the force balance Equation (15) and their boundary conditions into the weak form:

$$\int_V \sigma_{\mathrm{ij}} \delta \varepsilon_{ij} dV + \int_{\partial V} \sigma^*{}_{ij}\mathrm{n}_j \delta v_i dS = 0 \tag{17}$$

In the FEM, the displacement field at an arbitrary point within the solid will be specified by interpolating between nodes of mesh through the shape function N. Substituting the interpolated fields into the virtual work equation, we find the following:

$$\int_V \sigma_{ij}(\Delta \varepsilon_{\mathrm{kl}}(\mathrm{u}_i^a)) \frac{\partial N^a}{\partial \mathrm{x}_j} dV + \int_V \frac{\partial \Delta \sigma_{ij}}{\partial \Delta \varepsilon_{\mathrm{kl}}} \frac{\partial N^{\mathrm{b}}}{\partial \mathrm{x}_i} \frac{\partial N^a}{\partial \mathrm{x}_j} \Delta u_i^a dV + \int_{\partial V} \sigma^*{}_{ij}\mathrm{n}_j N^a dS = 0 \tag{18}$$

We can therefore rewrite the virtual work equation in a system of linear equations. The material-related parameter pertaining to the constitutive models, i.e.,

$$D_{ijkl}^{ep} = \partial \Delta \sigma_{ij} / \partial \Delta \varepsilon_{kl}, \tag{19}$$

is implemented in ABAQUS 6.13 finite element software through a UMAT subroutine [47–49]. In the implicit integration scheme, applying the backward Euler method to the flow Equation (13) and FV evolution Equation (14), we get the following:

$$\sigma_{DP,(n+1)}^{trial} - \sigma_{DP,(n)} - \Delta\sigma_{DP} - (Kab + 3G)\frac{1}{\sqrt{\frac{2}{9}b^2 + 1}} 2\Delta t \exp(-\frac{1}{\overline{v}_{f(n)} + \Delta\overline{v}_f}) \sin\mathrm{h}(\frac{\sigma_{DP(n)} + \Delta\sigma_{DP}}{\sigma_0}) = 0 \tag{20}$$

$$\Delta\overline{v}_f - \frac{\Delta t}{\alpha} \exp(-\frac{1}{\overline{v}_{f(n)} + \Delta\overline{v}_f})[\frac{1}{(\overline{v}_{f(n)} + \Delta\overline{v}_f)\beta S}(\cosh(\frac{\sigma_{DP(n)} + \Delta\sigma_{DP}}{\sigma_0}) - 1) - \frac{1}{n_D} + \kappa \nabla^2 \overline{v}_{f(n)}] = 0 \tag{21}$$

The initial guess solution $(\Delta\overline{v}_f{}^0, \Delta\sigma_{DP}{}^0)$ is taken as the solution from the previous time step. In the Newton–Raphson method, the solution $(\Delta\overline{v}_f{}^k, \Delta\sigma_{DP}{}^k)$ at the *n*th step will be updated for the calculation $(\Delta\overline{v}_f{}^{k+1}, \Delta\sigma_{DP}{}^{k+1})$ in the next step. The trial stress tensor is $\sigma_{DP,(n+1)}^{trial}$. The material stiffness matrix D_{ijkl}^{ep} can also be updated. By solving the system of linear equations of FV evolution and the strain and stress under a given external load, we can obtain the mechanical properties of BDMGCs containing various particles and initial stress. The material properties of bulk $Zr_{41.25}Ti_{13.75}Ni_{10}Cu_{12.5}Be_{22.5}$ MG are used [48]. The model systems have a total of 30,000 regular elements and the periodic boundary conditions in our FEM. Plane strain tensile loadings are applied with an effective strain rate of 0.1/s.

3. Results

The tensile stress–strain relationships of all systems with various particle characteristics and varying residual/initial stresses are shown in Figure 2. In the following, we will categorize 21 types of samples based on three statistical FV distributions of the matrix, incorporating three types of residual stress and two types of particles. For example, "MA-ρ_p0.02-P5" indicates that the BDMGC system comprises matrix A and hard particles (ρ_p = 0.02) subjected to a hydrostatic pressure of P = 5 GPa. Meanwhile, "MA-ρ_p0.02" signifies that the system is devoid of any initial stress. In general terms, compared with the MG matrix, the hard reinforcement increases strength, while the soft one results in a reduction in strength. The results corroborate that an FV density of 0.02 (or harder particles) enhances the strength of the composite (as illustrated by the blue lines in Figure 2). Conversely, softer particles with an FV density of 0.10 lead to a reduction in composite strength (as illustrated by the red lines in Figure 2) as compared to that in the monolithic sample (as represented by the black lines in Figure 2). This is evidenced by the changes observed in mechanical strength such as yield stress and peak stress. The yield stress is determined by the 0.2% strain offset method. The peak stress is taken at the first occurrence of the maximum stress. The flow stress is the asymptotic stress value at the large deformation strain. In Figure 2, the flow stresses of all systems with varying initial stresses remain the same as compared to the one without residual stress. The increased residual stresses (containing the compressive and tensile stresses) lead to a greater reduction in peak stress. As illustrated in Figure 2, the decrease in peak stress is alleviated as the heterogeneity of the matrix increases, which corresponds to the statistical variance of FV [49].

The standard deviations (SDs) of matrices A, B, and C are 0.002, 0.0115, and 0.0183, respectively. Comparing the peak stress of the samples "MA-ρ_p0.02-P5" and "MA-ρ_p0.02", it is clearly observed that the peak stress decreases from 2.0836 GPa to 1.7463 GPa. For the samples "MB-ρ_p0.02-P5" and "MB-ρ_p0.02", their peak stress decreases from 1.560 GPa to 1.4850 GPa. The peak stress of samples "MC-ρ_p0.02-P5" and "MC-ρ_p0.02" decreases from 1.7066 GPa to 1.6806 GPa. The yield stress reductions for samples "MA-ρ_p0.02", "MB-ρ_p0.02", and "MC-ρ_p0.02" are 0.4362 GPa, 0.07745 GPa, and 0.003976 GPa, respectively. As shown in Figure 3a, one can see a reduction in the strength (yield and peak stresses) of all samples as the standard deviation of the matrix increases. As particles are subjected to a hydrostatic pressure of $\boldsymbol{P}$ = 1GPa, the yield stress reductions for samples "MA-ρ_p0.02", "MB-ρ_p0.02", and "MC-ρ_p0.02" are 0.03776 GPa, 0.01965 GPa, and 0.01063 GPa, respectively. This trend becomes increasingly pronounced as the residual stress increases. For the system with soft particles (ρ_p = 0.1), the trend in strength reduction resembles that in hard particles, as shown in Figure 3. As particles are subjected to a hydrostatic pressure of $\boldsymbol{P}$ = 5 GPa, the reductions in peak (yield) stress for samples "MA-ρ_p0.02", "MB-ρ_p0.02", and "MC-ρ_p0.02" are 0.3144 GPa, 0.08725 GPa, and 0.04078 GPa (0.34089 GPa, 0.06675 GPa, and −0.02845 GPa), respectively.

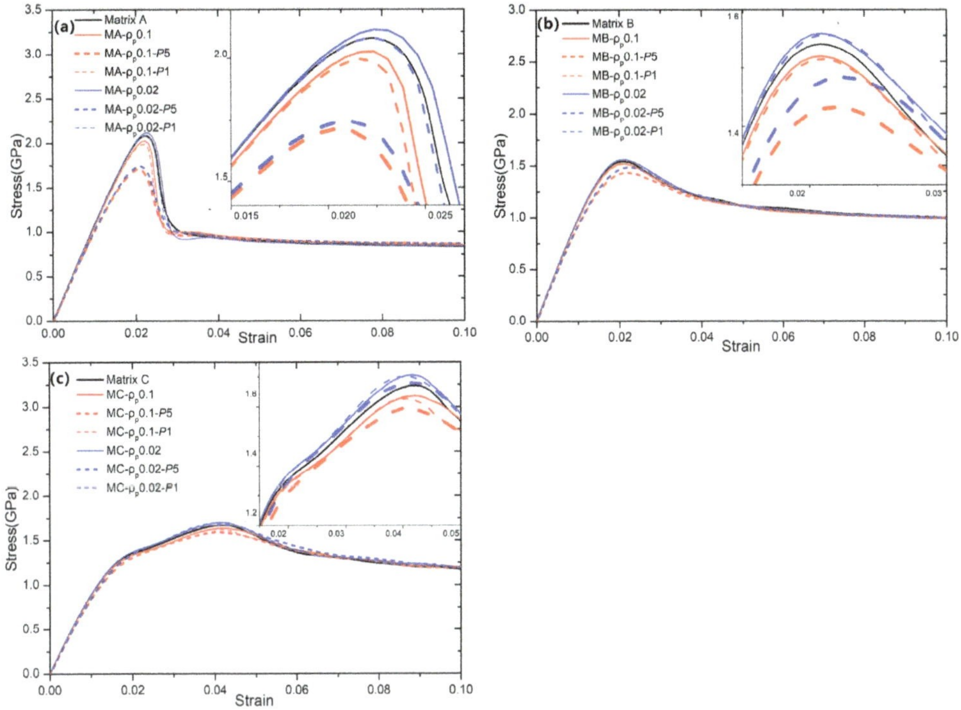

Figure 2. The stress–strain curves for the systems MA (**a**), MB (**b**), and MC (**c**) with soft particles ρ_p = 0.10 (red line) and hard particles ρ_p = 0.02 (blue line). The dashed and solid lines indicate systems with and without initial residual stresses, respectively. The black line denoting "Matrix" represents the monolithic MG.

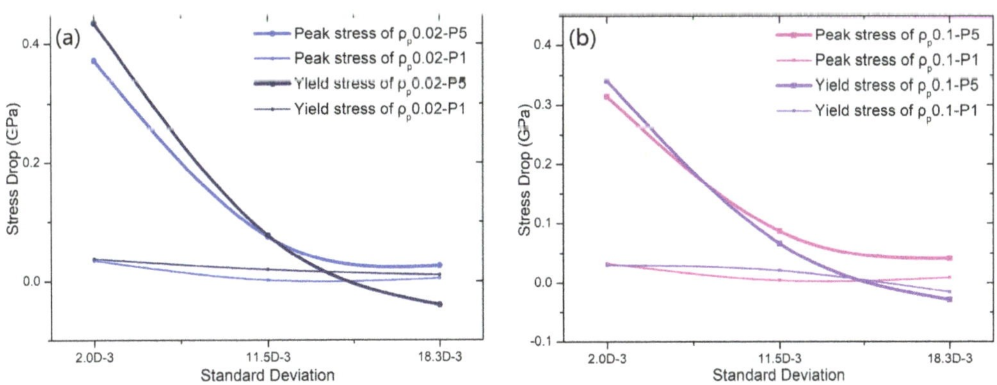

Figure 3. The mechanical strength reductions (yield stress and peak stress) in BDMGC samples with hard particles (ρ_p = 0.02) (**a**) and soft particles (ρ_p = 0.10) (**b**), influenced by initial stresses, are observed in relation to matrix heterogeneity (standard deviation of matrix).

4. Discussion

The presence of initial stresses (initial compressive and tensile stresses) contributes to a reduction in strength (peak stress), which suggests the emergence of new deformation mechanisms. This synergistic effects of residual/initial stress and heterogeneous microstructure on mechanical performance will be discussed later. To investigate the new deformation mechanism, the spatial FV distributions, strain, and stress of all systems under various applied strain states are examined to analyze the behaviors of local shear, as shown in Figures 4 and 5. In Figure 4, the samples "MA-ρ_p0.02" and "MA-ρ_p0.1" exhibit deformation in the absence of initial stress, while the deformation of samples "MA-ρ_p0.02-*P*5" and "MA-ρ_p0.1-*P*5" containing initial stresses is shown in Figure 5. Figure 4a shows that for the sample "MA-ρ_p0.02" containing hard particles, SBs usually initiate at the matrix and then propagate into mature SBs along the direction of the maximum effective stress (Figure 4b). The location for the initiation of SBs is random due to the random distribution of particles. The stoppage and obstacles of SBs occur when they encounter hard particles, leading to an increase in strength. During the process of the propagation of SBs, some but not all SBs will meet, and they are subsequently stopped by hard particles (Figure 4c,d). A higher fraction of particles increases the likelihood of encountering obstacles for SBs. For the sample "MA-ρ_p0.1" containing soft particles (Figure 4e), deformation first happens in the soft particle zone. Minor SBs will quickly initiate and nucleate in glassy droplets. However, they are trapped and restricted on the droplets (Figure 4f). The main SBs usually initiate at the middle region between two soft particles whose center line is about 45° off the loading axis (Figure 4g). The SBs will propagate and pass through soft particles. Several main SBs will connect and then traverse through the cross-section, as shown in Figure 4h.

In the following, we present a detailed examination of BDMGCs with initial stresses. The initial stresses and FV distribution of samples "MA-ρ_p0.02-*P*5" and "MA-ρ_p0.1-*P*5" are illustrated in Figure 5a,e, respectively. For sample "MA-ρ_p0.02-*P*5", the initiation point of SBs is at the interfaces between the matrix and hard particles (Figure 5b). The total strain (0.74%) of SB nucleation is significantly smaller than that (2.46%) of the sample without initial stresses. With further deformation, the nucleus of SBs will propagate and develop into mature SBs along the direction of the maximum effective stress (Figure 5c). The SBs will intersect and interconnect with each other (Figure 5d). For sample "MA-ρ_p0.1-*P*5", deformation first occurs in the soft particle zone under low strain (Figure 5f). However, the initial stress can expand the soft zone (Figure 5g), and then minor SBs will propagate along the direction of the maximum effective stress in the matrix (Figure 5h). In samples "MA-ρ_p0.02-*P*5" and "MA-ρ_p0.1-*P*5", since the initiation point and direction of SBs are identical, mature SBs appear nearly indistinguishable from one another.

For the systems with a tough matrix (larger statistical variance), the initial stress and FV distribution of samples "MC-ρ_p0.02-*P*5" and "MC-ρ_p0.1-*P*5" are illustrated in Figure 6a,e, respectively. For sample "MC-ρ_p0.02-*P*5", the initiation point of SBs is at the interfaces between the matrix and hard particles, as shown in Figure 6b. As deformation continues, the deformed regions are still restricted to their original location, while new deformed areas with higher FV values emerge elsewhere. Lots of localized deformation zones form into minor SBs in Figure 6c. The deformation bands in matrix C appear irregular and jagged, unlike the smooth and straight ones in matrix A. Short minor SBs begin to intertwine and linkup, and this then leads to the formation and propagation of mature SBs across the sample, as shown in Figure 6d. For sample "MC-ρ_p0.1-*P*5", deformation first occurs in the soft particle zone under low strain. However, minor SBs in the soft zone will be restricted by the toughened matrix in Figure 6f. For the MC system with a larger range of FV distribution, the points with a large FV will also be good for the initiation of minor SBs. The initial residual stress cannot result in the easy expansion of minor SBs, as shown in Figure 6f. The nuclei of minor SBs will propagate as rugged, zigzag SBs, with a few maturing and propagating along the direction of the maximum effective stress in the matrix (Figure 6h).

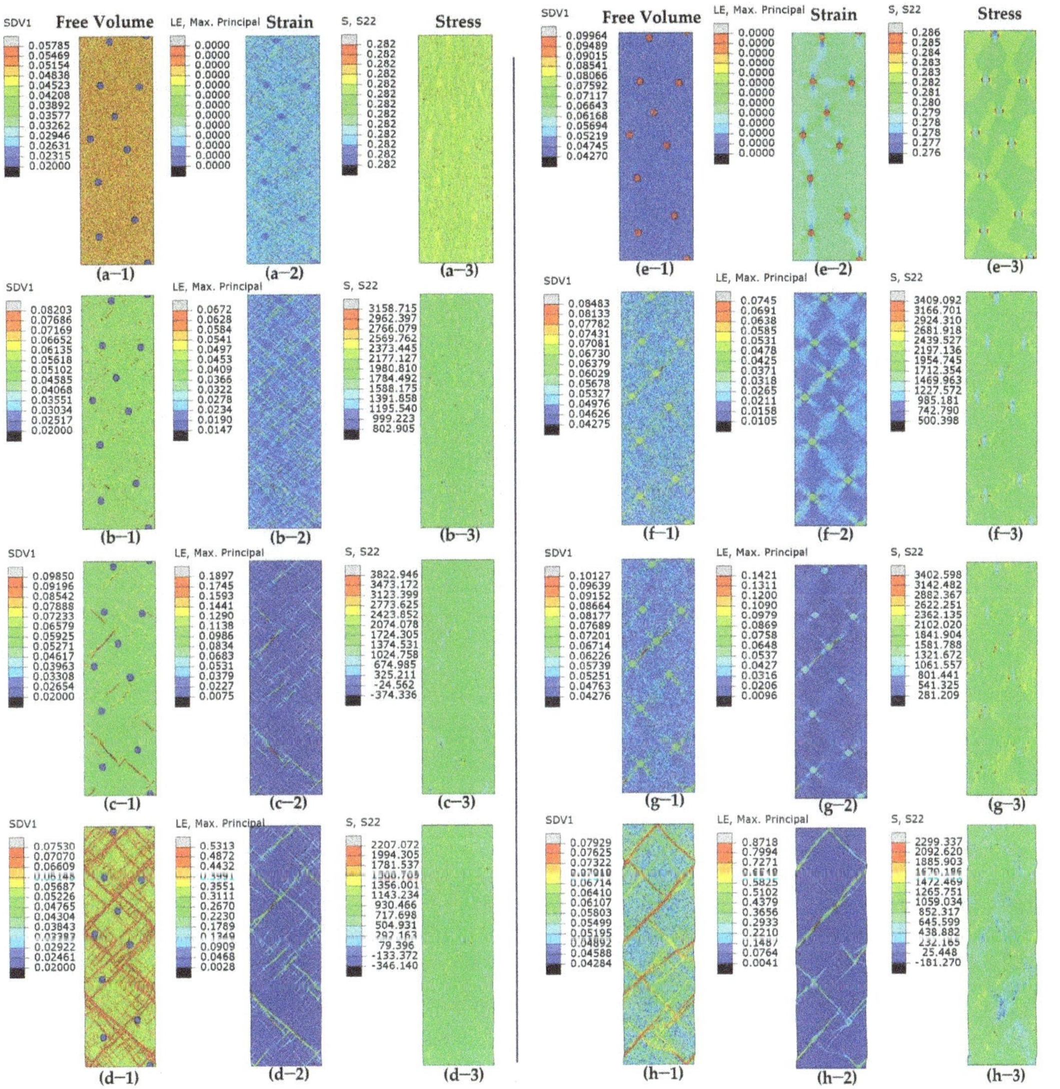

Figure 4. Contour plots for free volume (-1), shear strain (-2), and shear stress (-3) for MA system with (**a–d**) hard particles (ρ_p = 0.02) or (**e–h**) soft particles (ρ_p = 0.1) under varying tensile strains. Color scheme is shown on left of each figure.

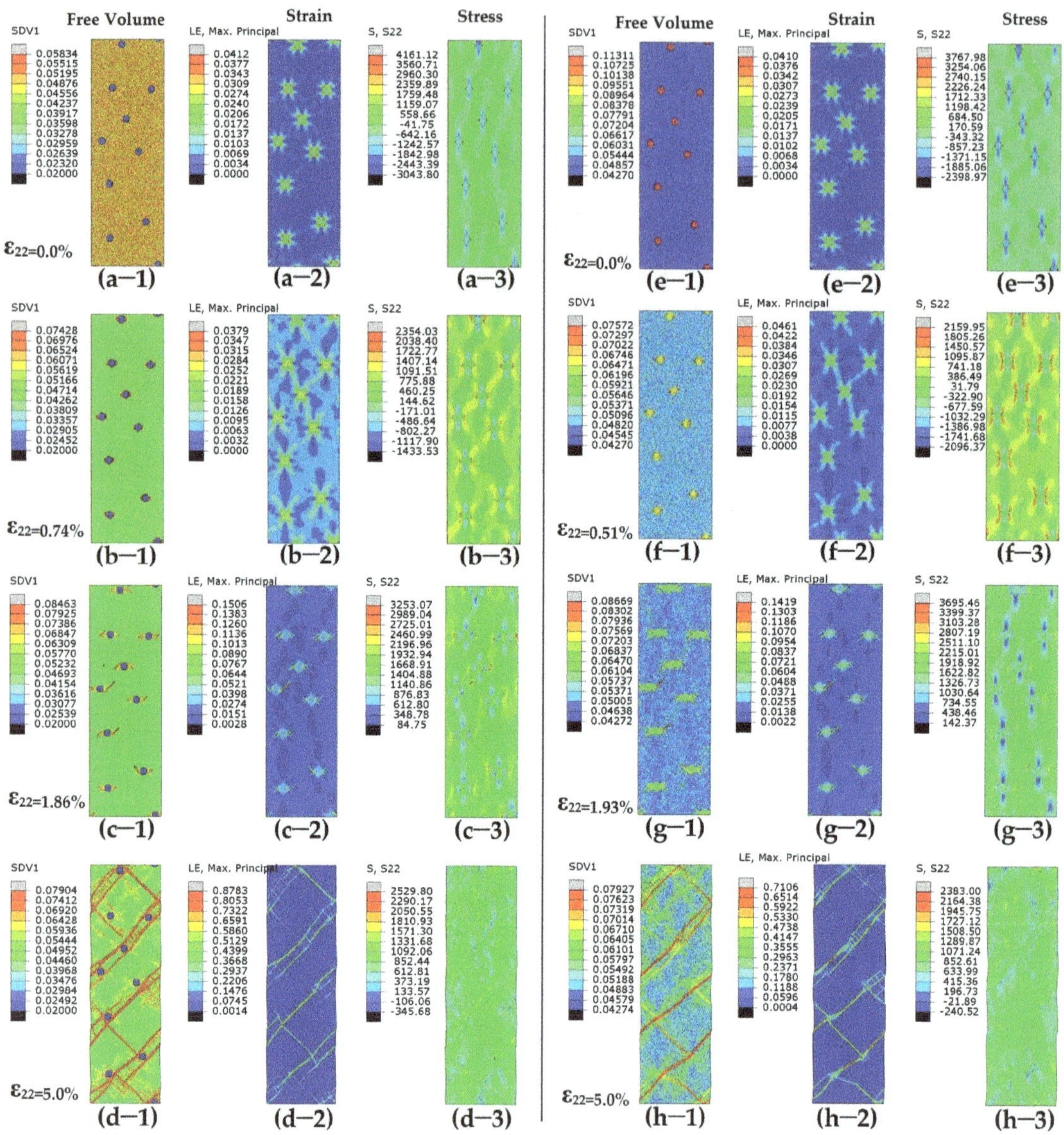

Figure 5. Contour plots for free volume (-1), shear strain (-2), and shear stress (-3) for MA system with (**a**–**d**) hard particles (ρ_p = 0.02) or (**e**–**h**) soft particles (ρ_p = 0.10) under varying tensile strains. MA system contains initial stresses before deformation. Color scheme is shown on left of each figure.

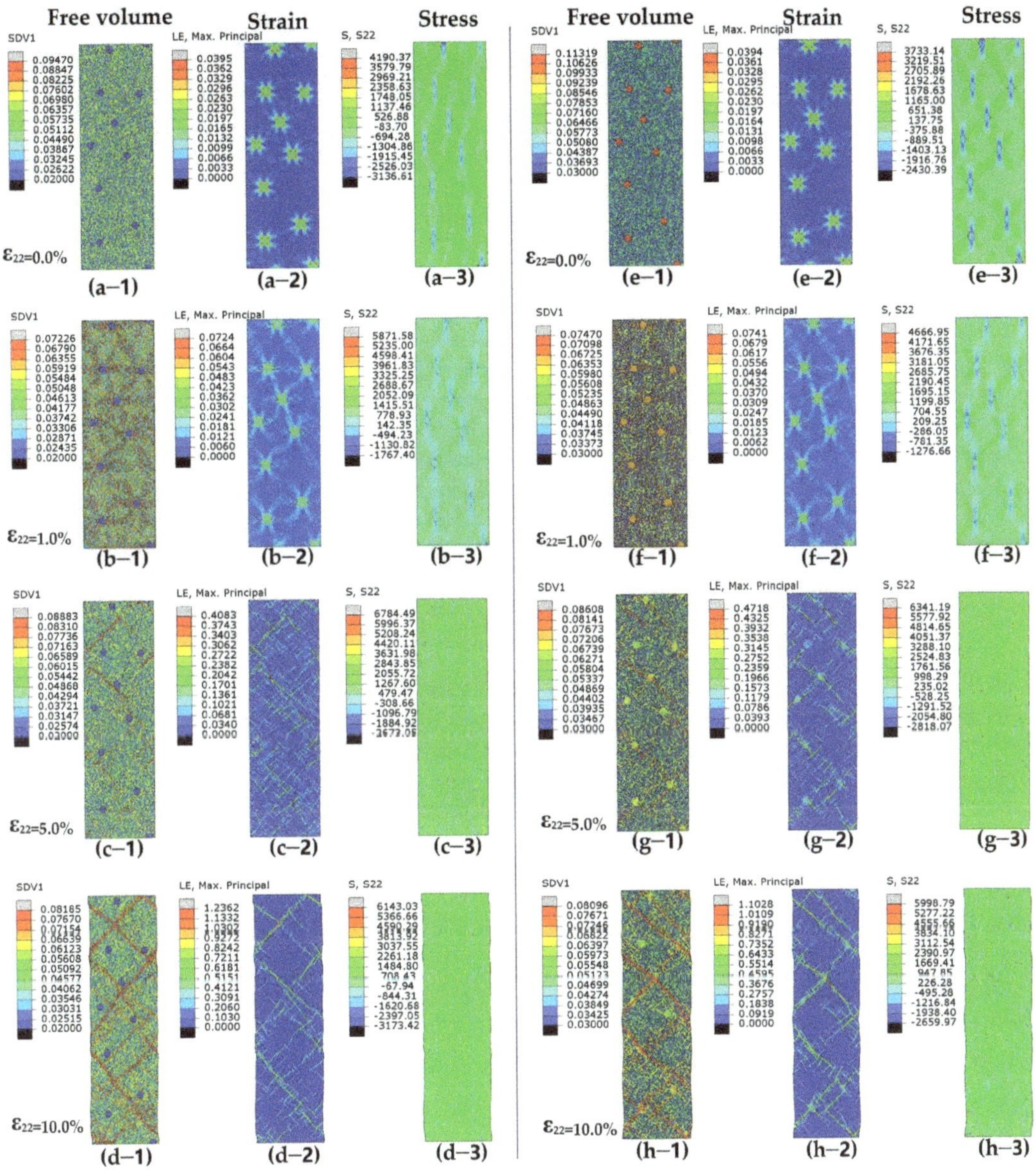

Figure 6. Contour plots for free volume (-1), shear strain (-2), and shear stress (-3) for MC system (**a**–**d**) with hard particles (ρ_p = 0.02) or (**e**–**h**) soft particles (ρ_p = 0.10) under varying tensile strains. MC system contains initial stresses before deformation. Color scheme is shown on left of each figure.

The peak stress of the samples "MA-ρ_p0.02-P5", "MA-ρ_p0.02-P1", and "MA-ρ_p0.02" shows a clear decrease as initial stresses (both compressive and tensile stresses) increase. Similarly, in systems MB and MC with soft or hard particles, the peak stress also decreases as the initial equivalent stresses increase. The standard deviation (SD) or heterogeneity of the matrix can enhance toughness and strength within a given type of matrix. The interplay between initial stresses and the significant heterogeneity of the matrix, coupled with the characteristics of particles, indicates the potential emergence of novel deformation mechanisms. The synthetic effect of initial spatial stresses and heterogeneity results from the SBs induced by temporal and spatial stress localization. The initial spatial distribution of the tensile stresses may put the disordered atoms in a high-energy state according to free volume theory, and consequently, small deformation can store enough energy and stress to trigger microscopic rearrangements, which contribute small local macroscopic shearing. The nucleation and propagation of SBs in the matrix are influenced by heterogeneity (the statistical variance of FV distribution). It is observed that shear banding initiates predominantly in regions of higher tensile stress and larger FV, and SBs grow along the direction of the maximum shear stress during deformation.

5. Conclusions

Using finite element modeling, we obtain the mechanical properties of BDMGCs with different spatial stresses and various properties of particle reinforcement, quantitatively investigating their influences on the mechanical properties and mechanisms of shear banding. The heterogeneous distribution of initial tensile stresses reduces the strength of BDMGCs, particularly in those containing a homogenous matrix. The spatial distribution of residual/initial stresses can tune and even determine the nucleation of SBs if the residual/initial tensile stress is high enough. We also show that the heterogeneous microstructure of BDMGCs can influence the propagation of SBs. By elucidating the micro-scale mechanics of BDMGC systems, we provide foundational insights for designing BDMGCs with spatial stresses and heterogeneous microstructures. The quantitative framework demonstrates that BDMGCs can be optimized by tuning heterogeneous stresses and microstructures.

Author Contributions: Y.W.: Writing—review and editing, Writing—original draft, Investigation, Formal analysis, Data curation, Funding acquisition. G.Z.: Writing—review and editing, Writing—original draft, Investigation, Formal analysis, Data curation. M.L.: Writing—review and editing, Validation, Supervision, Resources, Project administration, Methodology, Investigation, Formal analysis, Conceptualization. All authors have read and agreed to the published version of the manuscript.

Funding: This research was funded by Fundamental Research Funds for the Central Universities (FRF-TP-20-028A1, FRF-BD-25-002, and No. FRF-BD-23-02) and the Fundamental Research Funds for the Central Universities and The Youth Teacher International Exchange & Growth Program (QNXM20210044) and the Research Grants Council of Hong Kong Special Administrative Region, China (grant number 15233823).

Data Availability Statement: The raw data supporting the conclusions of this article will be made available by the authors on request.

Acknowledgments: Y.W. Wang thanks his daughter JIALV for mental support.

Conflicts of Interest: The authors declare no conflicts of interest.

References

1. Klement, W.I.; Willens, R.H.; Duwez, P. Non-crystalline structure in solidified gold-silicon alloys. *Nature* **1960**, *187*, 869–870. [CrossRef]
2. Kumar, G.; Rector, D.; Conner, R.D.; Schroers, J. Embrittlement of Zr-based bulk metallic glasses. *Acta Mater.* **2009**, *57*, 3572. [CrossRef]
3. Scudino, S.; Jerliu, B.; Pauly, S. Ductile bulk metallic glasses produced through designed heterogeneities. *Scr. Mater.* **2011**, *65*, 815–818. [CrossRef]
4. Demetriou, M.D.; Launey, M.E.; Garrett, G.; Schramm, G.P.; Hofmann, D.C.; Johnson, W.L.; Ritchie, R.O. A damage-tolerant glass. *Nat. Mater.* **2011**, *10*, 123–128. [CrossRef] [PubMed]
5. He, Q.; Shang, J.K.; Ma, E.; Xu, J. Crack-resistance curve of a Zr-Ti-Cu-Al bulk metallic glass with extraordinary fracture toughness. *Acta Mater.* **2012**, *60*, 4940–4949. [CrossRef]
6. Choi-Yima, H.; Johnson, W.L. Bulk metallic glass matrix composites. *Appl. Phys. Lett.* **1997**, *71*, 3808–3810. [CrossRef]

7. Hofmann, D.C.; Suh, J.Y.; Wiest, A.; Duan, G.; Lind, M.L.; Demetriou, M.D.; Johnson, W.L. Designing metallic glass matrix composites with high toughness and tensile ductility. *Nature* **2008**, *451*, 1085–1089. [CrossRef]
8. Hays, C.C.; Kim, C.P.; Johnson, W.L. Microstructure Controlled Shear Band Pattern Formation and Enhanced Plasticity of Bulk Metallic Glasses Containing in situ Formed Ductile Phase Dendrite Dispersions. *Phys. Rev. Lett.* **2000**, *84*, 2901–2904. [CrossRef]
9. Conner, R.D.; Dandliker, R.B.; Johnson, W.L. Mechanical properties of tungsten and steel fiber reinforced $Zr_{41.25}Ti_{13.75}Cu_{12.5}Ni_{10}Be_{22.5}$ metallic glass matrix composites. *Acta Mater.* **1998**, *46*, 6089–6102. [CrossRef]
10. Zhang, H.F.; Li, H.; Wang, A.M.; Fu, H.M.; Ding, B.Z.; Hu, Z.Q. Synthesis and characteristics of 80vol.% tungsten (W) fibre/Zr based metallic glass composite. *Intermetallics* **2009**, *17*, 1070–1077. [CrossRef]
11. Das, J.; Tang, M.B.; Kim, K.B.; Theissmann, R.; Baier, F.; Wang, W.H.; Eckert, J. "Work-hardenable" ductile bulk metallic glass. *Phys. Rev. Lett.* **2005**, *94*, 205501. [CrossRef] [PubMed]
12. Sun, Y.F.; Wei, B.C.; Wang, Y.R.; Li, W.H.; Cheung, T.L.; Shek, C.H. Plasticity-improved Zr-Cu-Al bulk metallic glass matrix composites containing martensite phase. *Appl. Phys. Lett.* **2005**, *87*, 051905. [CrossRef]
13. Hofmann, D.C. Shape memory bulk metallic glass composites. *Science* **2010**, *329*, 1294–1295. [CrossRef] [PubMed]
14. Qiao, J.; Jia, H.; Liaw, P.K. Metallic glass matrix composites. *Mat. Sci. Eng. R* **2016**, *100*, 1–69. [CrossRef]
15. Trexler, M.M.; Thadhani, N.N. Mechanical properties of bulk metallic glasses. *Prog. Mater. Sci.* **2010**, *55*, 759–839. [CrossRef]
16. Li, X.; Li, G.; Ma, J.; Cao, Y.; Xu, Y.; Ming, W. Progress in the preparation, forming and machining of metallic glasses. *J. Manuf. Process.* **2024**, *117*, 244–277. [CrossRef]
17. Sohrabi, S.; Fu, J.; Li, L.; Zhang, Y.; Li, X.; Sun, F.; Ma, J.; Wang, W.H. Manufacturing of metallic glass components: Processes, structures, and properties. *Prog. Mater. Sci.* **2024**, *144*, 101283. [CrossRef]
18. Zhou, F.; Du, C.; Cheng, C.; Xu, L.; Du, Z.; Gao, G.; Liu, A.; Fu, H. Penetration performance and fragmentation mechanism behind target of tungsten fibre/zirconium-based bulk metallic glass matrix composite rod. *Int. J. Refract. Met. Hard Mater.* **2023**, *112*, 106160. [CrossRef]
19. Lu, H.; Zhang, Z.; Tang, Y.; Zhou, H. Unravelling the relation between free volume gradient and shear band deflection induced extra plasticity in metallic glasses. *J. Mech. Phys. Solids* **2024**, *192*, 105806. [CrossRef]
20. Chang, H.J.; Yook, W.; Park, E.S.; Kyeong, J.S.; Kim, D.H. Synthesis of metallic glass composites using phase separation phenomena. *Acta Mater.* **2010**, *58*, 2483–2491. [CrossRef]
21. Wang, Y.W.; Zheng, G.P.; Li, M. Orientation effects on strengthening mechanism of network-structured metallic glass composites and nanoglasses. *J. Non-Cryst. Solids* **2025**, *666*, 123702. [CrossRef]
22. Wang, Y.W.; Zheng, G.P.; Li, M. Extraordinary enhancement of the toughness and plasticity of multilayered metallic glass composites with gradient heterogeneous interfaces. *Philos. Mag. Lett.* **2025**, *105*, 2544113. [CrossRef]
23. Ke, H.B.; Zeng, J.F.; Liu, C.T.; Yang, Y. Structure heterogeneity in metallic glass: Modeling and experiment. *J. Mater. Sci. Technol.* **2014**, *30*, 560–565. [CrossRef]
24. Greer, A.L.; Cheng, Y.Q.; Ma, E. Shear bands in metallic glasses. *Mat. Sci. Eng. R* **2013**, *74*, 71–132. [CrossRef]
25. Mear, F.; Vaughan, G.; Yavari, A.; Greer, A.L. Residual-stress distribution in shot-peened metallic-glass plate. *Philos. Mag. Lett.* **2008**, *88*, 757–766. [CrossRef]
26. Cheng, Y.; Pang, S.; Chen, C.; Zhang, T. Tailoring residual stress to achieve large plasticity in $Zr_{55}Al_{10}Ni_5Cu_{30}$ bulk metallic glass. *J. Alloys Compd.* **2017**, *690*, 176–181. [CrossRef]
27. Lu, J.; Ravichandrana, G. Pressure-dependent flow behavior of $Zr_{41.2}Ti_{13.8}Cu_{12.5}Ni_{10}Be_{22.5}$ bulk metallic glass. *J. Mater. Res.* **2003**, *18*, 2039–2049. [CrossRef]
28. Xiao, Y.; Wu, Y.; Liu, Z.; Wu, H.; Lü, Z.P. Effects of cooling rates on the mechanical properties of a Ti-based bulk metallic glass. *Sci. China* **2010**, *53*, 394–398. [CrossRef]
29. Scudino, S.; Surreddi, K.; Khoshkhoo, M.; Sakaliyska, M.; Wang, G.; Eckert, J. Improved Room Temperature Plasticity of $Zr_{41.2}Ti_{13.8}Cu_{12.5}Ni_{10}Be_{22.5}$ Bulk Metallic Glass by Channel-Die Compression. *Adv. Eng. Mater.* **2010**, *12*, 1123–1125. [CrossRef]
30. Flores, K.M.; Dauskardt, R.H. Mean stress effects on flow localization and failure in a bulk metallic glass. *Acta Mater.* **2001**, *49*, 2527–2537. [CrossRef]
31. Aben, H.; Anton, J.; Ois, M.; Viswanathan, K.; Chandrasekar, S.; Chaudhri, M.M. On the extraordinary strength of Prince Rupert's drops. *Appl. Phys. Lett.* **2016**, *109*, 231903. [CrossRef]
32. Launey, M.E.; Busch, R.; Kruzic, J.J. Effects of free volume changes and residual stresses on the fatigue and fracture behavior of a Zr-Ti-Ni-Cu-Be bulk metallic glass. *Acta Mater.* **2008**, *56*, 500–510. [CrossRef]
33. Zhang, X.; Xue, Y.; Zhang, H.; Fu, H.; Wang, Z.; Nie, Z.; Wang, L. Thermal Residual Stresses in W Fibers/Zr-based Metallic Glass Composites by High-energy Synchrotron X-ray Diffraction. *J. Mater. Sci. Technol.* **2015**, *31*, 159–163. [CrossRef]
34. Guo, Y.B.; Barkey, M. FE-simulation of the effects of machining-induced residual stress profile on rolling contact of hard machined components. *Int. J. Mech. Sci.* **2004**, *46*, 371–388. [CrossRef]
35. Majzoobi, G.H.; Azizi, R.; Nia, A. A three-dimensional simulation of shot peening process using multiple shot impacts. *J. Mater. Process. Technol.* **2005**, *164*, 1226–1234. [CrossRef]

36. Zhang, Y.; Wang, W.H.; Greer, A.L. Making metallic glasses plastic by control of residual stress. *Nature* **2006**, *5*, 857–860. [CrossRef]
37. Scudino, S.; Surreddi, K.B.; Eckert, J. Mechanical properties of cold-rolled $Zr_{60}Ti_5Ag_5Cu_{12.5}Ni_{10}Al_{7.5}$ metallic glass. *Phys. Status Solidi A* **2010**, *207*, 1118–1121. [CrossRef]
38. Joo, S.; Pi, D.; Setyawan, A.; Kato, H.; Janecek, M.; Kim, Y.; Lee, S.; Kim, H. Work-Hardening Induced Tensile Ductility of Bulk Metallic Glasses via High-Pressure Torsion. *Sci. Rep.* **2015**, *5*, 9660. [CrossRef]
39. Chen, Y.H.; Huang, J.C.; Wang, L.; Nieh, T.G. Effect of residual stresses on nanoindentation creep behavior of Zr-based bulk metallic glasses. *Intermetallics* **2013**, *41*, 58–62. [CrossRef]
40. Nieh, T.G.; Yang, Y.; Lu, J.; Liu, C.T. Effect of surface modifications on shear banding and plasticity in metallic glasses: An overview. *Prog. Nat. Sci.* **2012**, *22*, 355–363. [CrossRef]
41. Shao, Y.; Zheng, W.; Guo, W.; Lü, S.; Wu, S. In situ Fe-rich particle reinforced Mg-based metallic glass matrix composites via dealloying in metallic melt. *Mater. Lett.* **2021**, *285*, 129165. [CrossRef]
42. Sun, L.; Jiang, M.Q.; Dai, L.H. Intrinsic correlation between dilatation and pressure sensitivity of plastic flow in metallic glasses. *Scr. Mater.* **2010**, *63*, 945–948. [CrossRef]
43. Zheng, G.P.; Shen, Y. Simulation of shear banding and crack propagation in bulk metallic glass matrix composites. *J. Alloys Comp.* **2011**, *509*, 136–140. [CrossRef]
44. Kim, D.H.; Kim, W.T.; Park, E.S.; Mattern, N.; Eckert, J. Phase separation in metallic glasses. *Prog. Mater. Sci.* **2013**, *58*, 1103–1172. [CrossRef]
45. Spaepen, F. A microscopic mechanism for steady state inhomogeneous flow in metallic glasses. *Acta Metall.* **1977**, *25*, 407–415. [CrossRef]
46. Argon, A. Plastic deformation in metallic glasses. *Acta Metall.* **1979**, *27*, 47–58. [CrossRef]
47. Gao, Y.F. An implicit finite element method for simulating inhomogeneous deformation and shear bands of amorphous alloys based on the free-volume model. *Modell. Simul. Mater. Sci. Eng.* **2006**, *14*, 1329–1345. [CrossRef]
48. Zhao, M.; Li, M. Interpreting the change in shear band inclination angle in metallic glasses. *Appl. Phys. Lett.* **2008**, *93*, 241906. [CrossRef]
49. Wang, Y.W.; Li, M.; Xu, J.W. Toughen and harden metallic glass through designing statistical heterogeneity. *Scr. Mater.* **2016**, *113*, 10–13. [CrossRef]
50. Wang, Y.W.; Li, M.; Xu, J.W. Free volume gradient effect on mechanical properties of metallic glasses. *Scr. Mater.* **2017**, *130*, 12–16. [CrossRef]
51. Wang, Y.W.; Gleiter, H.; Li, M. From patterning heterogeneity to nanoglass: A new approach to harden and toughen metallic glasses. *MRS Bull.* **2023**, *48*, 56–67. [CrossRef]
52. Wang, Y.W.; Zheng, G.P.; Li, M. Manage local deformation by patterning structural heterogeneity: Controlling toughness in honeycomb patterned metallic glass composites. *Mater. Des.* **2025**, *253*, 113846. [CrossRef]
53. Bower, A.F. *Applied Mechanics of Solids*, 1st ed.; CRC Press: Boca Raton, FL, USA, 2010; pp. 213–229.

nanomaterials

MDPI

Article

A New Method for Accelerated Aging of Nanoparticles to Assess the Colloidal Stability of Albumin-Coated Magnetic Nanoparticles

Boris Nikolaev [1], Ludmila Yakovleva [1], Viacheslav Fedorov [1,2,3], Natalia Yudintceva [1,2], Daria Tarasova [1], Elizaveta Perepelitsa [4], Anastasia Dmitrieva [1], Maksim Sulatsky [1], Sivaprakash Srinivasan [5], Shirish H. Sonawane [5], Anusha Srivastava [6], Sharad Gupta [6], Avinash Sonawane [6], Stephanie E. Combs [7] and Maxim Shevtsov [1,2,7,*]

1 Laboratory of Biomedical Nanotechnologies, Institute of Cytology of the Russian Academy of Sciences (RAS), 194064 Saint Petersburg, Russia; nikolaevhpb@gmail.com (B.N.); yakluda5@gmail.com (L.Y.); gotosethxen@hotmail.com (V.F.); yudintceva@mail.ru (N.Y.); dariyataras0va@yandex.ru (D.T.); dmitriyeva03@bk.ru (A.D.); m_sulatsky@mail.ru (M.S.)
2 Personalized Medicine Centre, Almazov National Medical Research Centre, 197341 Saint Petersburg, Russia
3 Department of Inorganic Chemistry and Biophysics, Saint-Petersburg State University of Veterinary Medicine, 196084 Saint-Petersburg, Russia
4 Research Center of Neurology, 125367 Moscow, Russia; ross8marquise@gmail.com
5 Department of Chemical Engineering, National Institute of Technology, Warangal 506004, Telangana State, India; ss21chrer06@student.nitw.ac.in (S.S.); shirish@nitw.ac.in (S.H.S.)
6 Department of Biosciences & Biomedical Engineering, Indian Institute of Technology Indore, Indore 453552, Madhya Pradesh, India; phd2201171001@iiti.ac.in (A.S.); shgupta@iiti.ac.in (S.G.); asonawane@iiti.ac.in (A.S.)
7 Department of Radiation Oncology, Technische Universität München (TUM), Klinikum Rechts der Isar, 81675 Munich, Germany; stephanie.combs@tum.de
* Correspondence: maxim.shevtsov@tum.de; Tel.: +49-173148882

Abstract: The colloidal long-storage stability of nanosized drugs is a crucial factor for pharmacology, as they require much time for robust estimation. The application of bioavailable magnetic nanosuspensions in theranostics is limited by incomplete information about their colloidal stability in the internal media of human organisms. A method for the accelerated temperature stress "aging" of magnetic nanosized suspensions is proposed for the rapid assessment and prediction of the colloidal stability over time of nanosized iron oxide suspensions stabilized by albumin HSA. Colloidal stability is assessed using dynamic light scattering (DLS), fluorescence spectroscopy, electrophoresis, and ion monitoring methods during short- and long-term storage. Rapid assessment is achieved by short high-temperature (70 °C) processing of carboxymethyl-dextran-coated nanosol in the presence of albumin. The role of albumin in the sustained stability of superparamagnetic iron oxide particles (SPIONs) was studied under conditions mimicking blood plasma (pH = 7.4) and endolysosomal cell compartments (pH = 5.5). According to the fluorescence quenching and DLS data, colloidal stability is ensured by the formation of an HSA corona on carboxymethyl-dextran-coated SPIONs and their process of clustering. In the presence of albumin, the colloidal stability of nanoparticles is shown to increase from 80 to 121 days at a storage temperature of 8 °C The prognostic shelf life of magnetic nanosol is estimated by calculating the Van't Hoff's relation for the rate of chemical reactions. The validity of using the Van't Hoff's rule is confirmed by the agreement of the calculated activation energy at 8 °C and 70 °C. The developed method of the accelerated aging of nanoparticles can not only be employed for the estimation of the shelf life of magnetic nanoparticles coated with HSA in vitro but also for assessing the stability of SPIONs applied in vivo.

 https://doi.org/10.3390/nano15070475

Keywords: metal oxide nanoparticles; superparamagnetic iron oxide nanoparticles; SPIONs; protein corona; nanoparticle aggregation; method of accelerated aging; nanoparticle stability

1. Introduction

Magnetic nanoparticles (MNPs) are nanoscale materials used in analytical chemistry and biosensors. They consist of a metal core and an outer shell, with the number of atoms varying from a few units to several thousand [1–4]. These nanoclusters, with their active surface and high sorption capacity, can interact with various biomolecules because of their small size. They also have ferro- and superparamagnetic properties, making them useful in precision medicine [5,6]. A variety of magnetic nanoparticles have been synthesized, including Co, Fe, Ni, iron oxides, and ferrites. Superparamagnetic iron oxide nanoparticles (SPIONs), with the formula Fe_3O_4, are particularly of interest for research and application in biomedicine because of their cost-effectiveness, stability, and biocompatibility [7–9]. Using various synthesis and processing methods, monodisperse suspensions and other magnetic nanoparticles with varying surface morphologies can be obtained [10–12]. For the biomedical application of nanoparticles, it is necessary that they fulfill a number of requirements, including the ability to form a stable colloidal system in aqueous solutions and other biocompatible solvents and to vary the parameters of the solution (e.g., salt concentration, pH, and temperature) within intervals determined by the specific purpose of the study. However, because of their high reactivity, there is practically no inert medium for nanoparticles. One of the key features of nanoparticles in solution is their tendency to aggregate; therefore, the practical use of nanoparticle solutions is associated with their stabilizing agents (coating the surface of the magnetic "core", adding stabilizers, selecting solvents, etc.) [13].

The stability of a colloidal solution includes both its aggregate and kinetic consistency. Aggregative stability is the immutability of particle sizes in a colloidal solution over time, preventing particles from sticking together, which is considered to be more important for system applications. Sedimentation stability refers to the uniform distribution of colloidal particles due to Brownian motion. Loss of aggregative stability leads to a coarse state, affecting sedimentation stability. Symptoms include turbidity, color change, and precipitation [14]. The attraction between nanoparticles will have a magnetic nature and change in line with the double electric layer dynamics. In this article, both aggregate and sedimental stabilities were investigated using dynamic light scattering, temperature stress tests, and turbidimetry.

Magnetic nanoparticle suspensions lose stability and monodispersity without stabilization, especially in the presence of coagulants. Steric stabilization occurs when surfactant molecules adsorb onto the surface of nanoparticles, preventing the formation of aggregates. This process creates a physical barrier preventing particles from sticking together, ensuring well-dispersed systems. Electrostatic stabilization occurs when charged particles repulse each other upon the formation of charges because of electrolyte ion adsorption or surface group dissociation [15]. The addition of multicharged ions with high adsorptive capacity contributes to the aggregation of nanoparticles due to changes in the surface charge and double electric layer. Therefore, only single-charged ions find use in electrostatic stabilization [16,17].

Steric stabilization is crucial for maintaining the stability of MNPs in biomedical applications. It relies on the density of the macromolecules on the surface of MNPs and the interaction of the stabilizing coating with the dispersion medium. Natural (dextran, alginate, chitosan, etc.) or synthetic (PEG, PVS, PVP, etc.) polymers are mainly used as stabilizing agents [18–21]. The effectiveness of steric stabilization depends on a homogeneous polymer coating and the presence of bridge effects, which can cause particle aggregation due to high-molecular-weight interactions [22]. Some polymer coatings are shown to decrease specific molecules' adsorption [23]. Dextran is one of the more widely accepted polymers for particle modification because of its biocompatibility, variable size of its chains [24]. Even though dextran molecules are shown to desorb from particle surfaces, crosslinking agents may be used for polymer anchoring [25].

A simple dextran coating can be switched to charged macromolecules or proteins to further stabilize suspensions [26]. However, this method is only effective at low ionic strength or pH, which differs from the isoelectric point of MNPs and proteins. The ionic strength of the medium determines the diffuse layer on the nanoparticles' surface, preventing interaction and causing aggregation [27]. To obtain stable iron oxide nanoparticles both electrostatic and steric stabilizations (example in Figure 1) are necessary, as the isoelectric point of MNPs is pH 6.8 [28]. Compounds like polystyrene sulfonates, polyaspartate, polyacrylamide, polyacrylic acid, and carboxymethylcellulose are used as stabilizers [29–31]. A carboxymethyl-dextran coating may offer the best opportunities for application, as it combines the biocompatibility of dextran with high reactivity for modification. MNPs coated with carboxymethyl dextran exhibit superparamagnetism and stability, potentially serving as contrast agents in MRI [32].

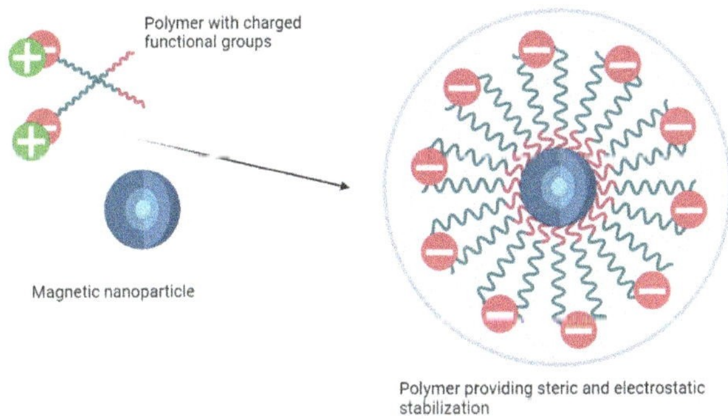

Figure 1. Electrosteric coating of nanoparticles.

Nanoparticles interact with cellular systems, which have varying media parameters that influence a magnetic suspension's stability. Nanoparticles adsorb macromolecules upon contact, creating a protective coating called a corona [33,34]. This corona formation depends on the particle surface's properties and can be controlled through modification, but then the size of the nanoparticle–corona complex also affects the biodistribution of the therapeutic composition [35]. Most up-to-date synthesis approaches, however, still achieve great applicability for nanoparticles [36].

A possible combination of steric and electrostatic stabilizations, as well as tackling unwanted protein sorption approaches, is coating or conjugating magnetite nanoparticles with BSA. Several commercially available nanoparticle preparations, as well as a number of agents undergoing preclinical research, have noted the use of albumin for stabilization [37,38]. For example, a magnetite–Au@BSA conjugate has been shown to retain a size distribution and morphology similar to pure magnetite nanoparticles [39]. Coating with BSA did not interfere with the superparamagnetism in a phantom MRI study. Similarly, BSA-TEOS-coated magnetite was used for four cycles of lysozyme extraction without changes in its physicochemical properties [40]. Many research groups have previously investigated SPION stability in the presence of serum proteins. In [41], the authors immobilized BSA on SPIONs intended for hyperthermia. Their findings suggest that BSA-SPION particles remain stable in water and PBS, and the passivation of the particle surface with protein prevents significant aggregation. Folic-acid-doped BSA also provides the possibility of the prolonged usage of SPIONs in targeted delivery [42], presenting high biocompatibility as well. Finally, the models presented in [43] emphasize the importance of balancing electrostatic and steric stabilizations to prevent particle clumping and surface ion chelation during in vivo application

While substantial research has been conducted, the question of the precise effect of albumin on the time intervals for using magnetite nanoparticles is still open to debate. In [44], the authors monitored the properties of stored SPIONs over a 12-week period, reporting substantial stability for dextran-coated preparations. As for an albumin corona, the protein coat lowers the unwanted uptake and acidic degradation of particles when assessed via an in vitro model [45]. In this study, the stability of MNPs in the presence of albumin was evaluated both under normal storage conditions and by modeling the use of particles in vivo. The goal was to formulate and offer a quick and available method of evaluating electrosterically coated SPIONs' shelf life. To do this, we evaluated the colloidal stability of nanoparticles that interfaced with BSA (to mimic a stabilizer) or HAS (to mimic blood serum) in media with various pHs. Many methods for assessing colloidal stability have been described, such as UV/VIS absorbance, FTIR, and geometry and charge monitoring, all of which are time-consuming [46]. In this paper, for the first time, we adapt an accelerated aging method for characterizing colloidal stability, which makes it possible to evaluate the stability of nanosuspensions both under storage conditions and to simulate the stability of particles when absorbed by cells or introduced into the body.

2. Materials and Methods

2.1. SPIONs Synthesis and Characterization

SPIONs were prepared by coprecipitation from Fe^{2+} and Fe^{3+} ($FeCl_3 \cdot 6H_2O$, $FeSO_4 \cdot 7H_2O$, Vekton, Saint-Petersburg, Russia) salt solutions with 1:2 (0.3 and 0.6 mol/L respectively) molar ratio in nitrogen media at pH 10 and 80 °C, adapted from the Massart method [47,48]. The precipitation was performed by the dropwise addition of NH_4OH (Vekton, Saint-Petersburg, Russia) to the iron salt solution under a nitrogen gaseous atmosphere with vigorous stirring. The magnetite formation follows the reaction shown below:

$$2Fe^{3+} + Fe^{2+} + 8OH^{-} = Fe_3O_4 + 4H_2O.$$

After the addition of reductant, the magnetite crystal formation was completed by continuous mixing at 80 °C for 8 min in inert media. The resulting magnetite crystals were washed with water on a magnetic rack until reaching the neutral pH of media and then weighted.

Nanoparticles were stabilized by coating biopolymer carboxymethyl dextran with a dextran-to-iron weight ratio of 3:1 (CMDx, 15 kDa, Sigma Aldrich, St. Louis, MO, USA) upon mixing the NP suspension with the polymer solution for 5 min. The resulting nanoparticles were then pulse-sonicated at 22 kHz and 1 kW output power (6 times for a 5 min interval with a 3 min rest) and purified by magnetic separation to remove large multidomain ferromagnetic particles. Further centrifuge separation, membrane filtration, and dialysis were applied to produce the final monodispersion.

The concentration of Fe^{+3} ions in the NP suspension was measured by the thiocyanate method [49]. Briefly, diluted SPION samples of 0.1 mL were incubated for 10 min at 80 °C in highly acidic (HNO_3, Vekton, Saint-Petersburg, Russia) media to convert all iron to Fe(III) form. Then, the volume of the sample was adjusted to 2.4 mL with water. The samples were then mixed with 0.8 mL of a 0.8 M solution of potassium thiocyanate (Lenreactiv,

Saint-Petersburg, Russia), and the Fe(III) formed a red complex with the thiocyanate which could be measured by molecular absorbance at 575 nm.

Determination of the carboxymethyl dextran concentration was by the anthrone method. This method is based on the breakdown of complex carbohydrates into monosaccharides in an acidic medium, followed by their dehydration and the formation of hydroxymethylfurfural, which forms a bluish-green complex compound upon reaction with anthrone. Then, 550 μL of distilled water, 50 μL of the diluted particle solution, and 3 mL of 0.05% anthrone sulfate (Sigma Aldrich, St. Louis, MO, USA) solution were mixed. The samples were heated in a boiling water bath for 2 min and then cooled, with the solution acquiring an emerald color, and the control sample remaining yellow–green. In the colored solutions, the dextran concentration was determined spectrophotometrically at a wavelength of 625 nm against the control with water. The concentration of carboxymethyl dextran was 2–3 g/mL. Before use, the MNP samples were dialyzed.

A transmission electron microscope, a Libra 120 (Carl Zeiss, Oberkochen, Germany), was used to produce the micrographs of the CMDx-SPION clusters with and without HSA. The samples were placed on copper grids coated with formvar/carbon films (Electron Microscopy Sciences, Hatfield, PA, USA) and stained with a 1% aqueous solution of uranyl acetate. The analysis of the particle size was performed using ImageJ software (National Institutes of Health, Bethesda, MD, USA; version 1.54d) to estimate the cluster areas and GraphPad Prism software (GraphPad Software, La Jolla, CA, USA; version 9.5.1.) to plot the distribution of cluster areas across samples. Each distribution included 70 clusters of varying sizes, selected randomly from multiple TEM images to ensure representativeness. The cluster areas were calculated in nm^2 and scaled by $\times 10^{-3}$ for graphical representation to enhance clarity.

The hydrodynamic size and zeta potential of the SPIONs were assessed via dynamic light scattering (DLS) using a Zetasizer Pro (Malvern Instruments, Malvern, UK). The zeta particle potential was determined by applying an electric field in a standard spectrophotometer cuvette. A suspension (3 mL volume) was placed in a glass cuvette and treated by coherent laser radiation (633 nm wavelength), and the power of the light field scattered by the sample was recorded at an angle of 176 degrees using a photoelectric multiplier, ADC, and computer. The autocorrelation function of the light intensity was analyzed according to the standard device protocol (cumulant procedures). An experimental autocorrelation function was processed as a sum of the exponentials and calculated by the translational diffusion relation based on Stokes law. To prevent multiple light scattering, the suspension was diluted up to 50 times to reach the needed rate of homogeneity in the cuvette. Each measurement was performed with three repetitions. The Zetasizer Pro uses a combination of laser Doppler velocimetry and phase analysis light scattering (PALS) in a patented technique called M3-PALS to measure the particle electrophoretic mobility. The Zetasizer Pro software (ZS Xplorer; version 3.00) produced a frequency spectrum from which the electrophoretic mobility and, hence, zeta potential was calculated (with built-in software), implementing the Smoluchovsky model and Henry equation. The data were interpreted in terms of hydrodynamic diameter, polydispersion index, and Zeta potential. The size of protein corona was estimated by DLS using a comparison of the control and MNPs after heating at 70 °C in the range of 20–30 nm.

2.2. Albumin Fluorescence

A Varioskan Lux (Thermo Scientific, Waltham, MA, USA) reader was used to measure the fluorescence intensity of the complex of studied proteins with magnetic nanoparticles. By registering the temperature dependence of the relative intensity of the fluorescence of the proteins with the addition of SPIONs and calculating the fluorescence quenching constant, it is possible to deal with the nature of the interaction of the SPIONs and proteins (albumin from bovine and human sera, Sigma). A set of samples was prepared from a suspension of SPIONs-CMDx nanoparticles with different concentrations (by the Fe^{3+} content), as follows: 0, 0.028, 0.056, 0.084, 0.112, 0.14, 0.168, 0.196, 0.224, and 0.252 μM in 0.1 M PBS (Biolot, Saint-Petersburg, Russia) at a pH = 7.4, 5.5, or 4.5. The acidity of the buffer was set to represent normal blood plasma of an endosome and lysosome interior [50]. The sample also contained 5 mM BSA or HSA to mimic blood serum [51].

Measurements of the albumin fluorescence were performed at an excitation wavelength of 280 nm, an emission wavelength of 340 nm. The measurements were carried out at 25 and 36 °C.

An evaluation of the fluorescence quenching constant was performed according to the formula for the relative fluorescence intensity (RFI), as follows:

$$\frac{F_0}{F} = 1 + k \times [Q]$$

where F_0 is the value of the protein fluorescence intensity (BSA or HSA);

F is the value of the fluorescence intensity of the protein complex (BSA or HSA) when nanoparticles are added;

k is the quenching constant of protein fluorescence (BSA or HSA), mol^{-1};

Q is the concentration of magnetic nanoparticles, μM.

2.3. SPIONs' Stress Test

For the rapid assessment of the colloidal stability of SPIONs in storage conditions and in the presence of serum albumin, the technique of thermally induced accelerated "aging" was used, since without heating, the degradation and aggregation of NPs occur rather slowly. The methodology is based on the work of Rabel et al. [52]. However, several modifications were introduced. An assessment of the colloidal stability and a stress test were performed, which reduced the testing time from 24 h, as proposed by M. Rabel, to 2 h. A temperature of 70 °C was chosen to reduce the testing time (maximize the speed of possible reactions) when heating the composition of the nanoparticles with albumin, without causing structural changes. The possibility of assessing the colloidal stability and stress testing in the presence of HSA is demonstrated. The applicability of the temperature-adjusted storage time calculation method is also demonstrated for the case of a protein crown made from HSA.

In general, 50 mL of 0.1 M PBS (pH 7.4, 5.5 or 4.5) containing 50 μg of SPIONs, by the Fe^{3+} content, was filtered (0.22 μm PVDF) onto a sterile glass. After filtration, the resulting suspension was stirred in a 1000 min^{-1} shaker for 30 s. Then, 1.2 mL of liquid was taken from the resulting solution into 12 Eppendorf-type test tubes for the stress-test experiment, as follows:

Test tubes with the SPION/protein suspension were placed in a water bath (ICC basic pro 20, IKA, Staufen, Germany), where samples were heat treated at 70 °C for 120 min. The samples were taken after 30, 60, and 120 min, with an additional control sample that did not undergo heating. Then, 500 μL of each suspension was taken to determine the aggregate colloidal stability of the SPION clusters by DLS. The remaining suspension was centrifuged at 12,000 min^{-1} for 5 min. Then, 100 μL of the supernatant was taken for the determination of the iron content.

Similar experiments were carried out at a concentration of 0.05, 0.5, and 1.0 mg/mL HSA in a 0.01 M PBS.

2.4. SPIONs' Age Test and Shelf Life

To verify the results obtained in the stress test, the SPION suspensions containing 0.5 mg/mL HSA were subjected to long-term storage.

Likewise for the stress test, an aqueous suspensions of SPIONs in 0.1M PBS containing 0.05, 0.5, or 1.0 mg/mL of protein was added to sterile Eppendorf-type tubes, which were stored at a temperature of 8 °C.

Samples were taken for evaluation after 1, 30, 90, and 120 days of storage, as follows: 500 µL of suspension for DLS; the remaining suspension was centrifuged at 12,000 min^{-1} 5 min, and 100 µL of the supernatant was taken for colorimetry.

To clarify the temperature coefficient, the results of the accelerated "aging" and long-term storage of nanoparticles with protein (HSA content of 0.5 mg/mL) were compared. To differentiate among the stages of nanoparticle aging, we evaluated regions of unidirectional (decrease-only and increase-only) changes in the parameters of the suspension. Combining the tendencies of the hydrodynamic diameter and Fe ion concentrations in both the stress and age tests, we determined an approximate shelf life for a magnetic nanosuspension at 70 °C. The shelf life for other temperatures was considered according to the following relationship [53,54]:

$$C = A^{\frac{t_{exp}-t}{10}} \times C_{exp}$$

where C is the shelf life of SPIONs at temperature t; C_{exp} is the shelf life of SPIONs at t_{exp} in the stress test; and A is the temperature coefficient.

The validity of using Van't Hoff's rule was confirmed by calculating the activation energy at the following two temperatures: 8 °C and 70 °C. The fact that the activation energy and temperature approximately satisfy the relationship, taking into account the Arrhenius rate constants, confirms that the Van't Hoff rule can be used in this case (when assessing the effect of temperature on the rate of change in the iron content in the supernatant), as follows:

$$E_A = \frac{RT^0}{10}$$

$$E_A = \frac{RT_1 \times T_2}{T_2 - T_1} \times \ln\frac{k_2}{k_1}$$

The data used in the calculation included R = 8.31; T_1 = 70; T_2 = 8; k_2 = 0.1583; and k_1 = 0.3159 for the unimodal stages of the suspension's aging.

2.5. SPIONs NMR

The 1H NMR spectra and proton relaxation times were measured on a Spinsolve 60 spectrometer (Magritek, Aachen, Germany). The samples, in an aqueous suspension, were placed in a cylinder ampoule with a 5 mm diameter. The spectra were recorded at an ambient temperature without rotating the sample. The TMS sample was used as an external standard (0.0 ppm) for measuring the chemical shift. When recording the spectra, the one-pulse program was used, the number of accumulations was 5, and the delay time between pulses varied in the range of 2–5 s. Relaxation times T_1 and T_2 were measured using inversion recovery and the Carr–Purcell–Meibom–Gilles sequence according to a common device protocol.

2.6. Turbidimetry

The kinetic stability of a magnetic nanoparticle suspension in the presence of BSA, the HSA was studied using a turbidimetric method. Turbidimetric data on the kinetic stability of the SPION/protein suspension (light transmission in a colloid media) were measured over 0–5 days at 8 °C. Light absorption, including the contribution of multiple static light scatterings from aggregates, was measured on a Multiscan photometer (Thermo Fisher,

Waltham, MA, USA) equipped with a laser operating at 405, 450, and 620 nm. For the optical density measurements, 96 plate cells were used. Light intensities were recorded at a 180° angle and 5 sec of pulse shaking after the preliminary vortexing of a 1 μL sample for 1 min. A general scheme of the experiment is shown in Figure 2.

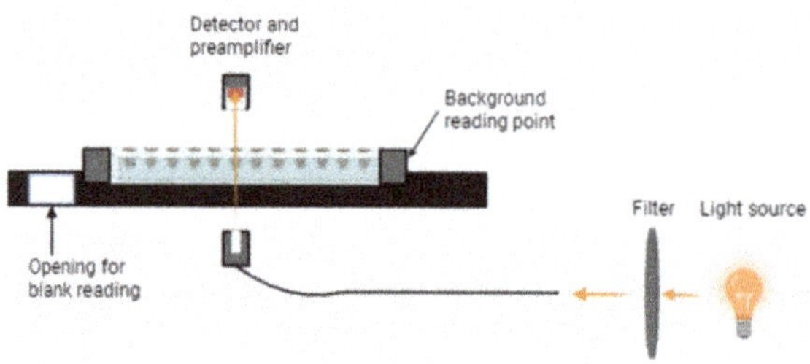

Figure 2. Turbidimetry experiment scheme.

A significant contribution to the weakening of the intensity of the light that passed through the analyzed dispersed system is made by the following two types of interaction of electromagnetic radiation with matter: light absorption and light scattering on particles. The law of change in the intensity of the incident light can be expressed as follows:

$$I = I_0 \times e^{(-\tau l)}$$

where I_0 is the intensity of the incident light;

τ is the turbidity;

l is the length of the path that passes by the light flux.

The ability of the studied system to scatter light is characterized by the turbidity coefficient, which is a product of the number of particles, n, in it and the scattering cross-section of particles, C:

$$\tau = n \times C$$

The scattering cross-section of particles is an integral of the scattered light in all directions relative to the center of the particle. This parameter is related to the geometric cross-section of the particle through the scattering coefficient, Q, as follows:

$$C = Q\pi d^2$$

where d is the particle diameter.

To avoid multiple light scattering, conditions of low concentrations were created, amounting to 50 μg Fe(III)/mL, at which each particle could be considered as an independent source of light scattering.

2.7. Statistics

The statistical evaluation was performed in Graphpad Prism software (version 9.5.1). For the statistical analysis of the parametric Student's *t*-test was employed for independent samples and parallel measurements. The significance level was equal to α = 0.05 for all tests. The confidence intervals were at the 95% level.

3. Results

3.1. Synthesized Nanoparticles' Characteristics

The initial studies of the physicochemical properties of the synthesized SPIONs show that the suspension consisted of highly dispersed ferrofluidic material with an ability to concentrate in a constant magnetic field and to later be resuspended immediately upon stirring or dilution. The TEM analysis revealed the formation of SPION clusters with varying areas, as previously described. To assess the distribution area of these clusters, both in the presence and absence of HSA, 70 randomly selected clusters were analyzed for each condition, with samples taken from multiple TEM images to ensure representativeness. The distribution plot (Figure 3A, right panel) demonstrates that the presence of HSA induced a shift in the cluster area, with the median value increasing from 738.5 nm^2 (without HSA) to 2422.1 nm^2 (with HSA). This corresponds to an approximately 3.3-fold increase in the median cluster area, indicating that albumin promotes SPION aggregation. The interquartile range (25% to 75%) also broadened, extending from 507.9 to 1313.5 nm^2 in the absence of HSA to 1453.1 to 4142.8 nm^2 in its presence, reflecting an increased variability in cluster sizes upon HSA's addition. These findings suggest that HSA plays a significant role in modulating SPION aggregation behavior (Figure 3A).

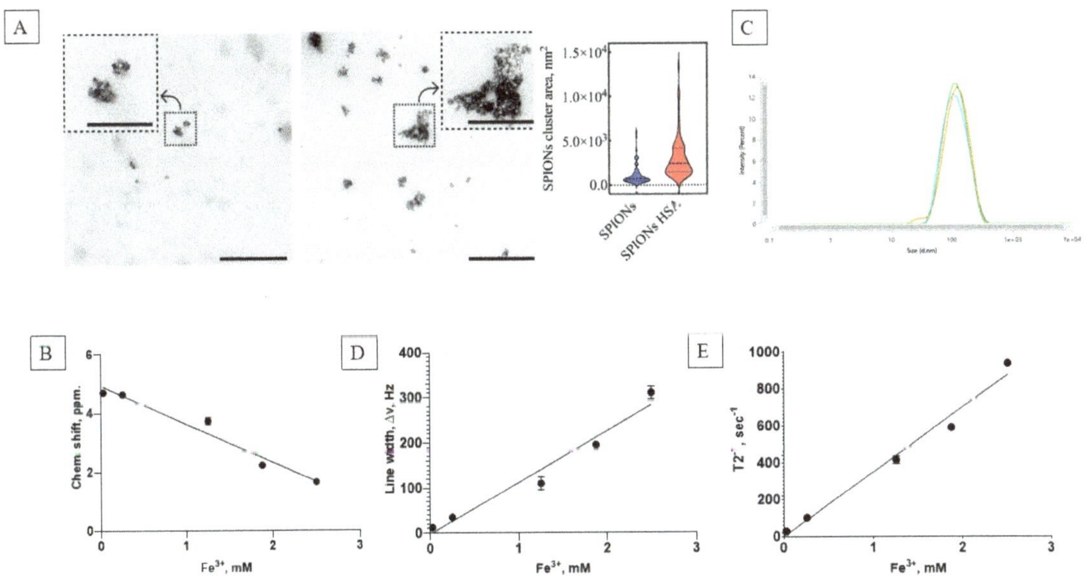

Figure 3. Physicochemical characterization of the magnetic nanoparticles. (**A**) TEM images of CMDx-SPION clusters without (left panel) and with HSA (center panel). The right panel presents the distribution of cluster areas (scaled by $\times 10^{-3}$ nm^2 for graphical clarity). Scale bars: 500 nm for the main images; 200 nm for insets. The violin plot shows the median cluster areas (dashed lines) and interquartile ranges (25% to 75%, dotted lines), illustrating the central tendency and variability in cluster sizes. The width of each distribution in the violin plot reflects the density of data points in different cluster areas, with broader sections indicating a higher concentration of values. (**B**) Chemical shift dependence on the SPION concentration in water. (**C**) Size distribution of the nanoparticles. (**D**) Line width dependence on the SPION concentration in water. (**E**) Relaxation time of the SPION water suspensions.

Synthesized suspensions had a high Fe ion content, as measured by the colorimetric method. The dynamics of the chemical shift (Figure 3B) and resonance line widening (Figure 3D) corresponded well with the superparamagnetic material. The NMR relaxometry studies show a linear dependence of the main magnetic parameters of the nanoparticle concentration, with a low T_2 relaxation time of ~1 ms and a high relaxation effectiveness (Figure 3E). The stable nanosuspension had a major dextran-coated nanoparticle fraction of 100 nm in size (Figure 3C).

3.2. Albumin Fluorescence Assessment

A Varioskan Lux reader was used to measure the fluorescence intensity of the complex formed between the studied proteins and magnetic nanoparticles. The temperature dependence of the relative fluorescence intensity of the albumin in the presence of SPIONs is used to determine the quenching constant for fluorescence.

As can be seen from the displayed plots (Figure 4A,B), the slope of the linear approximation for the fluorescence quenching differs non-significantly depending on the temperature in accordance with a static mechanism for quenching albumin fluorescence [55]. This suggests a simple protein corona formation between albumin and SPIONs, as a stable low-fluorescence complex between particle surface and protein is formed upon interaction without the dependence of fluorescent amino acids in a deeper protein structure. We also detect C_{50} for the protein quenching at a SPION concentration of 0.252 μM, corresponding to approximately half of the protein particle population in a suspension with a corona coat.

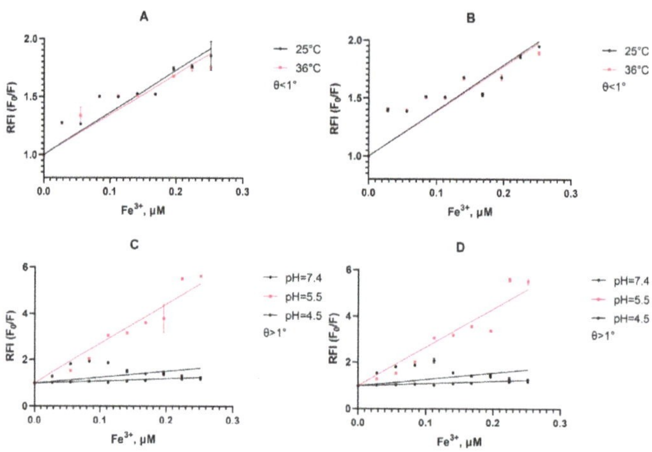

E

t, °C	Maximum of quenching constant, M^{-1}	
	BSA	HSA
25	3.72·10^6	3.62·10^6
36	3.52·10^6	3.58·10^6

Figure 4. Analysis of the albumin–nanoparticle interactions according to the dependence of the albumin fluorescence intensity (340 nm) on the Fe^{+3} content in SPIONs: (**A**,**B**) fluorescence intensity of the SPIONs with BSA and HSA, respectively, in PBS; (**C**,**D**) fluorescence intensity of the SPIONs with BSA and HSA, respectively, at a set pH; (**E**) calculated quenching constants.

As we traced the stability of this protein corona (Figure 4C,D), the quenching of the fluorescence intensity of the HSA is pH-dependent. Thus, at pH 5.5, the relative fluorescence intensity increases quite rapidly with an increase in the concentration of nanoparticles compared to other pHs. At pH 4.5, fluorescence quenching practically did not occur, which may be due to the proximity of this pH to the isoelectric point of the HSA (pI 4.7). Additionally, the similarity of the quenching constants may suggest that, in both albumin types, a fluorescent complex with a similar emission lifetime formed with no rapid dynamics in molecule conformation [56].This characteristic is shown both at 25 and 36 °C, and indicates an instability of the protein corona upon SPION uptake in cells, as endosomes and lysosomes have a lower pH.

3.3. Changes in SPION Parameters upon Albumin Incubation

The size and charge distributions of aqueous suspensions of SPIONs in the presence of albumin are illustrated in Figure 5. Based on the DLS analysis, the first peak in the figure can most likely be attributed to albumin nanoparticles with the size of ~10 nm. The observed polydisperse fractions belong to aggregates of protein in the sample media. The prepared SPIONs, however, showed a monodisperse character, with a 107 nm average size (by peak intensity) and a PDi of 0.170. The majority of the particle population had a surface charge of −28 mV. Upon incubation with albumin and protein corona formation, we observed that the size of nanoparticles inherited a multimodal distribution of albumin, with the main peak being close to 200 nm. This result was further affiliated by a shift in the polydispersity index of 0.17 (SPIONs) to 0.93 (SPIONs/albumin). At the same time, a protein corona seemed to neutralize most of the negative surface charge of SPIONs.

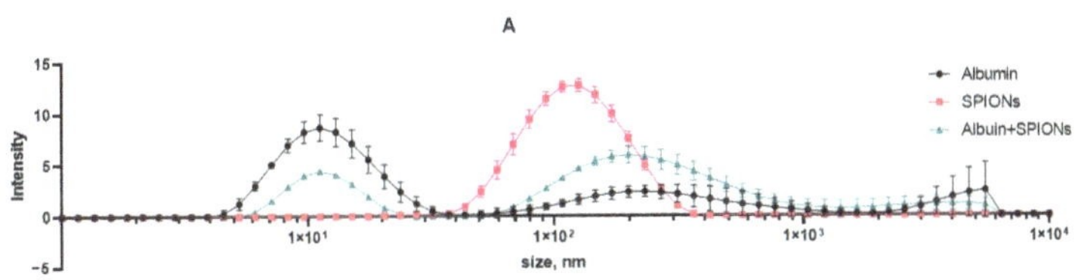

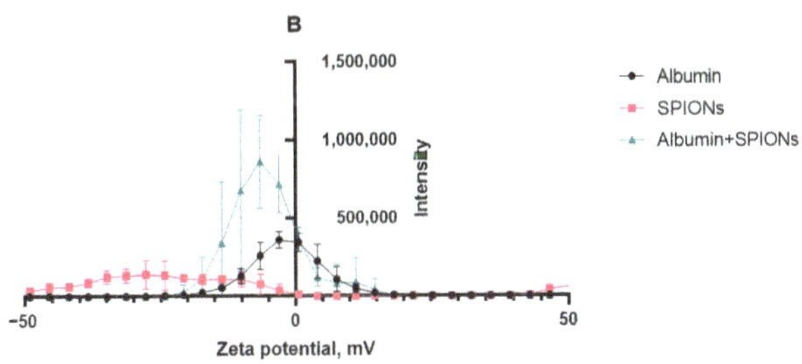

Figure 5. DLS results for SPIONs upon incubation with albumin: (**A**) size distribution; (**B**) zeta-potential distribution.

3.4. Turbidimetric Analysis of the SPION Suspension

With constant parameters for the number of particles and the path length of the light beam, the change in the turbidity coefficient's value is caused by a change in the scattering cross-section of the particles. Moreover, the turbidity value should increase proportionally to the increase in the effective hydrodynamic radius of the particles upon their aggregation. Pure albumin, here, exhibited no change in behavior depending on the concentration (overlapping data at 0.5 and 0.05 μg). Proceeding from a very low index change during the experiment (from 0.52 to 0.51, ~2% for SPIONs with 0.5 mg/mL albumin), we can assume the kinetic stability of the corona-coated nanoparticles with no significant cluster growth (Figure 6). Thus, the recording turbidity coefficient allows for an evaluation of the dispersion of the system, and in the kinetic monitoring mode, it is possible to track changes in the studied system over time.

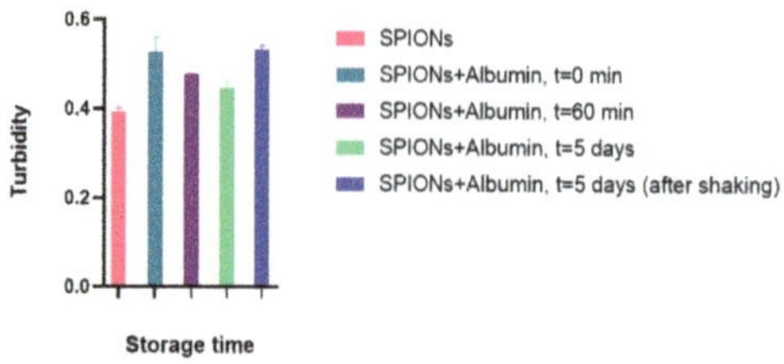

Figure 6. Turbidity of the magnetic suspension of SPION/albumin at a light absorbance of 405 nm after a 2 h treatment at an ambient temperature of 22–25 °C after 5 days of storage.

3.5. Particle Aging Dynamics

Monitoring the properties of SPIONs during accelerated aging in the presence of albumin at 70 °C, we obtained a set of dynamics of interest, as presented in Figure 7.

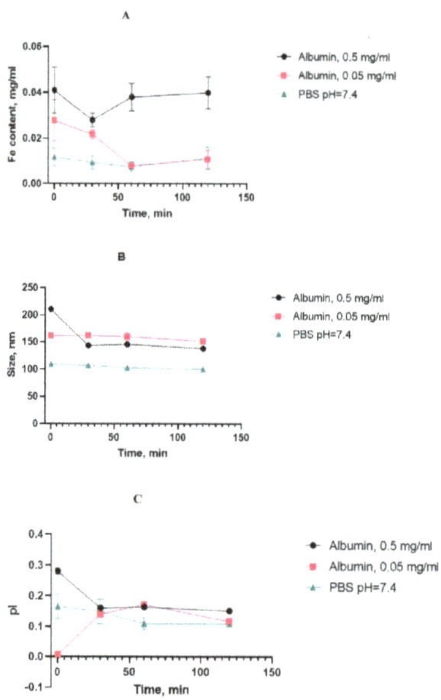

Figure 7. Physicochemical parameters of the SPIONs in the stress test: (**A**) Fe^{3+} content; (**B**) hydrodynamic diameter; (**C**) polydispersity index. n = 3 (for each parameter).

We initially chose the iron content in the supernatant after the suspension's centrifugation as a "stress marker" to quantify the amount of iron degrading from nanoclusters upon aging. With the increased presence of HSA, the concentration of iron in the supernatant did not increase with the heating time compared to the control point (without heating), and it can be concluded that the sedimentation stability of the system was maintained under the experimental conditions (Figure 8).

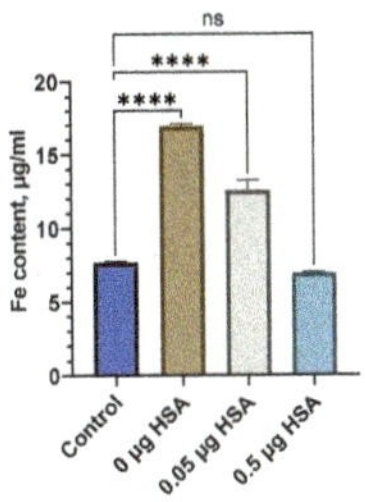

Figure 8. Iron content in the SPION sedimentation test after heating with HSA. ****—$p < 0.0001$; ns—non-significant.

The protein corona between albumin and SPIONs can be calculated as the difference between the sizes with and without albumin after heating for 120 min. A layer of approx. 20–30 nm was formed by the static mechanism.

As for the aggregation stability, the obtained results indicate that the SPION suspension underwent no significant aggregate formation during the course of the stress test. The decreasing tendency in the growth of the cluster diameter can be attributed to a decrease in the double electric layer and degradation of the dextran coat. Since there is also a decrease in the polydispersity index, it can be concluded that the particle size distribution becomes more uniform and the system becomes monodisperse, hence less prone to aggregation. On the basis of the high rate change in the physical properties of the magnetic nanosol at an HSA concentration of 0.5 mg/mL, this SPION/HSA suspension was chosen for long-term storage. The size, polydispersity index, and iron content in the supernatant were obtained after 4 months of storage of the magnetic nanoparticles (Figure 9).

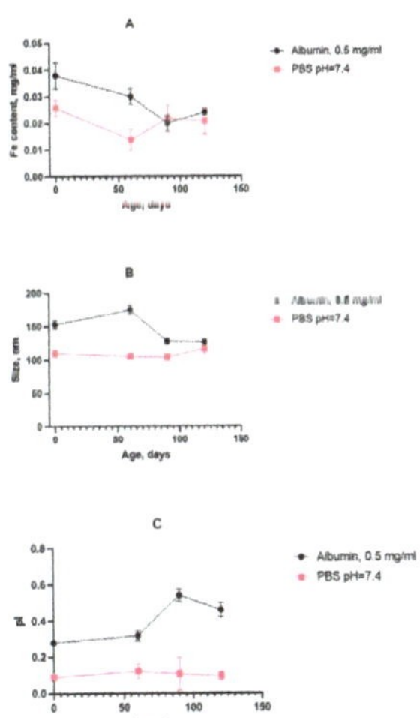

Figure 9. Properties of SPIONs during prolonged storage: (**A**) Fe^{3+} content; (**B**) hydrodynamic diameter; (**C**) polydispersity index. n = 3 (for each parameter).

The graph shows that the iron concentration did not exceed the initial value over the storage period, and it also reached a plateau, which confirms the sedimentation stability of the system. The particle size seemed to increase slightly during the measurement period, but then decreased and stabilized. This indicates the absence of the formation of large aggregates. The dynamics of the time evolution correspond to the stress tests. However, in the case of the prolonged storage of the SPION suspension with albumin, there were features of significant shifts in the polydispersity index, which may indicate uneven protein adsorption on the surface of nanoparticles.

During the storage period's analysis and calculation, based on the results obtained in the process of the "accelerated aging of MNPs" (2 h at 70 °C) or accelerated "aging" with a stress test, it was possible to estimate the time during which controlled quality indicators (size, PDI, and Fe(III)) remained within acceptable limits. An approximate relationship of the Van't Hoff's rule of about a 2–4-fold increase in the rates of chemical reactions with an increase in temperature by 10 °C was used to predict this time. For the initial determination, a coefficient of 2.5 was used as the common case for the shelf-life calculations [57,58]. The experimental value of the temperature coefficient was determined based on the data from the accelerated "aging" at 70 °C and a pH of 7.4 with the stress test and long-term storage with a stress test (Figures 8 and 10). The data on the linear approximation of the unidirectional changes in the iron concentration in the supernatants were compared (Figure 10A–D).

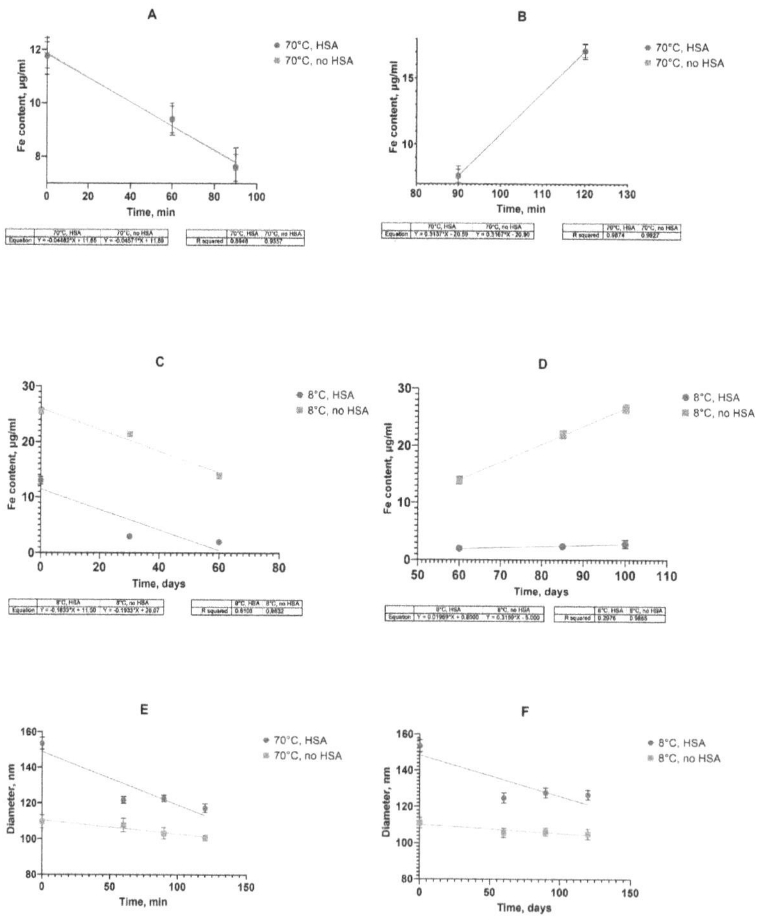

Figure 10. Storage stability of SPIONs in the dispersion during long cold storage (8 °C) and after short temperature stress action (70 °C) in the presence albumin: (**A**,**B**) Fe^{3+} release in degradation after temperature stress; (**C**,**D**) Fe^{3+} release in degradation during long storage at 8 °C; (**E**,**F**) nanocluster size dynamics for short stress and long storage, respectively. Control sample—no albumin added.

The figure shows the linear approximation data for the change in the iron concentration of the supernatant. The time for a 25% decrease in the iron concentration during the first stage was calculated using the linear approximation data on the dependence of the iron content on the supernatant over time. The time taken for the concentration to decrease to 9 μg/mL was 38.9 min. In accordance with Van't Hoff's rule using a coefficient of 2.5, the forecast for the time to achieve a similar concentration at 8 °C was calculated, which was 7.9 days. The experimental determination of the time to achieve a 25% decrease in the iron concentration in the supernatant in accordance with the linear approximation data during storage at 8 °C was 32.9 days. In accordance with the obtained experimental value, a coefficient of 2.5 was adjusted to 3.15 by the selection method (2.6–10.1, 2.8–16, 3.0–24.5, and 3.15–33.1). The temperature coefficient for the second stage was similarly checked with a 25% increase in the concentration and its value was determined to be 3.15. Thus, it was shown that it is possible to use the adjusted temperature coefficient of 3.15 to calculate the shelf life of a suspension of nanoparticles coated with carboxymethyl dextran based on the data for the entire period of the accelerated "aging" of two hours followed by a stress test. The shelf life of a suspension with known characteristics at 8 °C was 102.4 days (3.4 months), at 25 °C, 4.7 days; at 37 °C, 1.2 days. In the case of the accelerated "storage" of particles in the presence of albumin with subsequent stress testing, since a decrease in the rate of iron release into the supernatant was observed depending on the concentration of the HSA (Figure 9), up to its stabilization at an HSA of 0.5 mg/mL. The conditions for calculating the temperature coefficient and its refinement were insufficient. It is possible that a longer observation period than that used in the experiment (3–4 months) is required to register changes in this case. However, a change in the size of nanoclusters in the presence of albumin both during accelerated "storage" and long-term storage at 8 °C was recorded (Figure 10E,F)

The temperature coefficient of the process was estimated and refined taking into account the data obtained for nanoparticles without albumin in the medium. The time for a 10% decrease in the iron concentration was calculated using a linear approximation of the dependence of the hydrodynamic diameter of nanoclusters on time during accelerated "storage". The time taken for the size to decrease to 138 nm was 37 min. In accordance with Van't Hoff and using a coefficient of 3.15, the forecast for the time to achieve a similar concentration at 8 °C was calculated as 31.6 days. The experimental determination of the time to achieve a 10% decrease in the iron concentration in the supernatant according to the linear approximation during storage at 8 °C was 120.5 days. In accordance with the obtained experimental value, the coefficient of 3.15 was adjusted by the selection method to A = 3.91 (3.8–101, 3.9–118.7, and 3.91–120.6). The shelf life of the suspension with known characteristics at 8 ^{0}C was 120.6 days (4 months); at 25 °C, 11.9 days; and at 37 °C, 2.3 days. To understand the influence of the HSA, the temperature dependence of the hydrodynamic diameter of the nanoclusters on the time at the two temperatures of 70 °C and 8 °C was determined in the absence of albumin using the algorithm described above, with A = 3.32.

The shelf life of a suspension with known characteristics without albumin at 8 °C was 80 days (2.7 months); at 25 °C, 10.3 days; at 37 °C, 2.5 days.

Thus, the approximate shelf life for the SPIONs without and with HSA present were 80 and 121 days, respectively, at 8 °C. Based on the data from accelerated "aging" for two hours at 70 °C, it became possible to predict the characteristics of the nanoclusters for a period of 3–4 months for various experimental or storage temperatures.

4. Discussion

This study investigated magnetic nanosuspensions, which are unstable systems that may be prone to degradation during storage. The study investigates a nanodispersion of magnetic iron oxide particles of SPION/albumin, focusing on the sedimentation and aggregation properties.

The initial characterization of the synthesized nanoparticles confirmed the yield of superparamagnetic clusters of less than 200 nm in size (Figure 3). These particles adhere to the usual desired parameters for theranostic nanocarriers. Further study revealed a stable protein corona formed from albumin's interaction with the magnetic nanoparticles. The fluorescence analysis revealed a static fluorescence quenching mechanism, forming a stable SPION/albumin complex (Figure 4). The study also found that nonradiative transfer does not affect the deep amino acid residues of Trp and Tyr inside the protein corona. These findings are similar to those for citrate BSA/SPIONs with a dense monolayer of protein [59]. The heated albumin itself contains monomers and oligomers or aggregates. In addition to the native protein, the heterogenous fraction presents higher sedimentation coefficients [60]. This may lead to different behavior by MNPs covered with CMDx during heating (colloid solution and colloid sol). The carboxymethyl dextran shell easily adsorbs albumin and forms a hybrid coat layer. As opposed to citrate SPIONs, the ensured coat stabilization was provided by H-H bonds between saccharide residues and magnetite Fe domains. The zeta potential of such carboxymethyl dextran SPIONs under the influence HSA take on a neutral value (Figure 5). The latter leads to weak adherence between magnetic nuclei and nonpacked metastable clusters in an HSA saturation state on the surface. The presented model of this protein–SPION interface can be viewed in Figure 11. This general scheme matches well with the results of the structural analysis of SPIONs modified by blood sera proteins with magnetite nuclei and albumin [61].

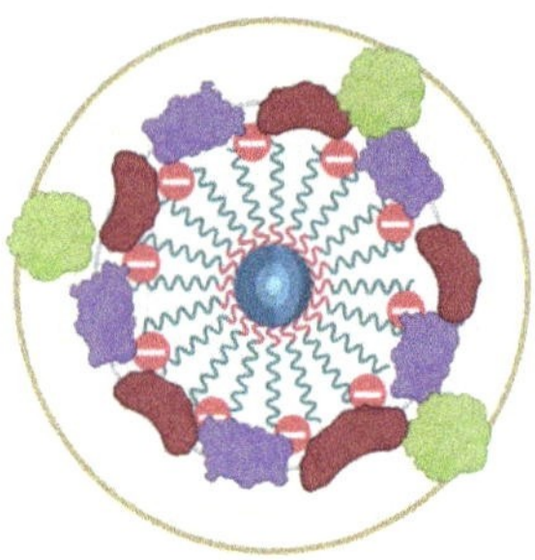

Figure 11. Schematic view of the sera protein (including albumin) corona around magnetic iron oxide nanoparticles stabilized by carboxymethyl dextran.

The obtained DLS results and estimated Fe ion content in the stable phase indicate that the adsorption of albumin on SPIONs may be accompanied by destabilization of magnetic flocculated clusters. According to the results of the DLS study, SPION/albumin nanoparticles coexist in the suspension as individual, separate, and stable particles and in the form of weakly adhered metastable clusters. These clusters are stabilized by van der Waals attraction and magnetic forces when placed in an external field. The multimodal fraction dispersion size function, as measured by light scattering, is evidence of such a structural arrangement. The size of the magnetic nanoparticles representing the dispersed phase of the colloid under study is comparable to the wavelength of the acting radiation, which can lead to light interference and a decrease in the scattering intensity, but the phase difference leading to this effect is completely compensated when registering a signal at an angle of zeros. This case is realized by applying the turbidimetric method of analysis. The turbidimetric approach allows us to use more simple phenomenological relations derived for particles whose size is smaller than the wavelength of light (Figure 6).

The bare SPIONs stabilized by carboxymethyl dextran were characterized in the monodispersed mode with a 100 nm range; The satisfactory stability of the suspension was achieved by the generation of a surface charge of −28 mV. The prepared albumin solution has a monomodal view, where the first peak in the light scattering plot is attached to albumin associates with a size of ~10 nm. In the concentrated protein solution, we observed a polymodal size dispersion with the appearance of nanoaggregated protein species, with an average size of 200–300 nm. Upon addition of albumin to the SPION suspension and following the formation of the protein corona, the size distribution function of the nanoparticles became multimodal, which resembled albumin's characteristics, with the main peak of the light scattering being typical for spheres with a size close to 200 nm. According to our findings, the protein albumin corona neutralizes most of the negative surface charge of the SPIONs but does not induce the falling out of magnetic material to a notable degree for a long period of time (Figure 8). The latter provide long-term colloid stability of SPION/HSA in storage (8 °C) for 6 months without phase separation (Figures 7 and 9).

The rapid temperature stress aging approach was tested for the case of magnetic nanosol stabilized by HSA protein. To reduce the amount of time taken to assess the colloidal stability of the magnetic nanoparticles in combination with human serum albumin (HSA) at a concentration of 0.5 mg/mL in a phosphate buffer from 6 months to several days, a heat-induced accelerated "aging" method was for the first time introduced for electrosterically stabilized magnetic nanoparticles (Figures 7 and 9). The accelerated stress-aging method was previously suggested as the cutting time for the evaluation of low-molecular pharmaceutical substances in solution [62,63]. The temperature stress-aging approach has not previously been applied in the case of the magnetic iron oxide nanoparticles with carboxymethyl dextran. The peculiar feature of the studied magnetic suspension is the usage of HSA compared to the usual BSA-only study [64]. The calculation of the storage time based on the temperature coefficient for compositions with nanoparticles as for pharmaceutical substances is proposed. An amendment was introduced to calculate the temperature coefficient based on the comparison of the actual and estimated storage times at a temperature of 8 °C. The initial stabilization of the magnetic nanoparticles was achieved by coating with carboxymethyl dextran, with the subsequent addition of titratable amounts of albumin. Albumin has a heat conformational transition that can be taken into account. The reliability of the results of the accelerated stress-aging was controlled comparing long-term storage of the magnetic suspension at 8 °C. To assess the aggregation's stability, the size and polydispersity index of the particles with protein were measured using dynamic light scattering and electrophoresis. The sedimentation colloidal stability of the suspensions was estimated by the iron (III) content in the supernatant obtained after sedimentation of the particles in the centrifugal field of the centrifuge. It was found that the general monitoring of the colloidal stability parameters of a magnetic suspension in the presence of albumin reveals a correlation between aggregation and sedimentation stability. After the long-term storage of the magnetic suspension sediment, some deposits consisted of metastable nanoclusters. The metastable nature of the deposit was confirmed by a mild shaking of the vessel containing dispersion. Shaking transforms the deposit into a stable magnetic nanosuspension with the initial DLS parameters of the DLS measurement monitored before storage (Figure 10). The distribution pattern of the nanoparticles in a complex with protein during the accelerated aging was influenced by the presence of albumin, a key factor in the production of a metastable magnetic composition. This stability of the suspension and SPION/albumin allows for the consideration of the magnetic complex as an independent carrier. This suggests that albumin could be used as an additive for nanosuspensions of carboxymethyl-dextran-coated SPIONs, as well as ensures that administered nanocarriers might not immediately aggregate in blood flow. A decrease in the stability of the magnetic complex of albumin when lowered to pH = 4 should be taken into account when nanoparticles are used to deliver therapeutic agents to target cells, as following the disintegration of the magnetic nanocarrier in the endolysosomes, the release of the drug into the cytoplasm occurs. The magnetic carrier based on carboxymethyl-dextran-coated magnetite nanoparticles retained the necessary stability for the targeted delivery of therapeutic agents during parenteral administration in the presence of serum albumin.

5. Conclusions

The method of accelerated temperature stress "aging" of nanosized iron oxide suspensions is efficient for the rapid assessment and prediction of colloidal stability over time of nanosized magnetic iron oxide suspensions stabilized by albumin. The express approach is achieved by applying a high stress temperature of 70 °C and using the albumin's influence on the stability of the iron oxide nanosol. According to the fluorescence quenching and DLS data, colloidal stability is ensured by the formation of an albumin corona and clustering of SPIONs/albumin. The formation of magnetic nanoparticle clusters with a protein corona affects the sedimentation stability of the colloidal system under physical influences (temperature and sedimentation by a centrifugal field). The temperature stress approach is appropriate for the prediction of the colloidal stability of nanosized magnetic iron oxide suspensions over long periods, which was shown by the assessment of the shelf life of iron oxide dispersions with a therapeutic iron content of 50 µg/mL and albumin in the range from 0.05 to 0.5 mg/mL.

Author Contributions: Conceptualization, B.N., L.Y., V.F. and M.S. (Maxim Shevtsov); methodology, B.N., L.Y., V.F., N.Y., D.T., E.P., A.D., M.S. (Maxim Sulatsky), S.S., S.H.S., A.S. (Anusha Srivastava), S.G., A.S. (Avinash Sonawane), S.E.C. and M.S. (Maxim Shevtsov); software, L.Y. and V.F.; validation, L.Y. and V.F.; formal analysis, L.Y.; investigation, B.N., L.Y., V.F., N.Y., D.T., E.P., A.D. and M.S. (Maxim Sulatsky), S.S., S.H.S., A.S. (Anusha Srivastava), S.G., A.S. (Avinash Sonawane), S.E.C. and M.S. (Maxim Shevtsov); resources, S.E.C. and M.S. (Maxim Shevtsov); data curation, B.N, L.Y., V.F. and M.S. (Maxim Shevtsov); writing—original draft preparation, B.N., L.Y., V.F., N.Y., D.T., E.P., A.D., M.S. (Maxim Sulatsky), S.S., S.H.S., A.S. (Anusha Srivastava), S.G., A.S. (Avinash Sonawane), S.E.C. and M.S. (Maxim Shevtsov); writing—review and editing, B.N., L.Y., V.F., N.Y., D.T., E.P., A.D., M.S. (Maxim Sulatsky), S.S., S.H.S., A.S. (Anusha Srivastava), S.G., A.S. (Avinash Sonawane), S.E.C. and M.S. (Maxim Shevtsov); visualization, V.F; supervision, M.S. (Maxim Shevtsov); project administration, M.S. (Maxim Shevtsov); funding acquisition, S.E.C. and M.S. (Maxim Shevtsov). All authors have read and agreed to the published version of the manuscript.

Funding: This work was financially supported by the Ministry of Science and Higher Education of the Russian Federation (Agreement No. 075-15-2022-301), under grant no. R44AR079960 from the National Institute of Arthritis and Musculoskeletal and Skin Diseases, NIH, and the Technische Universität München (TUM) within the DFG funding program Open Access Publishing. Professor Avinash Sonawane was supported by the Department of Science and Technology, Government of India, through an INT/RUS/RFBR/380 grant. Professor Shirish H. Sonawane (NIT Warangal India) was supported by the Ministry of Education (MoE), Government of India, through the SPARC (Scheme for Promotion of Academic and Research Collaboration) Project, under grant no. SPARC/2019-2020/P2420/SL.

Data Availability Statement: The datasets used and/or analyzed during the current study are available from the corresponding author Maxim Shevtsov upon reasonable request.

Conflicts of Interest: All authors declare no conflicts of interest.

Abbreviations

MNP	Magnetic nanoparticle
SPIONs	Superparamagnetic iron oxide nanoparticles
BSA	Bovine serum albumin
HSA	Human serum albumin
CMDx	Carboxymethyl dextran
DLS	Dynamic light scattering
TEM	Transmission electron microscopy
PBS	Phosphate-buffered saline
RFI	Relative fluorescence intensity

References

1. Gareev, K.G.; Grouzdev, D.S.; Koziaeva, V.V.; Sitkov, N.O.; Gao, H.; Zimina, T.M.; Shevtsov, M. Biomimetic Nanomaterials: Diversity, Technology, and Biomedical Applications. *Nanomaterials* **2022**, *12*, 2485. [CrossRef] [PubMed]
2. Zhou, Y.; Liu, R.; Shevtsov, M.; Gao, H. When imaging meets size-transformable nanosystems. *Adv. Drug Deliv. Rev.* **2022**, *183*, 114176. [CrossRef] [PubMed]
3. Shevtsov, M.; Kaesler, S.; Posch, C.; Multhoff, G.; Biedermann, T. Magnetic nanoparticles in theranostics of malignant melanoma. *EJNMMI Res.* **2021**, *11*, 127. [CrossRef]
4. Han, H.; Yao, Y.; Robinson, R.D. Interplay between Chemical Transformations and Atomic Structure in Nanocrystals and Nanoclusters. *Acc. Chem. Res.* **2021**, *54*, 509–519. [CrossRef]
5. Chatterjee, K.; Sarkar, S.; Jagajjanani Rao, K.; Paria, S. Core/shell nanoparticles in biomedical applications. *Adv. Colloid Interface Sci.* **2014**, *209*, 8–39. [CrossRef]
6. Wu, W.; He, Q.; Jiang, C. Magnetic iron oxide nanoparticles: Synthesis and surface functionalization strategies. *Nanoscale Res. Lett.* **2008**, *3*, 397–415. [CrossRef]
7. Gogoi, B.; Das, U. Review on Magnetism in Nanomaterials and Superparamagnetism. In *Selected Progresses in Modern Physics*; Springer: Singapore, 2021; pp. 303–312.
8. Yaseen, W.; Ullah, N.; Xie, M.; Wei, W.; Xu, Y.; Zahid, M.; Oluigbo, C.J.; Yusuf, B.A.; Xie, J. Cobalt–Iron nanoparticles encapsulated in mesoporous carbon nanosheets: A one-pot synthesis of highly stable electrocatalysts for overall water splitting. *Int. J. Hydrog. Energy* **2021**, *46*, 5234–5249.
9. Anjum, S.; Tufail, R.; Rashid, K.; Zia, R.; Riaz, S. Effect of cobalt doping on crystallinity, stability, magnetic and optical properties of magnetic iron oxide nano-particles. *J. Magn. Magn. Mater.* **2017**, *432*, 198–207.
10. Kievit, F.M.; Zhang, M. Surface engineering of iron oxide nanoparticles for targeted cancer therapy. *Acc. Chem. Res.* **2011**, *44*, 853–862. [CrossRef]
11. Alphandéry, E. Iron oxide nanoparticles for therapeutic applications. *Drug Discov. Today* **2020**, *25*, 141–149.
12. Guardia, P.; Labarta, A.; Batlle, X. Tuning the size, the shape, and the magnetic properties of iron oxide nanoparticles. *J. Phys. Chem. C* **2011**, *115*, 390–396. [CrossRef]
13. Wang, Y.X.; Xuan, S.; Port, M.; Idee, J.M. Recent advances in superparamagnetic iron oxide nanoparticles for cellular imaging and targeted therapy research. *Curr. Pharm. Des.* **2013**, *19*, 6575–6593. [CrossRef] [PubMed]
14. Phan, H.T.; Haes, A.J. What Does Nanoparticle Stability Mean? *J. Phys. Chem. C* **2019**, *123*, 16495–16507. [CrossRef]
15. Song, T.; Gao, F.; Guo, S.; Zhang, Y.; Li, S.; You, H.; Du, Y. A review of the role and mechanism of surfactants in the morphology control of metal nanoparticles. *Nanoscale* **2021**, *13*, 3895–3910. [CrossRef]
16. Rouster, P.; Pavlovic, M.; Szilagyi, I. Destabilization of Titania Nanosheet Suspensions by Inorganic Salts: Hofmeister Series and Schulze-Hardy Rule. *J. Phys. Chem. B* **2017**, *121*, 6749–6758. [CrossRef]
17. Kotsmar, C.; Yoon, K.Y.; Yu, H.; Ryoo, S.Y.; Barth, J.; Shao, S.; Prodanovic, M.; Milner, T.E.; Bryant, S.L.; Huh, C. Stable citrate-coated iron oxide superparamagnetic nanoclusters at high salinity. *Ind. Eng. Chem. Res.* **2010**, *49*, 12435–12443.
18. Elbasuney, S. Sustainable steric stabilization of colloidal titania nanoparticles. *Appl. Surf. Sci.* **2017**, *409*, 438–447. [CrossRef]
19. Guzman, M.; Arcos, M.; Dille, J.; Rousse, C.; Godet, S.; Malet, L. Effect of the Concentration and the Type of Dispersant on the Synthesis of Copper Oxide Nanoparticles and Their Potential Antimicrobial Applications. *ACS Omega* **2021**, *6*, 18576–18590. [CrossRef]
20. Vinodhini, V.; Krishnamoorthi, C. Effect of dispersants on cytotoxic properties of magnetic nanoparticles: A review. *Polym. Bull.* **2022**, *79*, 8143–8192.
21. Vijayakanth, V.; Chintagumpala, K. A review on an effect of dispersant type and medium viscosity on magnetic hyperthermia of nanoparticles. *Polym. Bull.* **2023**, *80*, 4737–4781.

22. Luo, L.; Nguyen, A.V. A review of principles and applications of magnetic flocculation to separate ultrafine magnetic particles. *Sep. Purif. Technol.* **2017**, *172*, 85–99.
23. Sabzi Dizajyekan, B.; Jafari, A.; Vafaie-Sefti, M.; Saber, R.; Fakhroueian, Z. Preparation of stable colloidal dispersion of surface modified Fe3O4 nanoparticles for magnetic heating applications. *Sci. Rep.* **2024**, *14*, 1296. [CrossRef] [PubMed]
24. Can, H.K.; Kavlak, S.; ParviziKhosroshahi, S.; Guner, A. Preparation, characterization and dynamical mechanical properties of dextran-coated iron oxide nanoparticles (DIONPs). *Artif. Cells Nanomed. Biotechnol.* **2018**, *46*, 421–431. [CrossRef]
25. Borny, R.; Lechleitner, T.; Schmiedinger, T.; Hermann, M.; Tessadri, R.; Redhammer, G.; Neumuller, J.; Kerjaschki, D.; Berzaczy, G.; Erman, G.; et al. Nucleophilic cross-linked, dextran coated iron oxide nanoparticles as basis for molecular imaging: Synthesis, characterization, visualization and comparison with previous product. *Contrast Media Mol. Imaging* **2015**, *10*, 18–27. [CrossRef] [PubMed]
26. Hui, B.H.; Salimi, M.N. Production of iron oxide nanoparticles by co-precipitation method with optimization studies of processing temperature, pH and stirring rate. *IOP Conf. Ser. Mater. Sci. Eng.* **2020**, *743*, 012036.
27. Chekli, L.; Phuntsho, S.; Roy, M.; Lombi, E.; Donner, E.; Shon, H.K. Assessing the aggregation behaviour of iron oxide nanoparticles under relevant environmental conditions using a multi-method approach. *Water Res.* **2013**, *47*, 4585–4599. [CrossRef] [PubMed]
28. Amiri, P.; Behin, J.; Ghanbariebad, S. Degradation of polyvinylpyrrolidone-coated iron oxide nanoparticles through ozonation: Steric to electrosteric repulsion and electrostatic interactions. *Colloids Surf. A Physicochem. Eng. Asp.* **2023**, *675*, 131995.
29. Lu, Y.; Ballauff, M. Spherical polyelectrolyte brushes as nanoreactors for the generation of metallic and oxidic nanoparticles: Synthesis and application in catalysis. *Prog. Polym. Sci.* **2016**, *59*, 86–104.
30. Dong, H.; Xie, Y.; Zeng, G.; Tang, L.; Liang, J.; He, Q.; Zhao, F.; Zeng, Y.; Wu, Y. The dual effects of carboxymethyl cellulose on the colloidal stability and toxicity of nanoscale zero-valent iron. *Chemosphere* **2016**, *144*, 1682–1689. [CrossRef]
31. Leonel, A.G.; Mansur, H.S.; Mansur, A.A.P.; Caires, A.; Carvalho, S.M.; Krambrock, K.; Outon, L.E.F.; Ardisson, J.D. Synthesis and characterization of iron oxide nanoparticles/carboxymethyl cellulose core-shell nanohybrids for killing cancer cells in vitro. *Int. J. Biol. Macromol.* **2019**, *132*, 677–691. [CrossRef]
32. Liu, G.; Hong, R.; Guo, L.; Li, Y.; Li, H. Preparation, characterization and MRI application of carboxymethyl dextran coated magnetic nanoparticles. *Appl. Surf. Sci.* **2011**, *257*, 6711–6717. [CrossRef]
33. Sakulkhu, U.; Mahmoudi, M.; Maurizi, L.; Salaklang, J.; Hofmann, H. Protein corona composition of superparamagnetic iron oxide nanoparticles with various physico-chemical properties and coatings. *Sci. Rep.* **2014**, *4*, 5020.
34. Monopoli, M.P.; Walczyk, D.; Campbell, A.; Elia, G.; Lynch, I.; Bombelli, F.B.; Dawson, K.A. Physical-chemical aspects of protein corona: Relevance to in vitro and in vivo biological impacts of nanoparticles. *J. Am. Chem. Soc.* **2011**, *133*, 2525–2534. [CrossRef]
35. Hoang, K.N.L.; Wheeler, K.E.; Murphy, C.J. Isolation Methods Influence the Protein Corona Composition on Gold-Coated Iron Oxide Nanoparticles. *Anal. Chem.* **2022**, *94*, 4737–4746. [CrossRef] [PubMed]
36. Ferreira-Filho, V.C.; Morais, B.; Vieira, B.J.C.; Waerenborgh, J.C.; Carmezim, M.J.; Tóth, C.N.; Même, S.; Lacerda, S.; Jaque, D.; Sousa, C.T.; et al. Influence of SPION Surface Coating on Magnetic Properties and Theranostic Profile. *Molecules* **2024**, *29*, 1824. [CrossRef]
37. Bychkova, A.V.; Yakunina, M.N.; Lopukhova, M.V.; Degtyarev, Y.N.; Motyakin, M.V.; Pokrovsky, V.S.; Kovarski, A.L.; Gorobets, M.G.; Retivov, V.M.; Khachatryan, D.S. Albumin-Functionalized Iron Oxide Nanoparticles for Theranostics: Engineering and Long-Term In Situ Imaging. *Pharmaceutics* **2022**, *14*, 2771. [CrossRef]
38. Yu, S.; Peralvarez-Marin, A.; Minelli, C.; Faraudo, J.; Roig, A.; Laromaine, A. Albumin-coated SPIONs: An experimental and theoretical evaluation of protein conformation, binding affinity and competition with serum proteins. *Nanoscale* **2016**, *8*, 14393–14405. [CrossRef]
39. Danafar, H.; Baghdadchi, Y.; Barsbay, M.; Ghaffarlou, M.; Mousazadeh, N.; Mohammadi, A. Synthesis of Fe_3O_4-Gold hybrid nanoparticles coated by bovine serum albumin as a contrast agent in MR imaging. *Heliyon* **2023**, *9*, e13874. [CrossRef]
40. Santos, M.G.; de Carvalho, D.T.; Caminitti, L.B.; de Lima, B.B.A.; Cavalcanti, M.H.d.S.; dos Santos, D.F.R.; Virtuoso, L.S.; Hirata, D.B.; Figueiredo, E.C. Use of magnetic Fe3O4 nanoparticles coated with bovine serum albumin for the separation of lysozyme from chicken egg white. *Food Chem.* **2021**, *353*, 129442. [CrossRef]
41. Gawali, S.L.; Shelar, S.B.; Gupta, J.; Barick, K.C.; Hassan, P.A. Immobilization of protein on Fe_3O_4 nanoparticles for magnetic hyperthermia application. *Int. J. Biol. Macromol.* **2021**, *166*, 851–860. [CrossRef]
42. Liu, Z.; Wang, X.; Zhang, C.; Lin, K.; Yang, J.; Zhang, Y.; Hao, J.; Tian, F. Folic acid-coupled bovine serum albumin-modified magnetic nanocomposites from quantum-sized Fe_3O_4 and layered double hydroxide for actively targeted delivery of 5-fluorouracil. *Int. J. Biol. Macromol.* **2024**, *256*, 128385. [CrossRef] [PubMed]
43. Mekseriwattana, W.; Srisuk, S.; Kriangsaksri, R.; Niamsiri, N.; Prapainop, K. The Impact of Serum Proteins and Surface Chemistry on Magnetic Nanoparticle Colloidal Stability and Cellular Uptake in Breast Cancer Cells. *AAPS PharmSciTech* **2019**, *20*, 55. [CrossRef] [PubMed]

44. Unterweger, H.; Dézsi, L.; Matuszak, J.; Janko, C.; Poettler, M.; Jordan, J.; Bäuerle, T.; Szebeni, J.; Fey, T.; Boccaccini, A.R.; et al. Dextran-coated superparamagnetic iron oxide nanoparticles for magnetic resonance imaging: Evaluation of size-dependent imaging properties, storage stability and safety. *Int. J. Nanomed.* **2018**, *13*, 1899–1915. [CrossRef] [PubMed]
45. Yu, S.-M.; Gonzalez-Moragas, L.; Milla, M.; Kolovou, A.; Santarella-Mellwig, R.; Schwab, Y.; Laromaine, A.; Roig, A. Bio-identity and fate of albumin-coated SPIONs evaluated in cells and by the C. elegans model. *Acta Biomater.* **2016**, *43*, 348–357. [CrossRef]
46. Segets, D.; Marczak, R.; Schafer, S.; Paula, C.; Gnichwitz, J.F.; Hirsch, A.; Peukert, W. Experimental and theoretical studies of the colloidal stability of nanoparticles-a general interpretation based on stability maps. *ACS Nano* **2011**, *5*, 4658–4669. [CrossRef]
47. Wei, L.; Hervé, M.; Edouard, P. Use of different rapid mixing devices for controlling the properties of magnetite nanoparticles produced by precipitation. *J. Cryst. Growth* **2012**, *342*, 21–27. [CrossRef]
48. Shevtsov, M.A.; Nikolaev, B.P.; Yakovleva, L.Y.; Parr, M.A.; Marchenko, Y.Y.; Eliseev, I.; Yudenko, A.; Dobrodumov, A.V.; Zlobina, O.; Zhakhov, A.; et al. 70-kDa heat shock protein coated magnetic nanocarriers as a nanovaccine for induction of anti-tumor immune response in experimental glioma. *J. Control. Release* **2015**, *220*, 329–340. [CrossRef]
49. Gojmerac Ivšić, A.; Tamhina, B. Extraction and formation of iron (III) thiocyanate complexes: Application for spectrophotometric determination of iron. *Croat. Chem. ACTA* **2003**, *76*, 323–328.
50. Gekle, M.; Silbernagl, S. Comparison of the buffer capacity of endocytotic vesicles, lysosomes and cytoplasm in cells derived from the proximal tubule of the kidney (opossum kidney cells). *Pflügers Arch.* **1995**, *429*, 452–454. [CrossRef]
51. Gremese, E.A.-O.; Bruno, D.A.-O.; Varriano, V.; Perniola, S.A.-O.; Petricca, L.; Ferraccioli, G. Serum Albumin Levels: A Biomarker to Be Repurposed in Different Disease Settings in Clinical Practice. *J. Clin. Med.* **2023**, *12*, 6017. [CrossRef]
52. Rabel, M.; Warncke, P.; Gruttner, C.; Bergemann, C.; Kurland, H.D.; Muller, R.; Dugandzic, V.; Thamm, J.; Muller, F.A.; Popp, J.; et al. Simulation of the long-term fate of superparamagnetic iron oxide-based nanoparticles using simulated biological fluids. *Nanomedicine* **2019**, *14*, 1681–1706. [CrossRef] [PubMed]
53. Available online: https://pharmacopoeia.ru/ofs-1-1-0009-15-sroki-godnosti-lekarstvennyh-sredstv/ (accessed on 20 August 2024).
54. Available online: https://pharmacopoeia.regmed.ru/pharmacopoeia/izdanie-15/1/1-1/stabilnost-i-sroki-godnosti-lekarstvennykh-sredstv (accessed on 20 August 2024).
55. Yang, Q.; Liang, J.; Han, H. Probing the Interaction of Magnetic Iron Oxide Nanoparticles with Bovine Serum Albumin by Spectroscopic Techniques. *J. Phys. Chem. B* **2009**, *113*, 10454–10458. [CrossRef] [PubMed]
56. Ivanov, K.L.; Lukzen, N.N.; Doktorov, A.B. On the time dependence of rate coefficients of irreversible reactions between reactants with anisotropic reactivity in liquid solutions. *J. Chem. Phys.* **2016**, *145*, 064104. [CrossRef]
57. Trang, L.T.H.; Huong, N.T.X.; Trang, V.T. Research for stability of food products and functional foods. *Vietnam. J. Food Control.* **2022**, *5*, 170–182.
58. Chen, W.-Y.; Huang, H.-M., Lin, C.-C., Lin, F.-Y.; Chan, Y.-C. Effect of temperature on hydrophobic interaction between proteins and hydrophobic adsorbents: Studies by isothermal titration calorimetry and the van't Hoff equation. *Langmuir* **2003**, *19*, 9395–9403.
59. Mekseriwattana, W.; Thiangtrongjit, T.; Reamtong, O.; Wongtrakoongate, P.; Katewongsa, K.P. Proteomic analysis reveals distinct protein corona compositions of citrate and riboflavin-coated SPIONs. *ACS Omega* **2022**, *7*, 37589–37599. [CrossRef]
60. Yamasaki, K.; Anraku, M. Stability of albumin and stabilization of albumin preparations. In *Albumin in Medicine: Pathological and Clinical Applications*; Springer: Berlin/Heidelberg, Germany, 2016; pp. 25–49.
61. Sekar, G.; Haldar, M.; Kumar, D.T.; Doss, C.G.P.; Mukherjee, A.; Chandrasekaran, N. Exploring the interaction between iron oxide nanoparticles (IONPs) and human serum albumin (HSA): Spectroscopic and docking studies. *J. Mol. Liq.* **2017**, *241*, 793–800.
62. Vilar, G.; Fernández-Rosas, E.; Puntes, V.; Jamier, V.; Aubouy, L.; Vázquez-Campos, S. Monitoring migration and transformation of nanomaterials in polymeric composites during accelerated aging. *J. Phys. Conf. Ser.* **2013**, *429*, 012044. [CrossRef]
63. Sayle, D.C.; Caddeo, F.; Morgan, L.M.; Neale, R.L.; Sayle, T.X.; Brambila, C.; Nutter, J.; Bhatta, U.; Ta, K.M.; Flitcroft, J.M. Aging mechanisms of nanoceria and pathways for preserving optimum morphology. *Nano Today* **2023**, *51*, 101916.
64. Bychkova, A.V.; Lopukhova, M.V.; Wasserman, L.A.; Degtyarev, Y.N.; Kovarski, A.L.; Chakraborti, S.; Mitkevich, V.A. The influence of pH and ionic strength on the interactions between human serum albumin and magnetic iron oxide nanoparticles. *Int. J. Biol. Macromol.* **2022**, *194*, 654–665. [CrossRef]

nanomaterials

MDPI

Communication

Observation of Linear Magnetoresistance in MoO_2

Yulong Su *, Zhibin He, Ruizheng Jiang and Jundong Zhang

Marine Engineering College, Dalian Maritime University, Dalian 116026, China; hezb999@dlmu.edu.cn (Z.H.); jrzdmu@dlmu.edu.cn (R.J.); zhjundong@dlmu.edu.cn (J.Z.)
* Correspondence: suyulong@dlmu.edu.cn

Abstract: Magnetoresistance, the change in resistance with applied magnetic fields, is crucial to the magnetic sensor technology. Linear magnetoresistance has been intensively studied in semimetals and semiconductors. However, the air-stable oxides with a large linear magnetoresistance are highly desirable but remain to be fully explored. In this paper, we report the direct observation of linear magnetoresistance in polycrystalline MoO_2 without any sign of saturation up to 7 T under 50 K. Interestingly, the linear magnetoresistance reaches as large as 1500% under 7 T at 2 K. The linear field dependence is in great contrast to the parabolic behavior observed in single-crystal MoO_2, probably due to phonon scattering near the grain boundaries. Our results pave the way to comprehending magneto-transport behavior in oxides and their potential applications in magnetic sensors.

Keywords: linear magnetoresistance; MoO_2; polycrystalline; magneto-transport

1. Introduction

Magnetoresistance refers to the variation in resistivity under the applied magnetic fields. The study of magnetoresistance has been stimulated by the magnetic sensor technology due to its accuracy and non-contact sensing capabilities, thus ensuring a reliable and robust performance even in harsh environmental conditions. The first magnetoresistive effect was discovered in 1856 by William Thomson [1]. Later, Albert Fert and Peter Grünberg discovered giant magnetoresistance in multilayers composed of ferromagnetic and non-magnetic conductive layers and were awarded the 2007 Nobel Prize in Physics, setting up the foundation for the study of spintronics. The magnetic sensors are practically used in magnetic position sensing [2,3], storage/recording devices [4–6], non-destructive monitoring [7], biosensing platforms [8,9], and so on. Nowadays, the demand for highly sensitive and cost-effective magnetoresistance sensors has been rising due to the growing concerns about further improving the data storage capacity/detection sensitivity of current magnetoresistance sensors and cutting down the total cost of fabrication/production, which leads to the expansion of the application area of magnetoresistance sensors. Specifically, with emerging nanotechnology, the aim is to develop novel nanomaterials that exhibit large resistance change, with minimum requirements on extreme conditions and large magnetic fields.

Usually, the magnetoresistance exhibits a quadratic dependence in the low magnetic field limit, while under strong magnetic fields, it saturates to a constant value [10], as shown in the following Equation (1):

$$\frac{\Delta\rho}{\rho} \propto \begin{cases} (\mu H)^2, \mu H < 1 \\ C, \mu H \geq 1 \end{cases} \quad (1)$$

where μ is the carriers' mobility, and H is the magnetic field.

This classical magnetoresistance is usually limited to an amplitude of a few percentages. Moreover, researchers found that the amplitude or the magnetic field dependence of the magnetoresistance can be modified by multiple factors, such as magnetic ordering [11–14], Dirac physics [15], Landau levels [16], or spatial inhomogeneities [17]. In particular, Dirac physics in topological insulators, semimetals, or spatial inhomogeneities can induce pronounced linear magnetoresistance behavior. Recently, linear magnetoresistance under high fields was observed in some Dirac or Weyl semimetals such as Bi [18], InSb [19], Cd_3As_2 [20], NbP [21], LaSb [22], and ZrSiS [23]; and compensated semimetals such as WTe_2 [24,25], $TaAs_2$, and $NbAs_2$ [26]. There have also been reports of linear magnetoresistance in layered charge density wave and spin density wave compounds, such as iron pnictides [27,28], 2H-$NbSe_2$, and 2H-$TaSe_2$ [29]. The non-saturating linear magnetoresistance is of great interest for the magnetic sensors with high resolution and memory reading/storage applications. In spite of the intensive study on semimetals and semiconductors, the air-stable oxides with large linear magnetoresistance are highly desirable but remain to be fully explored. In this study, we focused on polycrystalline Molybdenum dioxide (MoO_2), an air-stable oxide and carried out a systematical study on its magnetic field-dependent resistivity.

MoO_2, owing to its monoclinic crystal structure (space group $P2_1/c$), is known as a useful anode material in Li-ion batteries due to its low electrical resistivity, high stability, large capacity, and high density associated with volume capacity [30,31]. Recently, crystalline MoO_2 was predicted to be a nodal-line semimetal when spin–orbital coupling is neglected [32]. A transport study revealed a quadratic field dependent magnetoresistance in single-crystalline MoO_2, and the maximum magnetoresistance reaches a value of $5.03 \times 10^4\%$ at 2 K and 9 T [33]. Substantially, two-dimensional MoO_2 nanoplates were prepared by chemical vapor deposition methods [34,35], which display a large linear magnetoresistance of up to 455% at 3 K and −9 T and a nonlinear Hall effect [35]. Here, we report the visualization of linear magnetoresistance behavior in polycrystalline MoO_2, in great contrast with the quadratic behavior in single crystals. The polycrystalline MoO_2 samples were synthesized by a facile technique spark plasma sintering. The linear magnetoresistance behavior is robust, ensuring the practical application. Thus, our experimental results pave the way to a better understanding of the magnetoresistance physics and the potential applications in magnetic sensors.

2. Materials and Methods

2.1. Synthesis of MoO_2 Sample

The MoO_2 samples were prepared by a spark plasma sintering (SPS) method. MoO_2 powders were grounded for 90 min, using a mortar. Then, the grounded powder was put into a graphite die and sintered by an SPS system (LABOX-650, SINTER LAND, Kanagawa, Japan) at 973 K for 2 h, under vacuum, with a uniaxial pressure of 50 MPa. The sample was subsequently cooled to room temperature naturally. The ingots obtained from sintering were 20.7 mm in diameter and 10 mm in thickness. The obtained pellets were polished and cut into bards with a specific shape and size for electric transport measurements.

2.2. Materials Characterizations

The MoO_2 sample was characterized by the X-ray diffraction (XRD) and scanning electron microscope (SEM) measurements. The XRD patterns were collected in SmartLab, with a Cu-K_α radiation source (λ = 1.5406 Å, Rigaku, Tokyo, Japan), with a 2 theta range of 10–80° at room temperature, with a scanning rate of 13.5 degree per min. Scanning electron microscopy (SEM) characterization was performed on an FEI QUANTA 200 FEG microscope (City of Hope, Duarte, CA, USA) at 20 kV.

2.3. Transport Measurements

The cleaved samples with a typical size of about $3.0 \times 1.0 \times 0.1$ mm^3 were used for transport measurements. The resistivity measurements were performed on a Quantum Design physical property measurement system (PPMS), with the highest magnetic field being 7 T. Standard four-probe resistivity (ρ_{xx}) measurements were carried out using a constant current mode. The magnetoresistance (MR) is defined as MR = $[\rho_{xx}(H, T) - \rho_{xx}(0, T)]/\rho_{xx}(0, T)$ and measured by applying the magnetic field, H, perpendicular to the sample. Electrical contacts were prepared by platinum wires and a silver paste. The Seebeck coefficient (S) was measured by the PPMS thermal transport option in continuous scanning mode, with a 0.2 K min^{-1} heating rate.

3. Results and Discussion

3.1. XRD and SEM Analysis of MoO_2 Samples

Figure 1a displays the schematic crystal structure of monoclinic MoO_2, which can also be considered to have a distorted rutile structure. Figure 1b shows the X-ray diffraction pattern of the prepared polycrystalline MoO_2 sample. All the characteristic peaks of the XRD pattern can be indexed into well-crystallized MoO_2, confirming its monoclinic crystal structure (PDF#32-0671) with the space group, $P2_1/c$; and lattice constants, a = 5.69 Å, b = 3.12 Å, c = 5.07 Å, and β = 112.78°. The grain size of MoO_2 was examined by the SEM in Supplementary Figure S1 and inset of Figure 2a, where the grain size ranges from 500 nm to 1.0 μm.

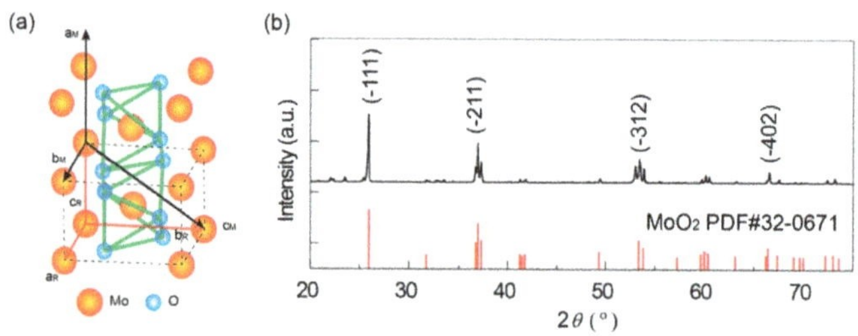

Figure 1. Crystal structure of MoO_2. (**a**) Schematic ball-and-stick model of monoclinic (M) MoO_2. The unit cell of the distorted rutile (R) structure is shown with dashed lines. (**b**) Powder XRD pattern of MoO_2.

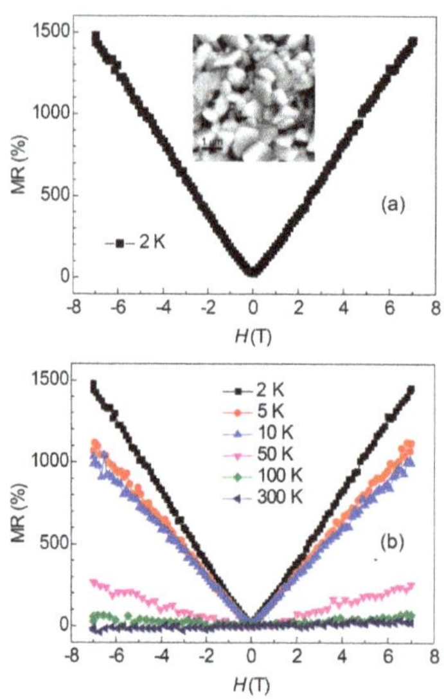

Figure 2. Magnetoresistance (MR) of polycrystalline MoO_2. (**a**) MR at 2 K. Here, MR is defined as MR = $[\rho_{xx}(H, T) - \rho_{xx}(0, T)]/\rho_{xx}(0, T) \times 100\%$. Inset: Scanning electron microscopy image of polycrystalline MoO_2. (**b**) MR at various temperatures ranging from 2 K to 300 K.

3.2. Magneto-Transport Properties of MoO_2 Samples

Figure 2 shows the temperature-dependent magnetoresistance by applying magnetic fields perpendicular to the sample. A positive, linear magnetic field dependence of the transverse magnetoresistance is visualized up to 7 T in Figure 2a. It is more obvious in the first-order derivative, $dMR(H)/dH$ (Supplementary Figure S2), that the linear dependence appears from 0.1 T. It is worth mentioning that the amplitude of the magnetoresistance is huge, reaching 1500% at H = 7 T and T = 2 K. Moreover, varying the temperature, T, from 2 K to 50 K dramatically affects the magnitude of MR but changes neither the sign nor the shape of MR-H curves, as shown in Figure 2b. It is obvious that the linear MR is strongly dependent on the temperature, T, and barely exists at a T above 50 K. The sensitivity of the magnitude of MR with T suggests the significant role of phonon scattering, which leads to a remarkable decrease with the increasing T. We need to mention that the linear magnetoresistance behavior is air-stable in polycrystalline MoO_2 and barely changes within three months.

To shed more light on the linear magnetoresistance behavior, we carried out a thermal transport measurement for the MoO_2 samples. Figure 3 displays the temperature-dependent resistivity, $\rho(T)$, and Seebeck coefficient, $S(T)$, curves. As shown in Figure 3a, the ρ-T curve of MoO_2 exhibits a metallic behavior. Above 100 K, the resistivity, ρ, increases with the increase in the temperature, T, while below 100 K, the resistivity, ρ, first increases and then decreases with a turnover at T = 50 K. It is different with single-crystal data of MoO_2, where the decrease in ρ below 50 K is absent probably due to electron scattering at the grain boundaries [33]. Figure 3b displays the temperature-dependent Seebeck coefficient, S, below 300 K. The Seebeck coefficient exhibits a non-monotopic temperature dependence, with the sign of the Seebeck coefficient changing from negative to positive with decreasing temperature. The negative Seebeck coefficient, S, below 300 K indicates electrons as the charge carrier in this system. However, there is a sign change from 16.7 K to 2 K and peaks at 10.8 K with a positive Seebeck coefficient.

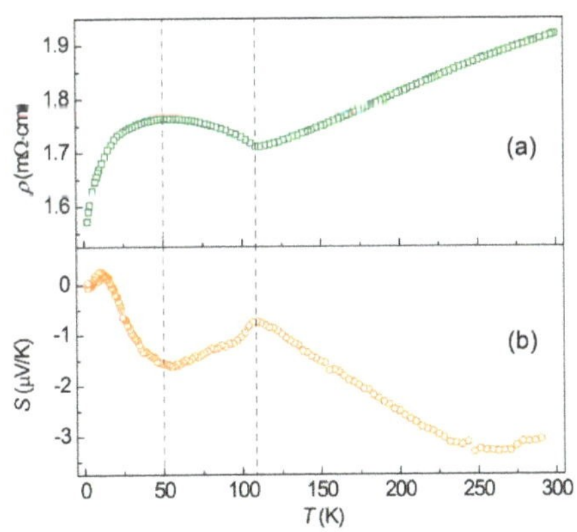

Figure 3. Temperature-dependent resistivity $\rho(T)$ (**a**) and Seebeck coefficient $S(T)$ (**b**) of polycrystalline MoO_2.

The Seebeck coefficient, S, can be described within a free electron model by Equation (2):

$$S(T) = -\frac{\pi^2}{3}\frac{k_B}{|e|}k_B T\left[\frac{N(0)}{n} + \frac{1}{\mu}\frac{d\mu}{dE_F}\right] \quad (2)$$

where E_F is the Fermi energy, and $N(0)$ is the density of states. The first term, n, determines the carrier type, while the second one can be qualitatively related to the variation in mobility and may change signs with temperature.

The Seebeck coefficient can also be calculated by Equation (3):

$$S = \frac{S_e b + S_h}{b + 1} \tag{3}$$

where S is the total Seebeck coefficient; S_e and S_h are the Seebeck coefficient of electrons and holes, respectively; and $b = \mu_e/\mu_h$ is the ratio of the mobilities of electrons and holes. It is known that the mobilities strongly depend on the temperature, as well as the crystallography, such as orientation and defects. Assuming identical charge carrier concentrations for holes and electrons, the sign change in the Seebeck coefficient happens when the mobility of holes becomes larger than the mobility of the electrons, which can occur because the mobility of electrons and holes is affected differently by surface scattering due to their different effective masses [36]. Similar explanations were given for the sign change in Bi, which suggested that the electron mobility becomes smaller than the hole mobility in the low temperature range, because the hole mobility was not significantly affected by boundary scattering [37].

Next, we discuss the possible origins for the linear magnetoresistance behavior, which is different from the parabolic magnetic field dependence in the MoO_2 single crystal. First, the observed quadratic magnetic field dependence in single-crystalline MoO_2 [33] can get out of the possible origin from Dirac physics, even though MoO_2 is predicted to be a nodal-line semimetal [34]. Moreover, the absence of linear magnetoresistance in the single crystal can also remove the possibility of magnetic ordering or Landau levels. Last but not least, non-saturating linear magnetoresistance can be comprehended semi-classically in the Parish and Littlewood framework [18] by a random resistor network model that mimics a disordered and strongly inhomogeneous conductor system. It has been evidenced that voids or inhomogeneities will cause linear magnetoresistance in disordered and strongly inhomogeneous semiconductors such as InSb [20]. In our polycrystalline MoO_2 samples, the grain size is of 0.5–1.0 μm, and the crystalline imperfections, especially grain boundaries, cannot be ignored. Thus, enhanced phonon scattering is expected near the grain boundaries [38,39]. We notice that the chemical vapor deposition-grown MoO_2 nanoplates (with a lateral size of ~10 μm and a thickness of 7.8–55.2 nm) also display a large linear magnetoresistance of up to 455% at 3 K and −9 T [35]. In two-dimensional films, the phonon–interface or phonon–boundary scattering increases as the film becomes thinner [40,41], especially when the sample size along the transport direction is much smaller than the mean free path of the phonon. Even though the grain boundaries are absent in those MoO_2 nanoplates [35], phonon–interface scattering is inevitable since the thickness range of 7.8–55.2 nm is considerably smaller than the phonon mean free path of the phonon, i.e., ~300 nm in silicon films [41]. Therefore, we believe that the grain boundary plays a remarkable role in the linear magnetoresistance behavior in polycrystal MoO_2, which is consistent with the sensitive temperature dependence, where the linear magnetoresistance disappears as temperature increases above 50 K (Figure 2b). More theoretical or experimental work in the future is needed to clarify the physical origin of this linear magnetoresistance in polycrystalline MoO_2.

4. Conclusions

In this paper, we report a linear magnetoresistance behavior in polycrystalline MoO_2 prepared by the spark plasma sintering technique. The linear magnetoresistance does not saturate up to 7 T under 50 K, with the measured magnitude reaching as large as ~1500% at T = 2 K and H = 7 T. Polycrystalline MoO_2 is air-stable and, thus, ensures a reliable and robust magnetoresistance performance even in harsh environmental conditions. Moreover, the facile synthesis of polycrystalline MoO_2, together with the feature of robust linear magnetoresistance, guarantees mass production and practical application. Our results provide valuable guidance towards the comprehension of magneto-transport behavior in oxides, as well as their potential applications in magnetic sensors.

Supplementary Materials: The following supporting information can be downloaded at https://www.mdpi.com/article/10.3390/nano14110915/s1, Figure S1. SEM images of polycrystalline MoO_2 samples; Figure S2. The magnetic field derivatives of MR at 2 K, 5 K, 10 K, and 50 K; Figure S3. R-T curves under 0 T and 7 T.

Author Contributions: Y.S., investigation, methodology, formal analysis, writing—original draft, and writing—review and editing; Z.H., writing—review and editing; R.J., writing—review and editing; J.Z., conceptualization, supervision, and writing—review and editing. All authors have read and agreed to the published version of the manuscript.

Funding: This work is supported by the China Ministry of Industry and Information Technology Project: Innovation Engineering of the Offshore LNG Equipment Industry Chain under Grant CBG3N21-2-7.

Data Availability Statement: Data will be made available upon request.

Conflicts of Interest: The authors declare that they have no known competing financial interests or personal relationships that could have appeared to influence the work reported in this paper.

References

1. Thomson, W. XIX. On the Electro-Dynamic Qualities of Metals:—Effects of Magnetization on the Electric Conductivity of Nickel and of Iron. *Proc. R. Soc. Lond.* **1997**, *8*, 546–550. [CrossRef]
2. Kartik, V.; Sebastian, A.; Tuma, T.; Pantazi, A.; Pozidis, H.; Sahoo, D.R. High-Bandwidth Nanopositioner with Magnetoresistance Based Position Sensing. *Mechatronics* **2012**, *22*, 295–301. [CrossRef]
3. Miyasaka, M.; Li, H.; Tay, K.V.; Phee, S.J. A Low-Cost, Point-of-Care Test for Confirmation of Nasogastric Tube Placement via Magnetic Field Tracking. *Sensors* **2021**, *21*, 4491. [CrossRef] [PubMed]
4. Chappert, C.; Fert, A.; Van Dau, F.N. The Emergence of Spin Electronics in Data Storage. *Nat. Mater.* **2007**, *6*, 813–823. [CrossRef]
5. Dowling, R.; Narkowicz, R.; Lenz, K.; Oelschlägel, A.; Lindner, J.; Kostylev, M. Resonance-Based Sensing of Magnetic Nanoparticles Using Microfluidic Devices with Ferromagnetic Antidot Nanostructures. *Nanomaterials* **2024**, *14*, 19. [CrossRef]
6. Marchal, N.; da Câmara Santa Clara Gomes, T.; Abreu Araujo, F.; Piraux, L. Giant Magnetoresistance and Magneto-Thermopower in 3D Interconnected Ni_xFe_{1-x}/Cu Multilayered Nanowire Networks. *Nanomaterials* **2021**, *11*, 1133. [CrossRef] [PubMed]
7. Hamia, R.; Cordier, C.; Dolabdjian, C. Eddy-Current Non-Destructive Testing System for the Determination of Crack Orientation. *NDT E Int.* **2014**, *61*, 24–28. [CrossRef]
8. Krishna, V.D.; Wu, K.; Perez, A.M.; Wang, J.-P. Giant Magnetoresistance-Based Biosensor for Detection of Influenza A Virus. *Front. Microbiol.* **2016**, *7*, 400. [CrossRef]
9. Rotundo, S.; Brizi, D.; Flori, A.; Giovannetti, G.; Menichetti, L.; Monorchio, A. Shaping and Focusing Magnetic Field in the Human Body: State-of-the Art and Promising Technologies. *Sensors* **2022**, *22*, 5132. [CrossRef]
10. Abrikosov, A.A. Quantum Magnetoresistance. *Phys. Rev. B* **1998**, *58*, 2788–2794. [CrossRef]
11. Jin, S.; Tiefel, T.H.; McCormack, M.; Fastnacht, R.A.; Ramesh, R.; Chen, L.H. Thousandfold Change in Resistivity in Magnetoresistive La Ca Mn O Films. *Science* **1994**, *264*, 413–415. [CrossRef] [PubMed]
12. Subramanian, M.A.; Toby, B.H.; Ramirez, A.P.; Marshall, W.J.; Sleight, A.W.; Kwei, G.H. Colossal Magnetoresistance Without Mn^{3+}/Mn^{4+} Double Exchange in the Stoichiometric Pyrochlore $Tl_2Mn_2O_7$. *Science* **1996**, *273*, 81–84. [CrossRef] [PubMed]
13. Röder, H. Lattice Effects in the Colossal-Magnetoresistance Manganites. *Phys. Rev. Lett.* **1996**, *76*, 1356–1359. [CrossRef] [PubMed]
14. Maksimovic, N.; Hayes, I.M.; Nagarajan, V.; Analytis, J.G.; Koshelev, A.E.; Singleton, J.; Lee, Y.; Schenkel, T. Magnetoresistance Scaling and the Origin of H-Linear Resistivity in $BaFe_2(As1\text{-}xPx)_2$. *Phys. Rev. X* **2020**, *10*, 041062. [CrossRef]
15. Zhang, W.; Yu, R.; Feng, W.; Yao, Y.; Weng, H.; Dai, X.; Fang, Z. Topological Aspect and Quantum Magnetoresistance of Beta-Ag_2Te. *Phys. Rev. Lett.* **2011**, *106*, 156808. [CrossRef] [PubMed]
16. Datta, S. *Electronic Transport in Mesoscopic Systems*; Cambridge Studies in Semiconductor Physics and Microelectronic Engineering; Cambridge University Press: Cambridge, UK, 1995; ISBN 978-0-521-59943-6.
17. Parish, M.M.; Littlewood, P.B. Non-Saturating Magnetoresistance in Heavily Disordered Semiconductors. *Nature* **2003**, *426*, 162–165. [CrossRef] [PubMed]
18. Yang, F.Y.; Liu, K.; Hong, K.; Reich, D.H.; Searson, P.C.; Chien, C.L. Large Magnetoresistance of Electrodeposited Single-Crystal Bismuth Thin Films. *Science* **1999**, *284*, 1335–1337.
19. Hu, J.; Rosenbaum, T.F. Classical and Quantum Routes to Linear Magnetoresistance. *Nat. Mater.* **2008**, *7*, 697–700. [CrossRef] [PubMed]
20. Liang, T.; Gibson, Q.; Ali, M.N.; Liu, M.; Cava, R.J.; Ong, N.P. Ultrahigh Mobility and Giant Magnetoresistance in the Dirac Semimetal Cd_3As_2. *Nat. Mater.* **2015**, *14*, 280–284. [CrossRef]
21. Shekhar, C.; Nayak, A.K.; Sun, Y.; Schmidt, M.; Nicklas, M.; Leermakers, I.; Zeitler, U.; Skourski, Y.; Wosnitza, J.; Liu, Z.; et al. Extremely Large Magnetoresistance and Ultrahigh Mobility in the Topological Weyl Semimetal Candidate NbP. *Nat. Phys.* **2015**, *11*, 645–649. [CrossRef]

22. Tafti, F.F.; Gibson, Q.D.; Kushwaha, S.K.; Haldolaarachchige, N.; Cava, R.J. Resistivity Plateau and Extreme Magnetoresistance in LaSb. *Nat. Phys.* **2016**, *12*, 272–277. [CrossRef]
23. Lv, Y.-Y.; Zhang, B.-B.; Li, X.; Yao, S.-H.; Chen, Y.B.; Zhou, J.; Zhang, S.-T.; Lu, M.-H.; Chen, Y.-F. Extremely Large and Significantly Anisotropic Magnetoresistance in ZrSiS Single Crystals. *Appl. Phys. Lett.* **2016**, *108*, 244101. [CrossRef]
24. Ali, M.N.; Xiong, J.; Flynn, S.; Tao, J.; Gibson, Q.D.; Schoop, L.M.; Liang, T.; Haldolaarachchige, N.; Hirschberger, M.; Ong, N.P.; et al. Large, Non-Saturating Magnetoresistance in WTe_2. *Nature* **2014**, *514*, 205–208. [CrossRef] [PubMed]
25. Adhikari, R.; Adhikari, S.; Faina, B.; Terschanski, M.; Bork, S.; Leimhofer, C.; Cinchetti, M.; Bonanni, A. Positive Magnetoresistance and Chiral Anomaly in Exfoliated Type-II Weyl Semimetal Td-WTe_2. *Nanomaterials* **2021**, *11*, 2755. [CrossRef] [PubMed]
26. Yuan, Z.; Lu, H.; Liu, Y.; Wang, J.; Jia, S. Large Magnetoresistance in Compensated Semimetals $TaAs_2$ and $NbAs_2$. *Phys. Rev. B* **2016**, *93*, 184405. [CrossRef]
27. Huynh, K.K.; Tanabe, Y.; Tanigaki, K. Both Electron and Hole Dirac Cone States in $BaFeAs_2$ Confirmed by Magnetoresistance. *Phys. Rev. Lett.* **2011**, *106*, 217004. [CrossRef] [PubMed]
28. Bhoi, D.; Mandal, P.; Choudhury, P.; Pandya, S.; Ganesan, V. Quantum Magnetoresistance of the PrFeAsO Oxypnictide. *Appl. Phys. Lett.* **2011**, *98*, 172105. [CrossRef]
29. Naito, M.; Tanaka, S. Galvanomagnetic Effects in the Charge-Density-Wave State of 2*H*-$NbSe_2$ and 2*H*-$TaSe_2$. *J. Phys. Soc. Jpn.* **1982**, *51*, 228–236. [CrossRef]
30. Zhou, L.; Wu, H.B.; Wang, Z.; Lou, X.W. Interconnected MoO_2 Nanocrystals with Carbon Nanocoating as High-Capacity Anode Materials for Lithium-Ion Batteries. *ACS Appl. Mater. Interfaces* **2011**, *3*, 4853–4857. [CrossRef]
31. Zhang, L.; Lin, H.; Zhai, L.; Nie, M.; Zhou, J.; Zhuo, S. Enhanced Supercapacitor Performance Based on 3D Porous Graphene with MoO_2 Nanoparticles. *J. Mater. Res.* **2017**, *32*, 292–300. [CrossRef]
32. Zhang, T.; Jiang, Y.; Song, Z.; Huang, H.; He, Y.; Fang, Z.; Weng, H.; Fang, C. Catalogue of Topological Electronic Materials. *Nature* **2019**, *566*, 475–479. [CrossRef]
33. Chen, Q.; Lou, Z.; Zhang, S.; Xu, B.; Zhou, Y.; Chen, H.; Chen, S.; Du, J.; Wang, H.; Yang, J.; et al. Large Magnetoresistance and Nonzero Berry Phase in the Nodal-Line Semimetal MoO_2. *Phys. Rev. B* **2020**, *102*, 165133. [CrossRef]
34. Apte, A.; Mozaffari, K.; Samghabadi, F.S.; Hachtel, J.A.; Chang, L.; Susarla, S.; Idrobo, J.C.; Moore, D.C.; Glavin, N.R.; Litvinov, D.; et al. 2D Electrets of Ultrathin MoO_2 with Apparent Piezoelectricity. *Adv. Mater.* **2020**, *32*, 2000006. [CrossRef]
35. Zhang, H.; Wu, Y.; Huang, Z.; Shen, X.; Li, B.; Zhang, Z.; Wu, R.; Wang, D.; Yi, C.; He, K.; et al. Synthesis of Two-Dimensional MoO_2 Nanoplates with Large Linear Magnetoresistance and Nonlinear Hall Effect. *Nano Lett.* **2023**, *23*, 2179–2186. [CrossRef] [PubMed]
36. Picht, O.; Müller, S.; Alber, I.; Rauber, M.; Lensch-Falk, J.; Medlin, D.L.; Neumann, R.; Toimil-Molares, M.E. Tuning the Geometrical and Crystallographic Characteristics of Bi_2Te_3 Nanowires by Electrodeposition in Ion-Track Membranes. *J. Phys. Chem. C* **2012**, *116*, 5367–5375. [CrossRef]
37. Nikolaeva, A.; Huber, T.E.; Gitsu, D.; Konopko, L. Diameter-Dependent Thermopower of Bismuth Nanowires. *Phys. Rev. B* **2008**, *77*, 035422. [CrossRef]
38. Rowe, D.M.; Shukla, V.S.; Savvides, N. Phonon Scattering at Grain Boundaries in Heavily Doped Fine-Grained Silicon–Germanium Alloys. *Nature* **1981**, *290*, 765–766. [CrossRef]
39. Sood, A.; Cheaito, R.; Bai, T.; Kwon, H.; Wang, Y.; Li, C.; Yates, L.; Bougher, T.; Graham, S.; Asheghi, M.; et al. Direct Visualization of Thermal Conductivity Suppression Due to Enhanced Phonon Scattering Near Individual Grain Boundaries. *Nano Lett.* **2018**, *18*, 3466–3472. [CrossRef] [PubMed]
40. Gu, X.; Wei, Y.; Yin, X.; Li, B.; Yang, R. Colloquium: Phononic Thermal Properties of Two-Dimensional Materials. *Rev. Mod. Phys.* **2018**, *90*, 041002. [CrossRef]
41. Ju, Y.S.; Goodson, K.E. Phonon Scattering in Silicon Films with Thickness of Order 100 Nm. *Appl. Phys. Lett.* **1999**, *74*, 3005–3007. [CrossRef]

nanomaterials

Article

Investigating the Mechanism of Rare-Earth Ion Incorporation into Glass–Ceramic Crystal Phases through Er^{3+} Ion Probe Characteristics

Zhixin Chen [1,†], Wenzhe Cui [1,†], Sijun Ren [2], Ju Yang [1], Jiayu Tian [1], Haitao Xia [1], Jiajing Shen [1] and Guozhong Ren [1,*]

[1] School of Physics, Changchun Normal University, Changchun 130032, China; qx202200244@stu.ccsfu.edu.cn (Z.C.)
[2] Faculty of Medicine, Dalian University of Technology, Dalian 116024, China
* Correspondence: renguozhong@ccsfu.edu.cn
† These authors contributed equally to this work.

Abstract: Exploring the intrinsic mechanisms of rare-earth ions entering the crystal phase has great significance for finely tuning the luminescent properties of glass–ceramics. Using Er^{3+} ions as a probe, X-ray diffraction was employed to precisely measure the crystallinity of SiO_2-PbF_2-Er_2O_3 glass–ceramics synthesized under various heat treatment conditions, confirming the occurrence of a rapid crystallization process. Additionally, by combining Judd–Ofelt theory with comprehensive analyses of absorption and fluorescence spectra, we calculated the relative proportions of Er^{3+} ions present in the crystal phase. We found that the crystallization process in the glass–ceramics and the incorporation of Er^{3+} ions into the crystal phase did not occur synchronously. This discovery provides new theoretical foundations and practical guidance for understanding the mechanism of rare-earth ion incorporation into crystal phases, which is significant for the development of functional materials with specific luminescent properties.

Keywords: glass–ceramics; Er^{3+} ions; Judd–Ofelt theory; crystal phase; probe

1. Introduction

Rare-earth-doped glass–ceramics have garnered significant attention due to their superior properties, including high strength, excellent oxidation resistance, good optical transmittance, adjustable thermal expansion coefficients, and low electrical conductivity. Their unique *f-f* transition spectroscopic characteristics, especially, spanning the visible to the near mid-infrared range [1], offer extensive application prospects in the defense, aerospace, electronics, power generation, and biomedical fields [2–5]. In glass–ceramics, rare-earth ions primarily reside within the crystal phase, and their concentration directly impacts the luminescent performance of the material. Therefore, understanding the specific mechanisms by which rare-earth ions enter the crystal phase is important.

Existing studies indicate that the doping concentration of rare-earth ions significantly influences the formation of the crystal phase, which in turn affects the luminescence intensity [6]. During the transformation from glass to glass–ceramics, the conditions of heat treatment are critical; excessively high temperatures can inhibit the formation of the crystal phase [7]. During this heat treatment, rare-earth ions migrate into the crystal phase. However, research on whether this migration process occurs synchronously with the crystallization of the glass is still limited.

The luminescence characteristics of rare-earth ions reveal information about the crystal field around them and highlight their unique value as probes for studying their environment. In recent years, the application of rare-earth ions as probes has increased, particularly with Eu^{3+} ions, where substantial progress has been made [8–10]. Similarly, Er^{3+} ions, which have similar chemical properties with Eu^{3+}, exhibit significant emissions in the near-infrared region and show remarkable luminescence in green (540 nm and 520 nm) and red (650 nm) wavelengths, applicable for upconversion luminescence in various matrices such as glass, glass–ceramics, single crystals, and nanocrystals [11,12]. The unique energy level structure of Er^{3+} ions allows them to achieve higher excited states through excited state absorption, cross relaxation, and energy transfer processes, thereby enhancing their potential as efficient upconversion luminescent materials and ion probes.

The phonon energy of the host matrix directly impacts the non-radiative relaxation rate of the excited states of rare-earth ions, influencing the efficiency of upconversion luminescence [13]. Selecting a matrix with lower phonon energy, such as fluorides, can help enhance the intensity of upconversion luminescence. By thermally treating fluoride precursor glasses, fluoro-oxide glass–ceramics composed of glass and crystal phases can be obtained [14–16]. In these fluoro-oxide glass–ceramics, fluoride crystals are uniformly distributed within the residual oxide glass matrix [17], combining the excellent optical properties of fluoride materials with the outstanding thermal stability and mechanical properties of oxide materials [18,19]. Fluoro-oxide glass–ceramics, containing low-phonon-energy fluoride crystals, can accommodate more rare-earth ions and provide superior optical performance [20]. During heat treatment, fluoride crystals nucleate and grow, while rare-earth ions in the glass preferentially migrate into these fluoride nanocrystals [21–24].

This study uses Er^{3+} as a probe to explore the mechanism by which rare-earth ions enter the crystal phase. Er^{3+}-doped SiO_2-PbF_2 glass–ceramics were prepared via a high-temperature solid-state synthesis method. X-ray diffraction (XRD), absorption spectra, fluorescence spectra, and Judd–Ofelt theory analysis were employed to investigate the effects of heat treatment duration on the entry of Er^{3+} ions into the glass-ceramic. By com-paring the degree of crystallinity with the content of rare-earth ions in the crystal phase, the study examines the asynchrony between the glass crystallization process and the process of dopant ions entering the crystal phase. This research provides a theoretical basis for optimizing heat treatment processes to improve the luminescence efficiency of rare-earth-doped glass–ceramics and expands the scope of applications for rare-earth ions as probes.

2. Materials and Methods

2.1. Materials and Sample Preparation

Glass and glass–ceramic samples were prepared using the high-temperature solid-state method. Raw materials were accurately weighed to a total of 15 g with a molar composition of $45SiO_2$-$55PbF_2$-$1Er_2O_3$. SiO_2 and PbF_2 were of analytical grade purity, while Er_2O_3 was 4N high purity. The weighed raw materials were ground in an agate mortar for one hour to ensure a uniform distribution of components. Subsequently, the mixed raw materials were placed in a crucible and then heated in a silicon carbide rod furnace at 930 °C for 15 min. Upon the completion of the melting process, the melt was immediately poured onto a preheated copper plate and covered with another copper plate to achieve rapid cooling and shaping, resulting in the glass sample.

A portion of the glass sample was placed in a crucible containing alumina powder and subjected to heat treatment in a silicon carbide rod furnace. After heat treatment, the samples were naturally cooled to room temperature. Both heat-treated samples and the original glass samples were collected for subsequent testing.

2.2. Characterization Techniques

The samples were analyzed by a 2500PC X-ray diffraction analyzer (Rigaku, Tokyo, Japan) with a step scanning speed of 0.25–1 s, a step length of 0.1, and a step angle of 2θ. The scanning range was 20°–60°. The glass transition temperature and crystallization temperature of the sample were obtained by using differential scanning calorimetry (SDT-2960: TA Instruments, New Castle, DE, USA). The measurements were carried out under a nitrogen atmosphere with a heating rate of 10 °C/min. The UV–visible absorption spectra of the samples were measured by a UV–visible photometer model UV-mini-1280 (Shimadzu, Tokyo, Japan) with a detection range of 200 nm–1200 nm. The excitation and emission spectra of the sample were obtained using a fluorescence spectrophotometer equipped with a 150 W xenon lamp (FL-4600: Hitachi Instrument, Tokyo, Japan). The excitation source for the spectrophotometer was a 980 nm semiconductor laser (IR-980/1–5000 mW: Raish Laser Optoelectronic Technology, Changchun, China).

3. Results

3.1. Characterization of the Degree of Crystallinity by XRD

Figure 1 shows the differential scanning calorimetry (DSC) curve of Er^{3+}-doped SiO_2-PbF_2 glass samples. Two critical characteristic temperatures are identified: the glass transition temperature (T_g) and the crystallization temperature (T_c), which are 390 °C and 430 °C. The glass transition temperature (T_g) marks the transformation of amorphous materials from a hard and brittle glass state to a more flexible, high-elastic state, typically represented by a distinct inflection point or step on the DSC curve. The crystallization temperature (T_c) corresponds to the temperature at which amorphous materials begin to crystallize and form an ordered structure, characterized by a significant exothermic peak on the DSC plot [25]. During the transformation of glass samples into glass–ceramics, the heat treatment temperature and time are critical for controlling the nucleation and growth of crystals. For example, crystal nuclei start forming at a heat treatment temperature of 460 °C, followed by gradual crystal growth with increased heat treatment time.

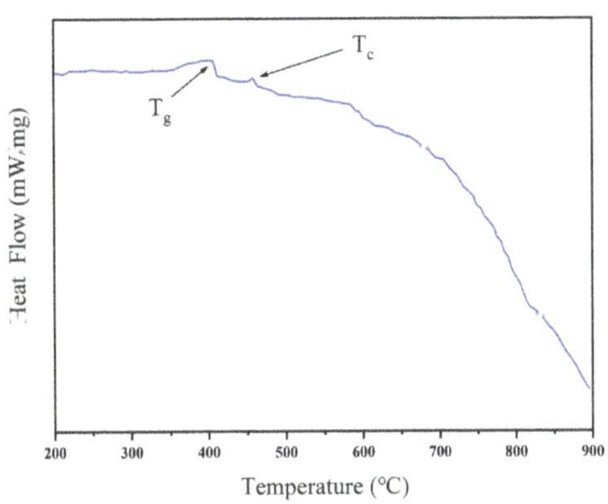

Figure 1. DSC pattern of Er^{3+}-doped glass sample.

The crystallization temperatures detected by DSC were used to heat treat the glass at 460 °C for 0.5 h, 1 h, 1.5 h, 2 h, 2.5 h and 3 h, respectively, to observe the entry of Er^{3+} into the glass–ceramics as the annealing time increased [26].

X-ray diffraction (XRD) is a powerful analytical technique that provides detailed information about the crystal structure of materials, including specific details about the samples being studied. In this work, XRD was utilized to identify the crystal phases that form in annealed glass.

Figure 2a shows the X-ray diffraction (XRD) patterns of the glass before heat treatment and of the glass–ceramics after different durations of heat treatment. The original glass sample exhibits an XRD pattern without prominent sharp peaks across all 2θ values, indicating its amorphous nature and the absence of crystal phases. After heat treatment at 460 °C for varying periods of 0.5 h, 1 h, 1.5 h, 2 h, 2.5 h, and 3 h, characteristic peaks indicative of crystal phase presence begin to appear in the XRD patterns, signaling the transformation of the glass samples into glass–ceramics. Four distinct diffraction peaks emerge, aligning closely with the peak positions in the standard card (PbF_2, ICDD PDF#06-0251), confirming PbF_2 as the crystal phase in the glass–ceramics. With extended heat treatment times, the positions of these four characteristic peaks remain stable, and no new diffraction peaks appear. The intensity of these characteristic peaks increases, suggesting an increase in the number of crystals. As shown in Figure 2a, the intensity of the diffraction peaks gradually increases with prolonged heat treatment time, indicating a relatively slow growth rate of PbF_2 crystal content.

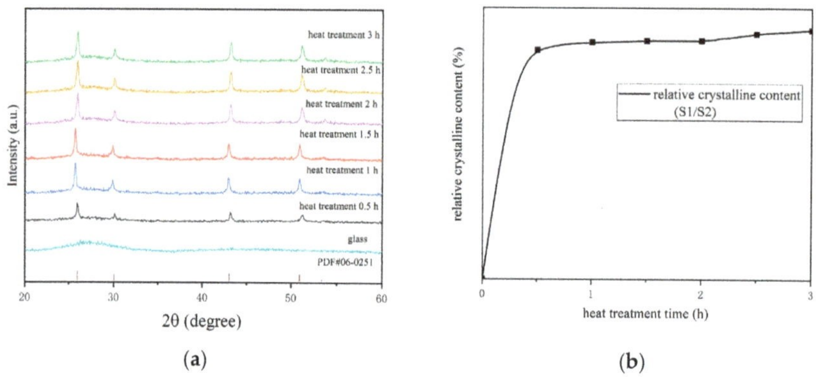

Figure 2. (**a**) X-ray diffraction patterns of glass and glass–ceramics; (**b**) trend curve of crystallinity over time.

The X-ray diffraction (XRD) patterns of the samples can be used to calculate the area of each diffraction peak. As shown in Table 1, by comparing the area S1 of the crystal phase diffraction peaks with the area S2 of the amorphous phase background diffraction, one can determine the relative crystallinity (S1/S2) of the samples subjected to different heat treatment durations. The results indicate that within the first 0.5 h, the relative crystal phase content rapidly increases to 29.6%. Over the subsequent period from 0.5 h to 3 h, the content further rises to 32.2%. This suggests that the rate of increase in the relative crystal phase content is most significant during the initial 0.5 h period.

Table 1. Diffraction peaks and glass peak areas.

S / Time (h)	Diffraction Peak S_1 (2θ = 26°)	Amorphous Halo S_2	Crystallinity (%)
0.5	87.1	294.7	29.6
1.0	159.4	520.5	30.6
1.5	152.1	493.2	30.8
2.0	178.6	579.7	30.8
2.5	198.7	626.6	31.7
3.0	189.0	585.9	32.2

As shown in Figure 2b, the ratio of the area of the crystal phase diffraction peaks to the area of the amorphous phase background gradually increases after heat treatment at 460 °C for periods of 0.5 h, 1 h, 1.5 h, 2 h, 2.5 h, and 3 h. This indicates that the content of PbF_2 crystals within the samples increases with prolonged heat treatment time, leading to an overall increase in crystallinity. However, this increase is very small, suggesting that the majority of the crystallization process is completed after 0.5 h of heat treatment. This finding confirms that the crystallization process in the glass–ceramics is rapid [27].

3.2. Calculation of J-O Parameters from Absorption Spectra

Judd-Ofelt (J-O) theory is a highly standardized and reliable analytical tool used to examine the spectral characteristics of rare-earth-doped materials. Developed independently by B.R. Judd and G.S. Ofelt in the 20th century, it has found widespread application in the qualitative analysis of rare-earth-ion-doped glasses and other materials. The J-O theory is particularly useful in characterizing radiative transitions in both doped solids and doped aqueous solutions. Its fundamental principle involves utilizing transitions between different energy levels of rare-earth ions to compute radiative parameters such as oscillator strengths, transition probabilities, fluorescence branching ratios, and radiative lifetimes from absorption spectra, within a certain degree of accuracy.

Figure 3 shows the absorption spectra of the glass and glass–ceramics over the wavelength range of 480 nm to 1000 nm, displaying six peaks corresponding to the transitions $^4S_{3/2} \rightarrow {}^4I_{15/2}$, $^4F_{7/2} \rightarrow {}^4I_{15/2}$, $^2H_{11/2} \rightarrow {}^4I_{15/2}$, $^4F_{9/2} \rightarrow {}^4I_{15/2}$, $^4I_{9/2} \rightarrow {}^4I_{15/2}$, $^4I_{11/2} \rightarrow {}^4I_{15/2}$. The corresponding peak wavelengths are 486 nm, 520 nm, 545 nm, 652 nm, 807 nm, and 972 nm, respectively. In the amorphous glass, the Er^{3+} ions are distributed within the amorphous structure, characterized by irregular arrangements of constituent ions. In glass–ceramics, due to the partial incorporation of Er^{3+} ions into the crystal phase, there is a change in the intensity of the peaks in the absorption spectra compared to those in the glass. This alteration reflects the different local environments experienced by the Er^{3+} ions in the amorphous versus the crystal regions, which in turn affects their electronic transitions and thus the observed spectral features.

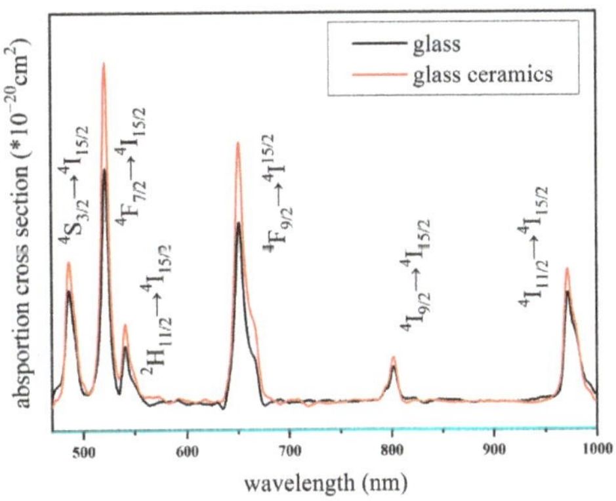

Figure 3. Absorption spectra of glass and glass–ceramics after treatment.

According to the J-O theoretical model, the experimental oscillator strength can be calculated by the following equation:

$$f_{exp} = \frac{mc^2}{N\pi e^2 \lambda^2} \int \alpha(\lambda) d\lambda \tag{1}$$

After further derivation, it can be concluded that:

$$\int \alpha(\lambda) d\lambda = \int \frac{\log_{10}(I_0(\lambda)/I(\lambda))}{d\log_{10} e} d\lambda = \frac{1}{0.43d} \int OD(\lambda) d\lambda \tag{2}$$

where $m = 9.1 \times 10^{-28}$ g is the electron mass, $e = 1.602 \times 10^{-19}$ C is the electron charge, $c = 2.998 \times 10^8$ m/s is the speed of light, d is the thickness of the sample, I_0 and I are the incident and effluent light intensities, $\alpha(\lambda)$ is the absorption coefficient, $OD(\lambda)$ is the optical density, and N is the doping concentration of rare-earth ions. The theoretical oscillator strength equation is as follows:

$$f_{cal}(J - J') = \frac{8\pi^2 mc}{3h(2j+1)\lambda} \cdot \chi_{ed} \cdot S_{ed} = \frac{8\pi^2 mc}{3h(2J+1)\lambda} \cdot \frac{(n^2+2)^2}{9n} \sum_{t=2,4,6} \Omega_t \left| \left\langle SLJ \middle\| U_{(t)} \middle\| S'L'J' \right\rangle \right|^2 \tag{3}$$

where $|<SLJ||U_{(t)}|||S'L'J'>|$ is the electric dipole transitions approximation matrix, and $U_{(t)}$ is the transition matrix element for rare-earth ions, independent of the matrix composition of the experimental samples. The transition matrix elements corresponding to Er^{3+} ions are shown in Table 2. χ_{ed} is the correction factor for the electric dipole radiative transitions, S_{ed} is the line intensity of the electric dipole, and Ω_t is the J-O intensity parameter.

Table 2. Er^{3+} transition matrix elements.

$U_{(\lambda=2,4,6)}$	$^4S_{3/2} \rightarrow {}^4I_{15/2}$	$^4F_{7/2} \rightarrow {}^4I_{15/2}$	$^2H_{11/2} \rightarrow {}^4I_{15/2}$	$^4F_{9/2} \rightarrow {}^4I_{15/2}$	$^4I_{9/2} \rightarrow {}^4I_{15/2}$	$^4I_{11/2} \rightarrow {}^4I_{15/2}$
$U_{(2)}$	0	0	0.7158	0	0	0.0276
$U_{(4)}$	0	0.1465	0.4138	0.5511	0.1587	0.0002
$U_{(6)}$	0.2225	0.6272	0.0927	0.4621	0.0072	0.3924

Assuming $f_{\exp} = f_{cal}$, one can obtain multiple sets of J-O intensity parameters Ω_t (t = 2, 4, 6). The determined value of Ω_t can be obtained by least squares fitting.

Let the scalar function be:

$$Y = \sum_{i=1}^{6}\{S_{ed}(i) - [\Omega_2 U(i,1) + \Omega_4 U(i,2) + \Omega_6 U(i,3)]\}^2 \tag{4}$$

where Ω_t (t = 2, 4, 6) is required to minimize the scalar function, and according to the method of least squares, a system of equations is established:

$$\frac{\partial Y}{\partial \Omega_2} = 0, \frac{\partial Y}{\partial \Omega_6} = 0, \frac{\partial Y}{\partial \Omega_6} = 0 \tag{5}$$

The Ω_2 of the glass is calculated to be 2.92, Ω_4 to be 1.44 and Ω_6 to be 0.91, all in units of 10^{-20} cm^2. Following the same methodology, the J-O strength parameters of the glass–ceramics heat-treated for different times are obtained as shown in Table 3:

Table 3. J-O strength parameters of Er^{3+}-doped glass–ceramics.

Sample	Ω_2 (10^{-20} cm^2)	Ω_4 (10^{-20} cm^2)	Ω_6 (10^{-20} cm^2)
0.5 h	2.31	1.24	0.84
1.0 h	1.42	0.94	0.75
1.5 h	1.27	0.89	0.73
2.0 h	1.11	0.84	0.72
2.5 h	1.03	0.81	0.70
3.0 h	0.93	0.78	0.70

Since the Ω_t parameter is influenced by the ligand field environment surrounding the trivalent lanthanide ions, it is especially suitable for evaluating the local environment of trivalent rare-earth ions in glass [28]. As shown in Table 3, this parameter undergoes significant changes before and after processing. For glass–ceramics prepared at the same temperature, even with different heat treatment durations, the variations in the parameters are evident. With increasing heat treatment time, the value of Ω_2 decreases, indicating a significant change in the microstructure around the trivalent Er^{3+} ions in the glass samples after heat treatment. This is because the covalency of the chemical bonds between trivalent lanthanide ions and their ligands strengthens as the Ω_2 value increases [29]. Given that the electronegativity of F^- ions is higher than that of O^{2-} ions, the covalency of Er-O bonds is stronger than that of Er-F bonds [30]. However, with increased heat treatment time, the decrease in the Ω_2 value suggests a weakening of the covalency of Er-O bonds and a relative strengthening of the covalency of Er-F bonds. This indicates that the Er^{3+} ions form crystal phases by bonding with F^- ions, signifying the migration of rare-earth ions from the glass phase to the crystal phase.

3.3. *Fluorescence Spectroscopy*

Figure 4a shows the fluorescence spectra of Er^{3+}-doped glass and glass–ceramic samples in the wavelength range of 400 nm to 700 nm. When excited by a 980 nm laser, the emission intensity of the glass sample is very low. However, when the glass is converted into a glass–ceramics through heat treatment, the emission intensity significantly increases. The spectrum displays four sharp peaks located at 410 nm, 520 nm, 545 nm, and 652 nm. Although the positions of the emission peaks remain unchanged, their intensity increases with the extension of heat treatment time, reaching a maximum at 3 h. This indicates that the luminescence of the glass–ceramics mainly originates from the Er^{3+} ions in the crystal phase, suggesting that the amount of rare-earth ions migrating from the glass phase to the crystal phase increases with prolonged heat treatment time.

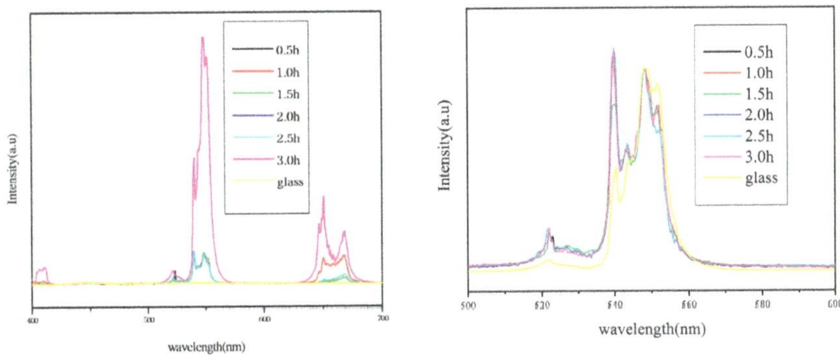

(a) (b)

Figure 4. (**a**) Fluorescence spectra of Er^{3+}-doped glass–ceramics. (**b**) Fluorescence spectrum after normalization at 545 nm.

Figure 4b presents the normalized fluorescence spectra at 545 nm, aimed at illustrating the presence of Er^{3+} ions in the crystal phase after heat treatment [31–34]. By normalizing the fluorescence spectra, it becomes possible to make clearer comparisons of spectral characteristics among different samples. The normalized spectral intensity of the glass sample is lower than others, a phenomenon associated with Stark level splitting. Analyzing the overall trend, from the relationship between fluorescence intensity (I) and spontaneous emission transition probability ($A_{JJ'}$):$I \propto A_{JJ'}$, it can be seen that the integrated area in Figure 4b correlates positively with the changes in the J-O intensity parameters, making quantitative analysis of the intensity parameters feasible.

Based on the calculated J-O intensity parameter Ω_t, the spontaneous radiative excursion chances between the J-J' energy levels can be calculated as follows:

$$A_{JJ'} = A_{ed} = \frac{64\pi^4 e^2}{3h(2J+1)\lambda^3} \cdot n^2 \chi_{ed} S_{ed} \tag{6}$$

According to statistical physics, the ratio of the number of particles in the ${}^2H_{11/2}$ and ${}^4S_{3/2}$ energy levels of which Er^{3+} conforms to Boltzmann statistics is given. For the system of nearly independent particles, the number of particles at different energy levels conforms to the Boltzmann distribution:

$$N_l = \omega_l e^{-\alpha-\beta} \tag{7}$$

The fluorescence intensity of the energy level leap is denoted by I:

$$I = A_{JJ'} \cdot N_l \tag{8}$$

Define $Ratio_{obs}$ as the fluorescence intensity ratio of 520 nm and 545 nm green luminescence produced by the upconversion process:

$$Ratio_{obs} = \frac{I_1}{I_2} \tag{9}$$

Associating the above equations and combining them with the J-O theory yields:

$$\begin{aligned} Ratio &= \frac{0.2225\Omega_6}{0.7158\Omega_2+0.4138\Omega_4+0.0927\Omega_6}\left(\frac{\lambda_{520nm}}{\lambda_{545nm}}\right)e^{-\frac{\Delta E}{kT}} \\ &= \frac{0.2225\Omega_6}{0.7158\Omega_2+0.4138\Omega_4+0.0927\Omega_6} \times 33.221 \end{aligned} \tag{10}$$

The temperature (T) is taken as room temperature. The excitation wavelength is 980 nm. The power is 0.5 W, and k is the Boltzmann constant.

From the fluorescence spectra, it is observed that the emission intensity of the glass is very low. After normalizing the fluorescence spectra, it was found that the peak area ratio at 520 nm and 545 nm is the same for both the glass and glass–ceramic samples; however, the fluorescence intensity of the glass sample is only one-thirtieth of the glass–ceramic sample. This is because the phonon energy of PbF_2 crystals is much lower than the glass matrix, reducing the non-radiative relaxation rate of Er^{3+} ions in PbF_2 crystals and enhancing the upconversion luminescence performance. Regardless of whether the rare-earth ions are located in the glass phase or the crystal phase, they can be detected through absorption spectra. However, the J-O parameters calculated from the absorption spectra are mixed, incorporating information from both the glass and crystal phases. Thus, when calculating the fluorescence intensity ratio of the green light emissions at 520 nm and 545 nm produced by upconversion processes, the result obtained using J-O parameters reflects the fluorescence intensity ratio of the entire sample, whereas the ratio derived from the peak areas in the fluorescence spectra represents only the fluorescence intensity ratio of the crystal phase.

Define $Ratio_{lum}$ as the theoretical value, which is the ratio calculated from the peak areas at 520 nm and 545 nm in the fluorescence spectra; and $Ratio_{obv}$ as the experimental value, which is obtained by substituting the J-O intensity parameters into Equation (10). According to the data in Table 4, between heat treatment durations of 0.5 h to 2.5 h, there is a significant difference between the theoretical and experimental values, indicating that the rare-earth ions are not only present in the crystal phase but also partly in the glass phase. As the heat treatment time increases, the rare-earth ions gradually migrate into the crystal phase, thereby reducing their presence in the glass phase. At 3 h, the theoretical and experimental values are closest, suggesting that the process of rare-earth ions entering the crystal phase is largely complete at this point.

Table 4. Theoretical and experimental values.

	0.5 h	**1.0 h**	**1.5 h**	**2.0 h**	**2.5 h**	**3.0 h**
$Ratio_{lum}$	4.98	5.02	5.01	4.96	4.99	5.02
$Ratio_{obv}$	2.77	3.76	4.01	4.41	4.55	4.91

3.4. Comparison of the Glass Crystallization Process with the Entry of Er^{3+} into the Crystal Phase

In the glass-ceramic samples, some of the Er^{3+} ions are distributed in the glass phase, while others are in the crystal phase. The process of rare-earth ions entering the crystal phase is largely completed at 3 h, the number of Er^{3+} ions in the crystal phase reaches its maximum value. Let the content of Er^{3+} ions in the crystal phase at 3 h be 1, and the content of Er^{3+} ions that entered the crystal phase before 3 h be X, with the remaining ions in the glass phase as $(1 - X)$. The J-O parameters for the glass are denoted as Ω_2', and those for the glass-ceramic are Ω_2''. The J-O parameter values obtained from the absorption spectra are substituted into Equation (11):

$$X \cdot \Omega_2'' + (1 - X) \cdot \Omega_2' = \Omega_2 \quad (11)$$

The results of the calculations are shown in Table 5:

Table 5. Ratio of the number of Er^{3+} ions entered into the crystal phase to the maximum number of Er^{3+} ions in the crystal phase.

Time	0.5 h	1.0 h	1.5 h	2.0 h	2.5 h	3.0 h
X	30.7%	75.4%	82.9%	90.9%	94.9%	100%

As shown in Table 5, the proportion of Er^{3+} ions in the crystal phase gradually increases with time. During the 0.5 h of heat treatment, the proportion of Er^{3+} ions in the crystal phase is 30.7% of the maximum, and by 3 h, this migration process is essentially complete. Compared to the glass crystallization process, the migration of rare-earth ions from the glass phase into the crystal phase is a slow process.

The crystallization process of glass–ceramics involves two stages: nucleation and growth. As shown in Figure 5a, when heat treatment is performed at the transformation temperature (T_g), the glass sample initially forms nuclei rapidly, followed by the growth of these nuclei. As the nuclei continue to grow, the crystallinity of the sample increases. Once the nuclei reach their maximum size, the crystallinity of the sample will also approach its maximum. However, in glass–ceramics, rare-earth ions are not entirely distributed within the crystal phase; some remain in the glass phase. When the number of rare-earth ions in the crystal phase reaches saturation, the rare-earth ions in the glass phase will no longer continue to migrate into the crystal phase. As illustrated in Figure 5b, Area A represents the state where Er^{3+} ions are entering the crystal phase but have not yet reached their maximum quantity; Area B represents the state where the quantity of Er^{3+} ions in the crystal phase has reached its maximum, meaning no additional Er^{3+} ions will enter the crystal phase. Transitioning from State A to State B takes approximately 3 h, which is a relatively slow process compared to the crystallization of the glass. This indicates that the crystallization of the glass and the migration of rare-earth ions into the crystal phase do not occur synchronously through heat treatment.

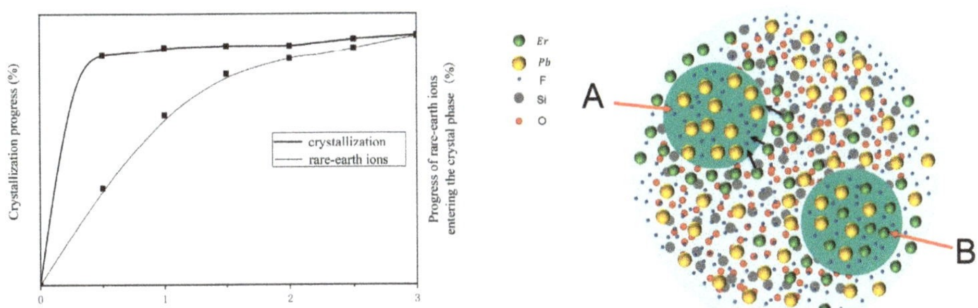

(a) (b)

Figure 5. (**a**) Curves of the variation of crystallization progress and the progress of rare-earth ions entering the crystal phase with heat treatment time. (**b**) Schematic distribution of Er^{3+} ions within the SiO_2-PbF_2 glass system.

4. Conclusions

In summary, we prepared glass-ceramic samples with a molar composition of 45SiO_2-55PbF_2-1Er_2O_3 using the high-temperature solid-state method. The differential scanning calorimetry (DSC) curve of the glass sample was analyzed to determine the optimal heat treatment temperature, which was found to be 460 °C. X-ray diffraction (XRD) analysis confirmed the presence of PbF_2 crystals in the samples after heat treatment. Based on the electric dipole transition characteristics of Er^{3+} ions, the Judd–Ofelt theory was applied to calculate the Judd–Ofelt parameters for the Er^{3+} ions. Fluorescence spectra showed that the emission intensity of the untreated glass sample was low under 980 nm excitation, indicating that upconversion fluorescence primarily originates from Er^{3+} ions within the crystal phase. By combining the Judd–Ofelt parameters with the normalized peak areas and intensity ratios of the emissions, we found that after 3 h of heat treatment, the rare-earth ions were essentially fully incorporated into the crystal phase. The results indicated that the incorporation of Er^{3+} ions into the crystal phase was a relatively slow process. In contrast to the XRD analysis, which showed that the transformation from glass to glass–ceramics was rapid, the incorporation of rare-earth ions into the crystal phase occurred progressively. During the heat treatment, the glass sample rapidly nucleated and grew, while the rare-earth ions gradually entered the crystal phase, suggesting that these two processes did not occur simultaneously. After crystallization was complete, the heat treatment time significantly influenced the concentration of rare-earth ions in the crystal phase, thereby determining the luminescence intensity of the material. This study, using Er^{3+} ions as a model, explored their environment within glass–ceramics, providing a reference for the application of other rare-earth ions in similar materials and broadening the scope of rare-earth ion applications. Moreover, this work has significant theoretical value for optimizing the luminescent properties of rare-earth-doped glass-ceramic materials.

Author Contributions: Conceptualization, Z.C. and W.C.; methodology, Z.C. and W.C.; software, S.R.; validation, S.R. and W.C.; formal analysis, J.Y.; investigation, J.Y. and J.T.; resources, Z.C.; data curation, Z.C. and J.T.; writing—original draft preparation, Z.C.; writing—review and editing, Z.C. and G.R.; visualization, H.X.; supervision, J.S.; project administration, G.R. All authors have read and agreed to the published version of the manuscript.

Funding: This research received no external funding.

Data Availability Statement: All data can be accessed upon request to the corresponding author.

Conflicts of Interest: The authors declare no conflicts of interest.

References

1. Atta, D.; Abdelghany, A.M.; Awad, W.M. Structural, optical properties, and laser spectroscopy of erbium-doped low-melting borate glasses. *Ceram. Int.* **2024**, *50*, 26528–26538. [CrossRef]
2. Chen, M.; Li, W.; Shen, M.; Zhu, S.; Wang, F. Glass–ceramic coatings on titanium alloys for high temperature oxidation protection: Oxidation kinetics and microstructure. *Corros. Sci.* **2013**, *74*, 178–186. [CrossRef]
3. Feng, M.; Jiang, C.; Chen, M.; Zhu, S.; Wang, F. A general strategy towards improving the strength and thermal shock resistance of glass-ceramics through microstructure regulation. *J. Mater. Sci. Technol.* **2022**, *120*, 139–149. [CrossRef]
4. Cheng, D.; Wei, C.; Huang, Y.; Zhang, Z.; Wang, D.; Liu, Z.; Newman, M.; Ma, T.; Chueh, Y.-H.; Zhang, X.; et al. Additive manufacturing of lithium aluminosilicate glass-ceramic/metal 3D electronic components via multiple material laser powder bed fusion. *Addit. Manuf.* **2022**, *49*, 102481. [CrossRef]
5. Lin, R.; Deng, C.; Li, X.; Liu, Y.; Zhang, M.; Qin, C.; Yao, Q.; Wang, L.; Wu, C. Copper-incorporated bioactive glass-ceramics inducing anti-inflammatory phenotype and regeneration of cartilage/bone interface. *Theranostics* **2019**, *9*, 6300–6313. [CrossRef]
6. Milewska, K.; Maciejewski, M.; Synak, A.; Łapiński, M.; Mielewczyk-Gryń, A.; Sadowski, W.; Kościelska, B. From Structure to Luminescent Properties of B_2O_3-Bi_2O_3-SrF_2 Glass and Glass-Ceramics Doped with Eu^{3+} Ions. *Materials* **2021**, *14*, 4490. [CrossRef]
7. Yongsheng, D.; Jie, M.; Yu, S.; Xuefeng, Z.; Hongxia, Z.; Hua, C.; Shunli, O.; Li, B. Crystallization characteristics and corrosion properties of slag glass-ceramic prepared from blast furnace slag containing rare earth. *J. Non-Cryst. Solids* **2020**, *532*, 119880. [CrossRef]
8. Freire, R.V.M.; Coelho, D.M.A.; Maciel, L.G.; Jesus, L.T.; Freire, R.O.; dos Anjos, J.V.; Junior, S.A. Luminescent Supramolecular Metallogels: Drug Loading and Eu(III) as Structural Probe. *Chem. Eur. J.* **2024**, *30*, 33. [CrossRef]

9. Cicconi, M.R.; Deng, H.; Otsuka, T.; Telakula Mahesh, A.; Khansur, N.H.; Hayakawa, T.; de Ligny, D. Photoluminescence Study of Undoped and Eu-Doped Alkali-Niobate Aluminosilicate Glasses and Glass-Ceramics. *Materials* **2024**, *17*, 2283. [CrossRef]
10. Yang, Y.; Li, D.; Qie, S.; Su, S.; Hu, M. Composite Eu@Cd-CP as a fluorescent probe for the detection of some food additives. *Spectrochim. Acta Part A Mol. Biomol. Spectrosc.* **2024**, *317*, 124401. [CrossRef]
11. Truong, N.M.P.; Sedano, M.; Durán, A.; Balda, R.; Pascual, M.J.; Klement, R. Er/Yb co-doped $LiYF_4$ transparent oxyfluoride glass-ceramics with up-conversion optical properties. *Ceram. Int.* **2023**, *49*, 41201–41209. [CrossRef]
12. Leontyev, A.V.; Nurtdinova, L.A.; Mityushkin, E.O.; Shmelev, A.G.; Zharkov, D.K.; Chuklanov, A.P.; Nurgazizov, N.I.; Nikiforov, V.G. Polarized luminescence in single upconversion $NaYbF_4$:Er rods. *New J. Chem.* **2024**, *48*, 14029–14038. [CrossRef]
13. Bungla, M.; Tyagi, M.; Ganguli, A.K.; Prasad, P.N. A $NaBiF_4$:Gd/Tb nanoscintillator for high-resolution X-ray imaging. *J. Mater. Chem. C* **2024**, *12*, 9595–9605. [CrossRef]
14. Liu, X.; Zhou, J.; Zhou, S.; Yue, Y.; Qiu, J. Transparent glass-ceramics functionalized by dispersed crystals. *Prog. Mater. Sci.* **2018**, *97*, 38–96. [CrossRef]
15. Dyussembekova, S.; Trusova, E.; Kichanov, S.; Podbolotov, K.; Kozlenko, D. A Study of PbF_2 Nanoparticles Crystallization Mechanism in Mixed Oxyde-Fluoride Glasses. *Ceramics* **2023**, *6*, 1508–1516. [CrossRef]
16. Kichanov, S.E.; Kozlenko, D.P.; Gorshkova, Y.E.; Rachkovskaya, G.E.; Zakharevich, G.B.; Savenko, B.N. Structural studies of nanoparticles doped with rare-earth ions in oxyfluoride lead-silicate glasses. *J. Nanoparticle Res.* **2018**, *20*, 54. [CrossRef]
17. Luo, J.; Xiao, Z.; Zeng, L.; Xu, J.; Liu, J.; Li, G.; Li, C.; He, H.; Tang, J. Germanate-based oxyfluoride transparent glass-ceramic embedded with Tm^{3+}:Ca_2YbF_7 nanocrystals for high-performance optical thermometer. *Ceram. Int.* **2023**, *49*, 4193–4203. [CrossRef]
18. He, Z.; Gao, Y.; Qiu, J. Effect of concentration of Er^{3+} on up-conversion luminescence in $NaYF_4$ embedded oxy-fluoride glass-ceramics under 1550 nm excitation. *Ceram. Int.* **2024**, *50*, 19302–19307. [CrossRef]
19. Shinozaki, K.; Ishii, Y.; Sukenaga, S.; Ohara, K. Ultrafast Nanocrystallization of BaF_2 in Oxyfluoride Glasses with Crystal-like Nanostructures: Implications for Upconversion Fiber Devices. *ACS Appl. Nano Mater.* **2022**, *5*, 4281–4292. [CrossRef]
20. Hu, F.; Chen, W.; Jiang, Y.; Song, W.; Wei, R.; Guo, H. Tm^{3+}-doped $Na_{0.5-x}Yb_{0.5+x}F_{2+2x}$ self-crystallization glass ceramics: Microstructure and optical thermometry properties. *J. Lumin.* **2019**, *214*, 116558. [CrossRef]
21. Babu, S.; Seshadri, M.; Balakrishna, A.; Reddy Prasad, V.; Ratnakaram, Y.C. Study of multicomponent fluoro-phosphate based glasses: Ho^{3+} as a luminescence center. *Phys. B Condens. Matter* **2015**, *479*, 26–34. [CrossRef]
22. Enríquez, E.; Berges, C.; Fuertes, V.; Gallego, A.; Naranjo, J.A.; Herranz, G.; Fernández, J.F. Ceramic Injection Moulding of engineered glass-ceramics: Boosting the rare-earth free photoluminescence. *Ceram. Int.* **2020**, *46*, 9334–9341. [CrossRef]
23. Marcondes, L.M.; Rodrigues, L.; Ramos da Cunha, C.; Gonçalves, R.R.; de Camargo, A.S.S.; Cassanjes, F.C.; Poirier, G.Y. Rare-earth ion doped niobium germanate glasses and glass-ceramics for optical device applications. *J. Lumin.* **2019**, *213*, 224–234. [CrossRef]
24. Jain, P.; Aida, T.; Motosuke, M. Fluorescence Anisotropy as a Temperature-Sensing Molecular Probe Using Fluorescein. *Micromachines* **2021**, *12*, 1109. [CrossRef] [PubMed]
25. Shánělová, J.; Honcová, P.; Málek, J.; Perejón, A.; Pérez-Maqueda, L.A. Direct comparison of surface crystal growth kinetics in chalcogenide glass measured by microscopy and DSC. *J. Am. Ceram. Soc.* **2023**, *106*, 6051–6061. [CrossRef]
26. Hu, Y.; Shen, Y.; Zhu, C.; Liu, S.; Liu, H.; Zhang, Y.; Yue, Y. Optical bandgap and luminescence in Er^{3+} doped oxyfluoro-germanate glass-ceramics. *J. Non-Cryst. Solids* **2021**, *555*, 120533. [CrossRef]
27. Wu, J.; Kermani, M.; Cao, L.; Wang, B.; Dai, Z.; Fu, L.; Hu, C.; Grasso, S. Rapid crystallization of $Li_{1.5}Al_{0.5}Ge_{1.5}(PO_4)_3$ glass ceramics via ultra-fast high-temperature sintering (UHS). *Int. J. Appl. Ceram. Technol.* **2023**, *20*, 2125–2130. [CrossRef]
28. Zaman, F.; Srisittipokakun, N.; Rooh, G.; Khattak, S.A.; Singkiburin, N.; Kim, H.J.; Sangwaranatee, N.; Kaewkhao, J. Investigation of Li_2O–Gd_2O_3–MO–B_2O_3–Nd_2O_3 (MO=Ba/Bi) glasses for laser applications by Judd–Oflet (J–O) theory. *J. Lumin.* **2019**, *215*, 116639. [CrossRef]
29. Liang, H.; Liu, S.; Zhang, P.; Lei, W.; Luo, Z.; Lu, A. Er_{3+}/Yb^{3+} co-doped SiO_2-Al_2O_3-CaO-CaF_2 glass: Structure, J-O analysis and fluorescent properties. *Mater. Sci. Eng. B* **2021**, *264*, 114919. [CrossRef]
30. Wang, F.; Tian, Y.; Jing, X.; Cai, M.; Zhang, J.; Xu, S. Effect of TeO_2 addition on thermal stabilities and 2.7 μm emission properties of fluoroaluminate–tellurite glass. *J. Quant. Spectrosc. Radiat. Transf.* **2015**, *165*, 93–101. [CrossRef]
31. Riahi, M.; Alizadeh, P.; Soltani, M. The impact of erbium ion on the crystallization process of yttrium fluorosilicate glass ceramics. *J. Non-Cryst. Solids* **2021**, *553*, 120499. [CrossRef]
32. Zhao, D.; Qiao, X.; Fan, X.; Wang, M. Local vibration around rare earth ions in SiO_2–PbF_2 glass and glass ceramics using Eu^{3+} probe. *Phys. B Condens. Matter* **2007**, *395*, 10–15. [CrossRef]
33. Keeling-Tucker, T.; Brennan, J.D. Fluorescent Probes as Reporters on the Local Structure and Dynamics in Sol–Gel-Derived Nanocomposite Materials. *Chem. Mater.* **2001**, *13*, 3331–3350. [CrossRef]
34. Tuyen, V.P.; Sengthong, B.; Quang, V.X.; Van Do, P.; Van Tuyen, H.; Xuan Hung, L.; Thanh, N.T.; Nogami, M.; Hayakawa, T.; Huy, B.T. Dy^{3+} ions as optical probes for studying structure of boro-tellurite glasses. *J. Lumin.* **2016**, *178*, 27–33. [CrossRef]

nanomaterials

MDPI

Article

Preparation and Characterization of Uniform and Controlled Silica Encapsulating on Lithium Yttrium Fluoride-Based Upconversion Nanoparticles

Yahya A. Alzahrani [1], Abdulmalik M. Alessa [2], Mona K. Almosaind [1], Rahaf S. Alarifi [1], Abdulaziz Alromaeh [3] and Masfer Alkahtani [1,*]

1 Future Energy Technologies Institute, King Abdulaziz City for Science and Technology (KACST), Riyadh 11442, Saudi Arabia; yalzhrani@kacst.edu.sa (Y.A.A.); mona.k.almosaind@gmail.com (M.K.A.); rahafalarifi11@gmail.com (R.S.A.)

2 Refining Technologies and Petrochemicals Institute, King Abdulaziz City for Science and Technology (KACST), Riyadh 11442, Saudi Arabia; aalessa@kacst.edu.sa

3 Microelectronics and Semiconductors Institute, King Abdulaziz City for Science and Technology (KACST), Riyadh 11442, Saudi Arabia; eng.abdulaziz2@gmail.com

* Correspondence: mqhtani@kacst.edu.sa; Tel.: +966-553322891

Abstract: In this work, we present an advancement in the encapsulation of lithium yttrium fluoride-based ($YLiF_4$:Yb,Er) upconversion nanocrystals (UCNPs) with silica (SiO_2) shells through a reverse microemulsion technique, achieving UCNPs@SiO_2 core/shell structures. Key parameters of this approach were optimized to eliminate the occurrence of core-free silica particles and ensure a controlled silica shell thickness growth on the UCNPs. The optimal conditions for this method were using 6 mg of UCNPs, 1.5 mL of Igepal CO-520, 0.25 mL of ammonia, and 50 μL of tetraethyl orthosilicate (TEOS), resulting in a uniform silica shell around UCNPs with a thickness of 8 nm. The optical characteristics of the silica-encased UCNPs were examined, confirming the retention of their intrinsic upconversion luminescence (UC). Furthermore, we developed a reliable strategy to avoid the coencapsulation of multiple UCNPs within a single silica shell. This approach led to a tenfold increase in the UC luminescence of the annealed particles compared to their nonannealed counterparts, under identical silica shell thickness and excitation conditions. This significant improvement addresses a critical challenge and amplifies the applicability of the resulting UCNPs@SiO_2 core/shell structures in various fields.

Keywords: lithium yttrium fluorides; upconversion nanoparticles; silica coating; recrystallization

1. Introduction

Rare earths-doped upconversion nanocrystals (UCNPs) have attracted significant interest for their exceptional chemical stability, significant Stokes shifts, narrow emission bandwidths, and resistance to photobleaching and photo blinking, coupled with their minimal toxicity [1–7]. These optical and chemical properties make them an important class of optical materials for biomedical studies [8,9], optical refrigeration for remote cooling [10–14], phosphor for display applications (such as LEDs) [15,16], and luminescent thermometers [4,17,18]. To optimally harness the unique advantages of UCNPs for bio applications and other quantum technologies such as optical lasing and laser refrigeration, significant efforts have been put into choosing and improving the host materials [19].

The ideal host material for these nanocrystals must be transparent and capable of incorporating optically active lanthanide ions. Yttrium-based hosts are particularly suitable for this purpose due to the comparable size and oxidation state of Y^{3+} ions. Among these, lithium yttrium fluoride ($YLiF_4$) is exceptional, meeting all these criteria while also being producible with superior optical and crystal quality [19,20]. Compared to $NaYF_4$, the most popular and efficient host material for lanthanide-doped NCs, $YLiF_4$ offers the advantage of a wide transparency range in the visible and near-infrared regions of the spectrum, which is essential for promising and efficient laser cooling [11,12,19–21]. Furthermore, lithium yttrium fluoride ($YLiF_4$) upconversion nanoparticles (UCNPs) are distinguished by their unique square bipyramidal shape, attributed to the fact that the tetragonal crystal phase's most stable crystal plane is the 011 plane. This results in eight surfaces with identical surface energies [19,20,22]. Such uniformity in surface energy allows for the easy and even application of various ligands or material functionalization onto the UCNPs, enhancing their suitability for biomedical uses [23,24].

To fully utilize the advantages of these promising YLF UCNPs in the above-mentioned applications, modifying their surface with a robust and controlled silica shell coating is required to enhance their colloidal and chemical robustness [25] and facilitate the attachment of ligands to the surface of YLF UCNPs [24–26], especially for bio applications. Furthermore, a silica coating on the surface of YLF UCNPS is highly effective for overgrowth of nanodiamond shells to produce a unique UCNPs@SiO_2@diamond shell for applications such as laser refrigeration and optical trapping in quantum and biological technologies.

Toward this goal, the reverse microemulsion method is a well-known technique for adding a silica layer to small and hydrophobic NPs [26]. This approach depends on key parameters, including the NPs shape and concentration, an amphiphilic surfactant (Igepal CO-520), along with ammonia and tetraethyl orthosilicate (TEOS). The silica coating process starts by forming tiny soap-like bubbles that attach to the particles, allowing their surface to be modified and then coated with a silica layer through a specific and controlled chemical reaction [26].

In this study, we aim to synthesize high-quality lanthanide-doped $YLiF_4$ UCNPs with an octahedral shape and to encapsulate them in precise and controlled silica shells. Also, we will introduce an effective method to avoid encapsulation of multiple UCNPs within a single silica shell with enhanced optical emission. We believe that this method will enhance the optical performance of the water-soluble, silica-coated UCNPs, making them very promising for future applications.

2. Materials and Methods

2.1. Materials

Ytterbium (III) chloride hexahydrate, yttrium (III) chloride hexahydrate, and erbium (III) chloride hexahydrate, which are classified as lanthanide elements or rare-earth chlorides, along with lithium hydroxide, ammonium fluoride, 28% concentrated ammonia, tetraethyl orthosilicate (TEOS), oleic acid, 1-octadecene, methanol, ethanol, cyclohexane, and deionized water, were sourced from Sigma-Aldrich, St. Louis, MO, USA. These chemicals were used as received, without undergoing any additional purification processes.

2.2. Synthesis of $YLiF_4$:Yb,Er (18/2%) UCNPs Nanocrystals

To synthesize $YLiF_4$:Yb,Er upconversion nanoparticles, start by dissolving $LnCl_3{\cdot}6H_2O$ (comprising 80% Y, 18% Yb, and 2% Er) in a mixture of oleic acid and 1-octadecene within a two-neck flask. Heat the mixture to 160 °C under argon for 40 min to achieve a clear solution, then cool to 60 °C. Following this, a prevortexed methanol solution containing $LiOH{\cdot}H_2O$ and NH_4F is injected into the flask. The mixture is maintained at 50 °C while stirring for nucleation, then heated to 160 °C to evaporate excess methanol and water. For crystallization, the temperature is raised to 300 °C and held for 1.5 h. Finally, the product is cooled, washed with ethanol, redispersed in cyclohexane, and stored at 4 °C for further use.

2.3. Preparation of $YLiF_4$:Yb,Er@SiO_2 UCNPs with Different UCNPs Contents

The production of $YLiF_4$:Yb,Er@SiO_2 upconversion nanoparticles (UCNPs) was conducted by adding reagents sequentially to a cyclohexane solution, held in a 50 mL round-bottomed flask at ambient temperature. The process began with mixing 30 mL of cyclohexane, 1.5 mL of Igepal CO-520, 0.25 mL of concentrated ammonia (28%), and 0.6 mL of a cyclohexane dispersion containing varying amounts (2.8 mg, 4 mg, 5 mg, 6 mg, and 7 mg) of $YLiF_4$:Yb,Er UCNPs. It is important to note that the experiment was performed separately for each specified amount of UCNPs. Following the addition of the UCNPs, the solution underwent ultrasonic treatment for 10 min to become transparent. The next step involved coating the UCNPs with silica shells, achieved by adjusting the concentration of the tetraethyl orthosilicate (TEOS) that was added, which ranged from 10 to 50 μL. After stirring the mixture for 24 h at room temperature, the $YLiF_4$:Yb,Er@SiO_2 UCNPs were separated via centrifugation. The nanoparticles were then washed with ethanol, using centrifugation at 6000 rpm for 10 min, and redispersed in 4 mL of ethanol through ultrasonication for 10 min, resulting in a colloidal solution.

2.4. *Preparation of $YLiF_4$:Yb,Er@SiO_2 UCNPs with Different Silica Shells*

The preparation of $YLiF_4$:Yb,Er@SiO_2 UCNPs began by sequentially adding chemicals to a cyclohexane solution within a 50 mL round-bottom flask at room temperature. The process started with the addition of 30 mL of cyclohexane, 1.5 mL of Igepal CO-520, 0.25 mL of 28% concentrated ammonia, and 0.6 mL of a cyclohexane dispersion containing 7 mg of $YLiF_4$:Yb,Er UCNPs. This mixture was then subjected to ultrasonic treatment for 10 min to form a clear solution. Next, to create $YLiF_4$:Yb,Er@SiO_2 UCNPs with various silica shell thicknesses, the concentrations of tetraethyl orthosilicate (TEOS) were varied between 10 and 50 μL, with each TEOS concentration tested in separate experiments. After adding TEOS, the mixture was stirred continuously for 24 h at room temperature. Finally, the $YLiF_4$:Yb,Er@SiO_2 UCNPs were separated using centrifugation, cleaned with ethanol by centrifuging at 6000 rpm for 10 min, and redispersed in 4 mL of ethanol through ultrasonication for another 10 min, producing a colloidal solution.

2.5. *Recrystallization of $YLiF_4$:Yb,Er UCNPs*

In a dual-necked 50 mL round-bottom flask, a precise mixture of 8 mL of oleic acid (OA) and 12 mL of 1-octadecene (ODE) was prepared. Into this mixture, 2 mL of previously synthesized core upconverting nanoparticles (UCNPs) were added under an argon (Ar) atmosphere, deliberately omitting the addition of lithium (Li), yttrium (Y), and fluorine (F) ions. The solution was then heated to 110 °C to evaporate the chloroform, with continuous stirring throughout the process. After 40 min, the temperature was increased to a specified heat-treatment level of 240 °C and maintained for 1.5 h. Upon completing the heat treatment, the samples were allowed to cool down to room temperature gradually. The cooled UCNPs underwent a thorough washing process with ethanol three times to remove impurities, followed by redispersion in 2 mL of cyclohexane to obtain the final product.

2.6. *Transmission Electron Microscope Imaging (TEM) of the Synthesized UCNPs@SiO_2*

TEM images of the synthesized UCNPs@SiO_2 were captured using a high-resolution JEOL 2010 200 kV TEM (JEOL Ltd., Tokyo, Japan). To prepare for TEM imaging, the TEM grids were subjected to plasma treatment. This crucial step eliminated organic residues and dust and also enhanced the grids' hydrophilicity, ensuring the sample solutions spread evenly. For the imaging process itself, we applied several drops of the synthesized UCNPs@SiO_2 NP solutions onto these treated grids. Then, to ensure the samples were adequately prepared for high-quality imaging, they were dried in a vacuum environment.

2.7. *Upconversion Emission Measurements of the Synthesized UCNPs@SiO_2*

We developed a specialized confocal laser scanning microscope tailored for the upconversion emission analysis of the synthesized UCNPs@SiO_2 NPs. This optical microscope consists of a 4f imaging system, a high magnification microscope objective (100×, NA = 0.8, model number: LMPlanFL N, Olympus, Tokyo, Japan), Thorlabs GVS 212 Galvano scanners (Newton, NJ, USA), a continuous wave (CW) 532 nm green laser (CNI lasers, Changchun, China), and a custom-made spectrometer.

Prior to optical characterizations of the synthesized UCNPs@SiO_2 NPs, we employed a drop-casting technique to evenly coat quartz substrates with the synthesized UCNPs and UCNPs@SiO_2 NPs, ensuring a consistent and thin layer of the samples for uniform optical measurements. These substrates were then mounted on the confocal microscope's x–y–z stage. The microscope's x–y scanner facilitated the optical scanning of the samples with the green laser and then the capturing of the emitted optical emission spectra from the UCNPs@SiO_2 NPs through the same objective lens. The spectra were analyzed using the home-made spectrometer equipped with a precise pinhole (Thorlabs, Newton, NJ, USA), an 1800-grooves/mm diffraction grating (Thorlabs, Newton, NJ, USA), achieving a high spectral resolution of 0.03 nm, a sensitive camera (Trius camera model SX-674, Starlight Xpress Ltd., Bottle Lane, UK), and a single photon counter (Hamamatsu photon counter model number H7155-21, Hamamatsu Photonics UK Limited, Welwyn Garden City, UK) for precise detection and quantification of emitted photons.

3. Results and Discussion

The $YLiF_4$:Yb,Er UCNPs used in this study were synthesized following a hydrothermal synthesis procedure previously reported in [18,19,21] and detailed in Section 2. A silica coating of the synthesized $YLiF_4$:Yb,Er UCNPs was performed using a reverse microemulsion method which was well reported and optimized for coating hydrophobic oleate-capped iron oxide (Fe_3O_4) nanoparticles [26]. This reverse microemulsion silica coating method is governed by the concentration of NPs, the surface polymer (Igebal CO-520), and the amount of ammonia and TEOS. For a successful silica coating of the synthesized $YLiF_4$:Yb,Er UCNPs, first, we varied the content of the synthesized UCNPs from 2.8 mg to 7 mg to avoid core-free silica nanoparticles as summarized in Table 1. The silica coating procedure used in this study is experimentally detailed in Section 2. The amount of Igepal CO-520 and ammonia was carefully chosen to be 1.5 mL and 0.25 mL, respectively, which was found to strongly control the size and number of the aqueous domains during the silica formation. These values were optimized for a successful silica coating around NPs after many carful experiments reported in [26]. The experimental results revealed a significant presence of core-free silica particles, with only a few UCNPs successfully encapsulated in silica shells when employing 2.8 mg and 4 mg of the synthesized UCNPs, as depicted in Figure 1a,b. Addressing this issue, we increased the UCNP concentration to 6 mg, resulting in a remarkable reduction in core-free silica particles as evidenced in Figure 1c. By further adjusting the UCNP amount to 7 mg for the silica coating process, we achieved an optimal one-to-one encapsulation ratio of UCNPs within the silica, demonstrating a precise match between the number of UCNPs and the aqueous domains, as illustrated in Figure 1d.

Prior to silica coating at different silica thicknesses, structural characterizations (morphology and size) of the synthesized $YLiF_4$:Yb,Er UCNPs were performed using a transmission electron microscope (TEM). For this, diluted drops of UCNPs (10×) in cyclohexane were placed on a carbon-based TEM grid. Figure 2a shows well-dispersed octahedral nanoparticles with an average size of 20 nm. Upon optimizing the UCNP concentration, we achieved a uniform coating of the synthesized $YLiF_4$:Yb,Er UCNPs with silica shells of varying thicknesses, as outlined in Table 2 and depicted in Figure 2b–e. The experimental methodology for this process is comprehensively described in Section 2. By maintaining a constant content of UCNPs at 7 mg, we systematically varied the tetraethyl orthosilicate (TEOS) volume from 10 μL to 50 μL. This adjustment, while keeping other parameters consistent as detailed in Table 2, allowed us to precisely control the silica shell thickness, which ranged from 3.2 nm to 8.1 nm. Notably, this method effectively prevented the formation of core-free silica particles, ensuring each UCNP was encapsulated within its own silica shell.

Table 1. Parameters for applying a SiO_2 coating to UCNPs at different concentrations.

Sample	UCNPs (mg)	Igepal CO-520 (mL)	Ammonia (mL)	TEOS (μL)	Silica Shell (nm)	Core-Free Silica (Yes/No)
1	2.8	1.5	0.25	30	6.5	yes
2	4	1.5	0.25	30	6.5	yes
3	5	1.5	0.25	30	6.5	yes
4	6	1.5	0.25	30	6.5	no
5	7	1.5	0.25	30	6.5	no

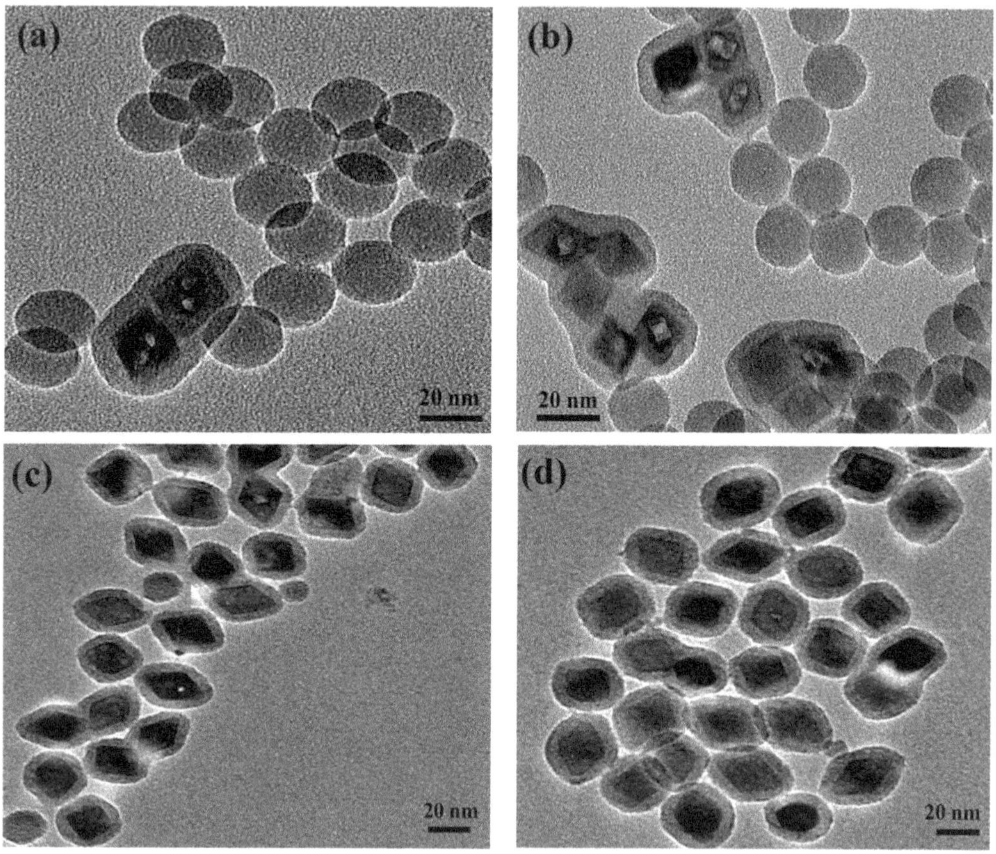

Figure 1. TEM images of UCNPs@SiO_2 NPs with different content of UCNPs (**a**) 2.8 mg, (**b**) 4 mg, (**c**) 6 mg, and (**d**) 7 mg.

Table 2. Parameters for fabricating UCNPs@SiO_2 with different TEOS Amounts.

Sample	UCNPs (mg)	Igepal CO-520 (mL)	Ammonia (mL)	TEOS (μL)	Silica Shell (nm)	Core-Free Silica (Yes/No)
6	7	1.5	0.25	10	3.2	no
7	7	1.5	0.25	20	4.5	no
8	7	1.5	0.25	30	6.2	no
9	7	1.5	0.25	40	7.2	no
10	7	1.5	0.25	50	8.1	no

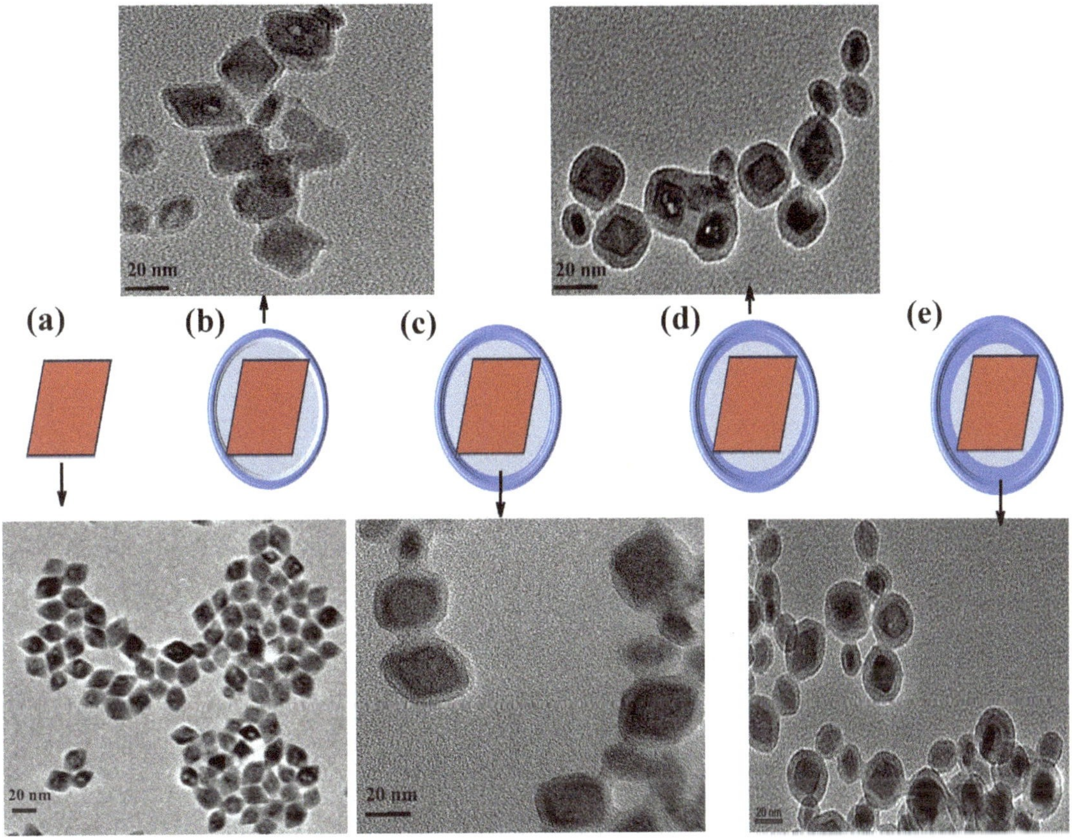

Figure 2. TEM images of UCNPs@SiO_2 with different TEOS amounts. (**a**) UCNPs before silica coating. (**b–e**) UCNPs@SiO_2 with different TEOS amounts ranging from 10 to 40 μL corresponding to silica coating thickness increases from 3.2 to 7.2 nm.

To better understand the silica shell formation on the synthesized UCNPs via the reverse microemulsion process, an illustration schematic was put together as shown in Figure 3. At the first step, Igepal CO-520, a type of detergent, naturally groups together in the cyclohexane solution to form micelles due to its hydrophilic groups. These micelles form because Igepal CO-520 molecules have both water-loving and water-repelling parts. In the oil-like environment of cyclohexane, the water-repelling parts stick out, driving the molecules to cluster together. When oleate-capped UCNPs are introduced into this mix in step number two, an interesting ligand exchange takes place. Some of the oleic acid on the nanoparticles' surface is replaced by Igepal CO-520, partially integrating the nanoparticles into the micelles. The addition of ammonia is the third and critical step, which causes the Igepal CO-520 micelles to absorb it and grow larger [25,26].

The Igepal CO-520 micelles will then be filled with ammonia upon its addition, and the micelle size is enlarged and forms a reverse microemulsion system. Subsequently in the fourth step, the added TEOS will hydrolyze at the oil/water interface and perform the ligand exchange with the Igepal CO-520 chemically absorbed on the UCNPs surface and will then transfer the UCNPs to the water phase. Finally in the last two steps, the hydrolyzed TEOS on the surface of the UCNPs undergoes a condensation process and forms silica shells.

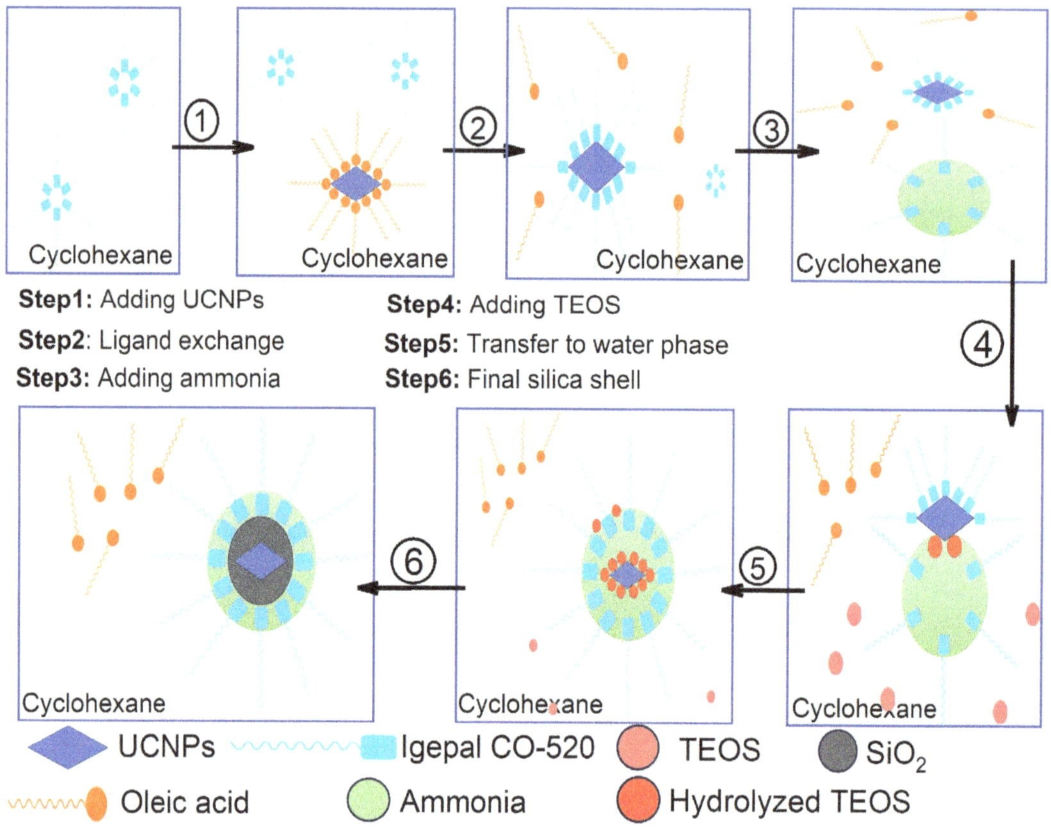

Figure 3. An illustration of the silica coating mechanism of silica shells around UCNPs.

The reverse microemulsion method is particularly advantageous for the synthesis of UCNPs@SiO_2 core-shells due to its exceptional control over shell thickness and uniformity. This method allows precise manipulation of silica deposition on UCNPs' cores by controlling the water-to-surfactant ratio and other reaction parameters within the microemulsion droplets. Unlike other synthesis techniques such as the Stöber process, which may produce larger and less uniform silica particles, reverse microemulsion facilitates the production of highly uniform silica shells at lower temperatures, preserving the quality of the sensitive cores. However, the technique has its drawbacks, such as the complex removal of surfactants from the final product, which can affect the purity and functionality of the silica shells. Moreover, scaling up this method for industrial production remains challenging due to the delicate balance of emulsion stability and the high cost of the surfactants and organic solvents used [26,27].

Comparatively, methods like sol–gel and hydrothermal are simpler and more cost-effective for producing silica particles at a high temperature, resulting in a lack of the fine control over shell morphology provided with reverse microemulsion [28].

Next, optical characterization of the synthesized UCNPs before and after silica coating was performed. For this purpose, a confocal scanning laser microscope was built and equipped with a near-infrared (NIR) laser (980 nm) and a custom-built spectrometer that covers a wide range of the spectrum from the visible (VIS) to NIR as illustrated in Figure 4a. The recorded emission spectra were filtered from the NIR excitation laser with a 750 nm short pass filter (SP filter). The optical microscope parts are detailed in Section 2. For optical testing, 1 μL of each sample (synthesized UCNPs and UCNPs@silica shells at different thicknesses) was spin-coated on a quartz slide to make a thin layer of the sample to avoid agglomeration during drying. The prepared substrates of the UCNP samples were then placed on the confocal microscope. Figure 4b shows the UC photoluminescence of the synthesized UCNPs before and after silica coating with two strong green emissions peaked at 520 nm and 550 nm corresponding to UC optical emission from both $^2H_{11/2}$ and $^4S_{3/2}$ excited states to the $^4I_{15/2}$ ground state. In addition, there was a weak red emission that peaked at 650 nm, corresponding to the UC optical transition from the $^4F_{9/2}$ excited states to the $^4I_{15/2}$ ground state of the upconverting ion Er^{3+} after absorption of two incident 980 nm photons [2,17,18] as illustrated in Figure 4b, inset.

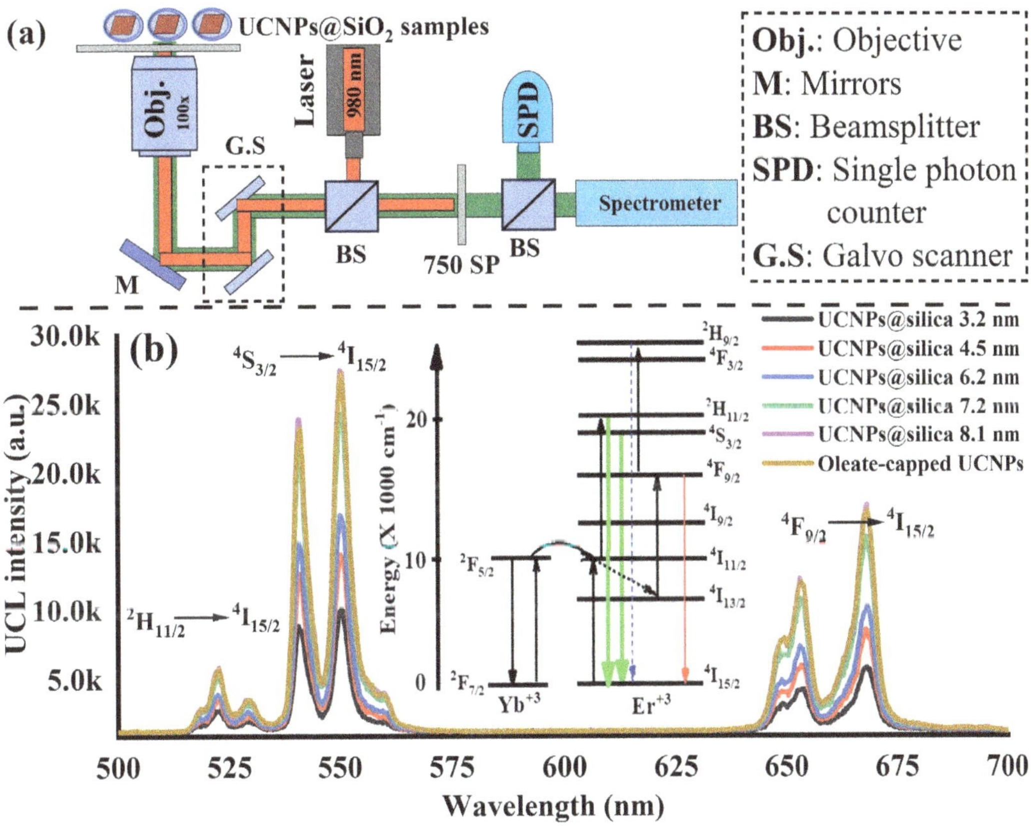

Figure 4. (**a**) An illustration of the custom-made optical microscope used for optical characterizations of UCNPs@SiO_2. The optical setup consists of an imaging system with 100× objective, 980 nm laser directed into the samples with a beamsplitter (BS) and a custom-made spectrometer to analyze the recorded optical spectrum from the examined samples. The recorded emission spectra were filtered from the excitation laser with a 750 nm short pass filter (SP filter). (**b**) Optical emission spectra of the synthesized UCNPs coated with and without silica shells. It can be seen in this figure how the silica coating shell thicknesses helps UCNPs retain the good UC emission when dispersed in water to be more compatible for practical applications. (**b, inset**) An illustration shows the subsequent upconversion process in rare-earth ion-doped $YLiF_4$ host crystals as they absorb near-infrared photons and emits visible photons.

In the analysis of the recorded optical spectra, we observed that the control UCNPs (oleate-capped) showed a bright UC photoluminescence due to a full coverage of the oleate group over their surface which prevents water penetration to quench the UC emission. Unlike the oleate-capped UCNPs, the UC photoluminescence was quenched and increased to retain their good upconversion luminescence efficiency as the silica shell thickness increases from 3 to 8 nm as shown in Figure 4b. This implies that the silica shell coating on UCNPs significantly enhances their upconversion emission efficiency by providing robust physical protection and reducing surface quenching, such as from water molecules. The silica layer acts as a barrier against environmental factors, preserving the UCNPs' core, while also isolating them from potential quenchers in the surrounding medium. This isolation minimizes nonradiative energy losses at the nanoparticle surface, ensuring higher upconversion efficiency by maintaining the excitation energy within the nanoparticles.

One significant challenge encountered during the growth of silica on upconversion nanoparticles (UCNPs) is the formation of nanoparticle aggregates under a common silica shell. For the nanoparticles to be effectively used in various applications, it is crucial that each UCNP is uniformly and individually coated with silica, with the dispersibility of the nanoparticles being a key factor. In some instances, we noted that UCNPs tended to cluster and form bundles before the application of the silica coating. This led to the production of aggregates with silica coatings as seen in our experiments as shown in Figure 5a. This aggregation is primarily due to the loss of oleic acid (OA) surface ligands during the cleaning process. It has been documented that OA can detach from the nanoparticle surface if UCNPs are excessively treated with ethanol and sonicated, leading to the particles clumping together [29]. To address this issue, Bian et al. developed a method to enhance the crystallinity of α-UCNPs by modifying the crystal edges through a combination of wet chemical treatment and thermal annealing in the presence of OA [29]. We hypothesized that this approach might also enhance the colloidal dispersibility of our UCNPs during silica coating and therefore produce single-core UCNPs@silica NPs.

Toward this goal, we conducted an experiment where UCNPs were annealed at high temperatures with an excess of OA as illustrated in Figure 5b. Experimental details of the UCNPs recrystallization can be found in Section 2. Results of the recrystallized UCNPs were compared with those of nonannealed UCNPs under identical concentrations and conditions during silica shell coating. OA post-treatment markedly improved the colloidal stability of the UCNPs in organic solvents and greatly enhanced the colloidal stability and uniformly of UCNPs. After that, we carried out a silica shell coating around UCNPs with one TEOS concentration (40 μL). Figure 5c shows that post annealing of the synthesized UCNPs in an excess of OA facilitated the successful growth of silica on their surface with single-core UCNPs@silica NPs. Furthermore, we also tested and compared the UC emission of the annealed and nonannealed UCNPs coated with the same thickness silica shell. We found that the UC luminance of the annealed particles enhanced $10\times$ time compared with nonannealed UCNPs at the same silica shell and excitation conditions as demonstrated in Figure 5d.

Finally, in contrast to earlier studies that primarily focused on silica encapsulating of commonly used sodium-based UCNPs without avoiding aggregation of the UCNPs during the silica encapsulation process [30–32], our obtained results introduce a refined encapsulation technique for $YLiF_4$:Yb,Er UCNPs with a barmaid-like shape with a precisely controlled silica (SiO_2) shell thickness. In addition, this reported method ensures individual UCNP encapsulation within silica shells, significantly boosting their optical properties, which is critical for their promising application in cutting-edge fields.

As a continuation of this work, we plan to expand on these advancements by developing diamond nanoshells around the synthesized UCNPs@SiO_2. This initiative aims to mitigate the current issues of background biofluorescence in bioimaging applications where diamond color centers are excited by visible light. Furthermore, we anticipate great potential in exploring Yb-doped YLF UCNPs coated with nanodiamond shells for applications such as laser refrigeration and optical trapping in quantum technologies.

These developments could revolutionize the functionality and application of UCNPs in advanced applications.

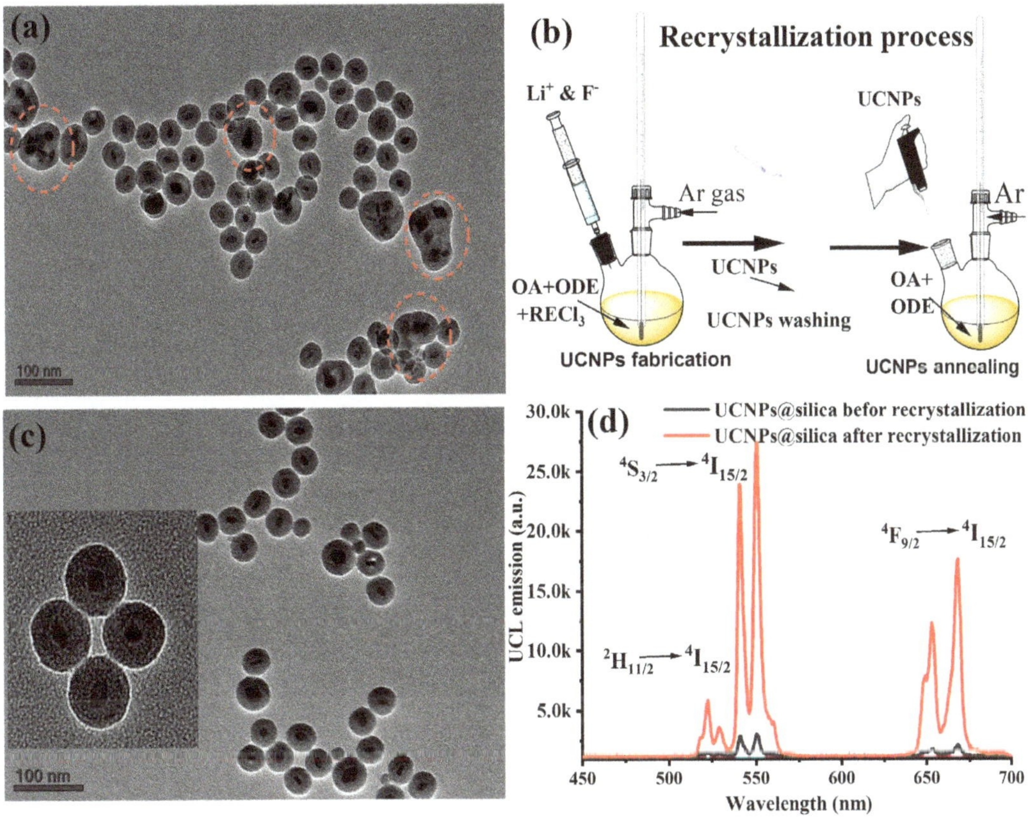

Figure 5. (**a**) Unwanted UCNP aggregates within a silica shell (highlighted in red dashed circles) during a common silica shell formation before the recrystallization process. (**b**) An illustration of the wet chemical annealing procedure used for UCNP recrystallization. (**c**) Post annealing of the synthesized UCNPs in an excess of OA facilitated the successful growth of silica on their surface with single-core UCNPs@silica NPs. (**d**) UC emission recorded from UCNPs@SiO_2 before and after recrystallization process sensing.

4. Conclusions

We have presented a precise encapsulation of $YLiF_4$:Yb,Er upconversion nanocrystals (UCNPs) with silica (SiO_2) shells using the reverse microemulsion method. The key parameters of this reported method were optimized to produce UCNPs@SiO_2 core/shell structures with adjustable shell thicknesses and ensure individual UCNP encapsulation. This approach successfully eliminates the occurrence of core-free silica particles and maintains the UCNPs' inherent upconversion luminescence, critical for their potential applications in many promising applications. Furthermore, our technique effectively prevents the coencapsulation of multiple UCNPs in a single silica shell, a significant achievement that enhances the functionality and applicability of these core/shell structures in advanced technological applications. This work not only contributes significantly to the field of nanomaterial encapsulation but also opens new pathways for utilizing upconversion nanocrystals in innovative ways.

Author Contributions: Conceptualization, M.A. and Y.A.A.; Data curation, Y.A.A., A.M.A., M.K.A., R.S.A. and A.A.; Formal analysis, M.A. and Y.A.A.; Funding acquisition, M.A.; Methodology, M.A. and Y.A.A.; Project administration, M.A.; Software, Y.A.A.; Validation, M.A. and Y.A.A.; Visualization, M.A.; Writing—original draft, M.A., A.A. and Y.A.A.; Writing—review and editing, Y.A.A. and M.A. All authors have read and agreed to the published version of the manuscript.

Funding: This work was funded by King Abdulaziz City for Science and Technology (KACST), Saudi Arabia.

Data Availability Statement: The data presented in this study are available on request from the corresponding author.

Conflicts of Interest: The authors declare no conflicts of interest.

References

1. Dubey, N.; Chandra, S. Upconversion nanoparticles: Recent strategies and mechanism based applications. *J. Rare Earths* **2022**, *40*, 1343–1359. [CrossRef]
2. Alkahtani, M.; Alsofyani, N.; Alfahd, A.; Almuqhim, A.A.; Almughem, F.A.; Alshehri, A.A.; Qasem, H.; Hemmer, P.R. Engineering Red-Enhanced and Biocompatible Upconversion Nanoparticles. *Nanomaterials* **2021**, *11*, 284. [CrossRef] [PubMed]
3. Liu, Q.; Zhang, Y.; Peng, C.S.; Yang, T.; Joubert, L.-M.; Chu, S. Single upconversion nanoparticle imaging at sub-10 W cm^{-2} irradiance. *Nat. Photonics* **2018**, *12*, 548–553. [CrossRef] [PubMed]
4. Siefe, C.; Mehlenbacher, R.D.; Peng, C.S.; Zhang, Y.; Fischer, S.; Lay, A.; McLellan, C.A.; Alivisatos, A.P.; Chu, S.; Dionne, J.A. Sub-20 nm Core–Shell–Shell Nanoparticles for Bright Upconversion and Enhanced Förster Resonant Energy Transfer. *J. Am. Chem. Soc.* **2019**, *141*, 16997–17005. [CrossRef] [PubMed]
5. Würth, C.; Fischer, S.; Grauel, B.; Alivisatos, A.P.; Resch-Genger, U. Quantum Yields, Surface Quenching, and Passivation Efficiency for Ultrasmall Core/Shell Upconverting Nanoparticles. *J. Am. Chem. Soc.* **2018**, *140*, 4922–4928. [CrossRef] [PubMed]
6. Wen, S.; Zhou, J.; Zheng, K.; Bednarkiewicz, A.; Liu, X.; Jin, D. Advances in highly doped upconversion nanoparticles. *Nat. Commun.* **2018**, *9*, 2415. [CrossRef] [PubMed]
7. Zhang, Y.; Wen, R.; Hu, J.; Guan, D.; Qiu, X.; Zhang, Y.; Kohane, D.S.; Liu, Q. Enhancement of single upconversion nanoparticle imaging by topologically segregated core-shell structure with inward energy migration. *Nat. Commun.* **2022**, *13*, 5927. [CrossRef] [PubMed]
8. Naher, H.S.; Al-Turaihi, B.A.H.; Mohammed, S.H.; Naser, S.M.; Albark, M.A.; Madlool, H.A.; Marzoog, H.A.M.A.; Jalil, A.T. Upconversion nanoparticles (UCNPs): Synthesis methods, imaging and cancer therapy. *J. Drug Deliv. Sci. Technol.* **2023**, *80*, 104175. [CrossRef]
9. Fan, Q.; Sun, C.; Hu, B.; Wang, Q. Recent advances of lanthanide nanomaterials in Tumor NIR fluorescence detection and treatment. *Mater. Today Bio* **2023**, *20*, 100646. [CrossRef] [PubMed]
10. Krishnaiah, K.V.; Ledemi, Y.; de Lima Filho, E.S.; Nemova, G.; Messaddeq, Y.; Kashyap, R. Development of Yb^{3+}-doped oxyfluoride glass-ceramics with low OH^- content containing CaF_2 nanocrystals for optical refrigeration. *Opt. Eng.* **2016**, *56*, 011103. [CrossRef]
11. Luntz-Martin, D.R.; Felsted, R.G.; Dadras, S.; Pauzauskie, P.J.; Vamivakas, A.N. Laser refrigeration of optically levitated sodium yttrium fluoride nanocrystals. *Opt. Lett.* **2021**, *46*, 3797–3800. [CrossRef] [PubMed]
12. Roder, P.B.; Smith, B.E.; Zhou, X.; Crane, M.J.; Pauzauskie, P.J. Laser refrigeration of hydrothermal nanocrystals in physiological media. *Proc. Natl. Acad. Sci. USA* **2015**, *112*, 15024–15029. [CrossRef] [PubMed]
13. Ortiz-Rivero, E.; Prorok, K.; Martín, I.R.; Lisiecki, R.; Haro-González, P.; Bednarkiewicz, A.; Jaque, D. Laser Refrigeration by an Ytterbium-Doped $NaYF_4$ Microspinner. *Small* **2021**, *17*, 2103122. [CrossRef] [PubMed]
14. Forberger, L.; Felsted, R.G.; Bard, A.B.; Luntz-Martin, D.R.; Vamivakas, A.N.; Pauzauskie, P.J. Synthesis and thermometry of NV^- nanodiamond alpha-$NaYF_4$ composite nanostructures. In *Optical Trapping and Optical Micromanipulation XIX*; SPIE: Bellingham, WA, USA, 2022; Volume 12198.
15. Jin, L.; Liu, Z.; Zhang, Y.; Wu, Y.; Liu, Y.; Deng, H.; Song, Q.; Xiao, S. Lanthanide-doped nanocrystals in high-Q microtoroids for stable on-chip white-light lasers. *Photonics Res.* **2022**, *10*, 1594–1601. [CrossRef]
16. Sarkar, D.; Ganguli, S.; Samanta, T.; Mahalingam, V. Design of Lanthanide-Doped Colloidal Nanocrystals: Applications as Phosphors, Sensors, and Photocatalysts. *Langmuir* **2019**, *35*, 6211–6230. [CrossRef] [PubMed]
17. Alkahtani, M.H.; Gomes, C.L.; Hemmer, P.R. Engineering water-tolerant core/shell upconversion nanoparticles for optical temperature sensing. *Opt. Lett.* **2017**, *42*, 2451–2454. [CrossRef] [PubMed]
18. Alzahrani, Y.A.; Alromaeh, A.; Alkahtani, M. Efficient Lithium-Based Upconversion Nanoparticles for Single-Particle Imaging and Temperature Sensing. *Materials* **2023**, *16*, 4354. [CrossRef] [PubMed]
19. Mulder, J.T.; Meijer, M.S.; van Blaaderen, J.J.; du Fossé, I.; Jenkinson, K.; Bals, S.; Manna, L.; Houtepen, A.J. Understanding and Preventing Photoluminescence Quenching to Achieve Unity Photoluminescence Quantum Yield in Yb:YLF Nanocrystals. *ACS Appl. Mater. Interfaces* **2023**, *15*, 3274–3286. [CrossRef] [PubMed]

20. Carl, F.; Birk, L.; Grauel, B.; Pons, M.; Würth, C.; Resch-Genger, U.; Haase, M. $LiYF_4$:Yb/$LiYF_4$ and $LiYF_4$:Yb,Er/$LiYF_4$ core/shell nanocrystals with luminescence decay times similar to YLF laser crystals and the upconversion quantum yield of the Yb,Er doped nanocrystals. *Nano Res.* **2021**, *14*, 797–806. [CrossRef]
21. Hong, A.R.; Kim, S.Y.; Cho, S.H.; Lee, K.; Jang, H.S. Facile synthesis of multicolor tunable ultrasmall $LiYF_4$:Yb,Tm,Er/$LiGdF_4$ core/shell upconversion nanophosphors with sub-10 nm size. *Dye. Pigment.* **2017**, *139*, 831–838. [CrossRef]
22. Littleford, T.E.; Jackson, R.A.; Read, M.S.D. An atomistic simulation study of the effect of dopants on the morphology of $YLiF_4$. *Phys. Status Solidi C* **2013**, *10*, 156–159. [CrossRef]
23. Rojas-Gutierrez, P.A.; DeWolf, C.; Capobianco, J.A. Formation of a Supported Lipid Bilayer on Faceted $LiYF_4$:Tm^{3+}/Yb^{3+} Upconversion Nanoparticles. *Part. Part. Syst. Charact.* **2016**, *33*, 865–870. [CrossRef]
24. Chien, H.-W.; Wu, C.-H.; Yang, C.-H.; Wang, T.-L. Multiple doping effect of $LiYF_4$:Yb^{3+}/Er^{3+}/Ho^{3+}/Tm^{3+}@$LiYF_4$:Yb^{3+} core/shell nanoparticles and its application in Hg^{2+} sensing detection. *J. Alloys Compd.* **2019**, *806*, 272–282. [CrossRef]
25. Hanske, C.; Sanz-Ortiz, M.N.; Liz-Marzán, L.M. Silica-Coated Plasmonic Metal Nanoparticles in Action. *Adv. Mater.* **2018**, *30*, e1707003. [CrossRef]
26. Ding, H.L.; Zhang, Y.X.; Wang, S.; Xu, J.M.; Xu, S.C.; Li, G.H. Fe_3O_4@SiO_2 Core/Shell Nanoparticles: The Silica Coating Regulations with a Single Core for Different Core Sizes and Shell Thicknesses. *Chem. Mater.* **2012**, *24*, 4572–4580. [CrossRef]
27. Solans, C.; Izquierdo, P.; Nolla, J.; Azemar, N.; Garcia-Celma, M.J. Nano-emulsions. *Curr. Opin. Colloid Interface Sci.* **2005**, *10*, 102–110. [CrossRef]
28. Milde, M.; Dembski, S.; Osvet, A.; Batentschuk, M.; Winnacker, A.; Sextl, G. Polymer-assisted sol–gel process for the preparation of photostimulable core/shell structured SiO_2/Zn_2SiO_4:Mn^{2+} particles. *Mater. Chem. Phys.* **2014**, *148*, 1055–1063. [CrossRef]
29. Bian, W.; Lin, Y.; Wang, T.; Yu, X.; Qiu, J.; Zhou, M.; Luo, H.; Yu, S.F.; Xu, X. Direct Identification of Surface Defects and Their Influence on the Optical Characteristics of Upconversion Nanoparticles. *ACS Nano* **2018**, *12*, 3623–3628. [CrossRef] [PubMed]
30. Alkahtani, M.; Alfahd, A.; Alsofyani, N.; Almuqhim, A.A.; Qassem, H.; Alshehri, A.A.; Almughem, F.A.; Hemmer, P. Photostable and Small YVO_4:Yb,Er Upconversion Nanoparticles in Water. *Nanomaterials* **2021**, *11*, 1535. [CrossRef] [PubMed]
31. Xu, S.; Yu, Y.; Gao, Y.; Zhang, Y.; Li, X.; Zhang, J.; Wang, Y.; Chen, B. Mesoporous silica coating $NaYF_4$:Yb,Er@$NaYF_4$ upconversion nanoparticles loaded with ruthenium(II) complex nanoparticles: Fluorometric sensing and cellular imaging of temperature by upconversion and of oxygen by downconversion. *Microchim. Acta* **2018**, *185*, 454. [CrossRef] [PubMed]
32. Kandoth, N.; Barman, S.; Chatterjee, A.; Sarkar, S.; Dey, A.K.; Pramanik, S.K.; Das, A. Photoactive Lanthanide-Based Upconverting Nanoclusters for Antimicrobial Applications. *Adv. Funct. Mater.* **2021**, *31*, 2104480. [CrossRef]

nanomaterials

MDPI

Article

Elastic Liposomes Containing Calcium/Magnesium Ferrite Nanoparticles Coupled with Gold Nanorods for Application in Photothermal Therapy

Ana Rita F. Pacheco [1,2], Ana Margarida Barros [1,2], Carlos O. Amorim [3], Vítor S. Amaral [4], Paulo J. G. Coutinho [1,2], Ana Rita O. Rodrigues [1,2] and Elisabete M. S. Castanheira [1,2,*]

1 Physics Centre of Minho and Porto Universities (CF-UM-UP), University of Minho, Campus de Gualtar, 4710-057 Braga, Portugal; anaritapacheco99@gmail.com (A.R.F.P.); pg42479@alunos.uminho.pt (A.M.B.)
2 Associate Laboratory LaPMET, Campus de Gualtar, 4710-057 Braga, Portugal
3 Physics Department and i3N, University of Aveiro, Campus de Santiago, 3810-193 Aveiro, Portugal
4 Physics Department and CICECO, University of Aveiro, Campus de Santiago, 3810-193 Aveiro, Portugal
* Correspondence: ecoutinho@fisica.uminho.pt

Abstract: This work reports on the design, development, and characterization of novel magneto-plasmonic elastic liposomes (MPELs) of DPPC:SP80 (85:15) containing $Mg_{0.75}Ca_{0.25}Fe_2O_4$ nanoparticles coupled with gold nanorods, for topical application of photothermal therapy (PTT). Both magnetic and plasmonic components were characterized regarding their structural, morphological, magnetic and photothermal properties. The magnetic nanoparticles display a cubic shape and a size (major axis) of 37 ± 3 nm, while the longitudinal and transverse sizes of the nanorods are 46 ± 7 nm and 12 ± 1.6 nm, respectively. A new methodology was employed to couple the magnetic and plasmonic nanostructures, using cysteine as bridge. The potential for photothermia was evaluated for the magnetic nanoparticles, gold nanorods and the coupled magnetic/plasmonic nanoparticles, which demonstrated a maximum temperature variation of 28.9 °C, 33.6 °C and 37.2 °C, respectively, during a 30 min NIR-laser irradiation of 1 mg/mL dispersions. Using fluorescence anisotropy studies, a phase transition temperature (T_m) of 35 °C was estimated for MPELs, which ensures an enhanced fluidity crucial for effective crossing of the skin layers. The photothermal potential of this novel nanostructure corresponds to a specific absorption rate (SAR) of 616.9 W/g and a maximum temperature increase of 33.5 °C. These findings point to the development of thermoelastic nanocarriers with suitable features to act as photothermal hyperthermia agents.

Keywords: calcium/magnesium ferrite; gold nanorods; elastic liposomes; photothermia

1. Introduction

In modern medicine, liposomes have emerged as ideal and effective nanocarriers. These well-studied biocompatible vesicles ensure the protection, stabilization and transport of different substances, both hydrophilic and hydrophobic, without changing their mechanism of action [1,2]. Considering the treatment of skin diseases, which represent one of the most frequent types of disorders all over the world [3], liposomes must be designed to ensure high penetration into the different layers of the skin, especially into its barrier, the stratum corneum (SC). For that purpose, elastic liposomes (ELs) are an excellent approach for topical applications due to their ultra-deformable properties [4–6]. Generally, ELs consist of an aqueous core surrounded by a double bilayer of phospholipids and single chain surfactants with a high radius of curvature, also known as edge activators (EAs) [1,7]. These are responsible for reducing the phase transition temperature and conferring an elastic nature to liposomes by modulating their interfacial tension and forcing the reorganization of the main phospholipid, increasing their flexibility. Thus, when a transdermal water gradient occurs, the high deformability of ELs allows them to pass spontaneously across the SC, without modifying their initial structure [7]. In fact, vesicles with this behavior manage to penetrate channels ten times smaller than their diameter, allowing them to reach the SC by the intercellular route [8]. Typical EAs include oleic acid, polysorbate (Tween), sorbitan (Span) and sodium cholate [9]. For the main phospholipid composition, lipids containing phosphatidylcholine with a T_m localized above the physiological temperature are usually included in the formulations. In this context, dipalmitoyl-phosphatidyl-choline (DPPC) shows great potential ($T_m \simeq 41$ °C) [10]. Below this value, liposomes are in a solid gel phase. However, with the temperature increase, they will attain a fluid liquid crystal state, and an enhanced permeability will be promoted, releasing their content [9]. Vesicles exhibiting this behavior are known as thermosensitive liposomes. Their conjugation with EAs leads to a reduction in the liposome's transition temperature. Considering topical applications, it is important that a system's T_m is adjusted and localized between 25 °C (room temperature) and 35 °C (average skin temperature [11]) for an improved permeation potential and stability, by optimizing the ingredients of elastic liposomes, as well as the lipid/EA ratio.

After ensuring topical administration through ELs, other strategies are required for skin disease therapy. On the one hand, the incorporation of magnetic nanoparticles (MNPs) into the nanoplatform is a beneficial solution, because with proper composition they can behave as multi-modal therapeutic agents (from diagnosis and magnetic targeting to drug delivery and hyperthermia mediators) [12]. To avoid toxicity issues, the biocompatibility of the MNPs must be ensured. Considering this, ferrites containing alkaline earth metals, such as Mg^{2+} and Ca^{2+} ions, are preferred for inclusion in the final formulation, as a result of their proved cell viability [13]. Promising previous results using magnesium doped ferrite nanoparticles have been reported, in particular for $Ca_{0.25}Mg_{0.75}Fe_2O_4$ MNPs. These have shown a superparamagnetic behavior and high thermal energy dissipation, when irradiated with near-infrared (NIR) light or by exposure to an alternating magnetic field [13–16]. To obtain enhanced properties, and especially a more efficient hyperthermia capability, shape anisotropic MNPs have proven to be a good option [17–20].

On the other hand, gold nanorods (AuNRs) can provide an even more pronounced photothermal conversion efficiency [21–24]. When irradiated by a particular frequency of light equal to the inherent frequency of the conductive electrons of AuNRs, these start to collectively oscillate at the nanorods surface, reaching their maximal amplitude of oscillation. This results in the free electrons' resonance, a phenomenon called localized surface plasmon resonance (LSPR) [24]. Since they show two LSPR bands (transverse and longitudinal), rod-shaped gold nanoparticles exhibit excellent light absorption in the NIR spectral region (650–1100 nm), which translates into suitable photothermal agents with an efficient body penetration and notable temperature rise at target tissues [25].

The incorporation of magnetic/plasmonic nanoparticles (MPNPs) in the aqueous interior of lipid vesicles, forming magneto-plasmonic elastic liposomes, is a promising approach with potential to treat topical disease conditions through hyperthermia. The combination of both magnetic and plasmonic components in a single system can maximize the cell damage by an abnormally localized heat under a NIR laser (plasmonic hyperthermia) [26,27]. Several magnetic/plasmonic nanostructures have been reported, most of them using ferrites as the magnetic component (e.g., iron oxide, manganese ferrite) and gold or carbon as the plasmonic element [28–30].

Herein, we propose a novel nanosystem capable of action as magnetic and plasmonic hyperthermia agent at low concentration for skin disease therapy. For this purpose, $Mg_{0.75}Ca_{0.25}Fe_2O_4$ nanocubes were coupled with Au nanorods and surrounded by a lipid bilayer of DPPC:SP80 (85:15) with elastic properties for improved penetration in the SC. The response of MPELs and their components to external stimuli was thoroughly examined, in order to determine their heating capability, along with investigating the impact of the interaction between MNPs and AuNRs. To the best of our knowledge, the coupling between Ca/Mg mixed ferrite nanocubes and Au nanorods has not previously been reported in the literature, as well as their encapsulation in elastic liposomes. Promising results were obtained for MPELs' future application in skin disease treatment by photothermal hyperthermia.

2. Materials and Methods

The chemicals iron(III) chloride hexahydrate, magnesium acetate tetrahydrate, calcium acetate hydrate, cis-9-octadecenoic acid (OA), tetrahydrofuran (THF), 1-octadecene, chloroform, hydrogen peroxide, sodium borohydride, dimethyl sulfoxide 99.9% (DMSO), gold(III) chloride solution, cetyltrimethylammonium bromide (CTAB), silver nitrate, hydrazine monohydrate, DL-cysteine 97%, imidazole hydrochloride, potassium phosphate monobasic, hydroquinone, sodium sulphate anhydrous, 1,1′-carbonyldiimidazole (CDI), sorbitan monostearate (SP60), sorbitan tristearate (SP65), sorbitan monooleate (SP80), sorbitan trioleate (SP85) and poly-oxyethylene-sorbitan monooleate (TW80) were purchased from Sigma-Aldrich (St. Louis, MO, USA). 1,2-Dipalmitoyl-*sn*-glycero-3-phosphocholine (DPPC) was purchased from Avanti Polar Lipids (Birmingham, AL, USA). Spectroscopic grade solvents and ultrapure water Milli-Q grade (MilliporeSigma, St. Louis, MO, USA) were used in all preparations.

2.1. Synthesis of $Mg_{0.75}Ca_{0.25}Fe_2O_4$ Nanocubes

Calcium/magnesium mixed ferrite nanocubes were prepared by an adapted protocol from Pacheco et al. [31], which involves the use of the surfactants octadecene and oleic acid as solvent and reducing agent, respectively. OA plays an important role in the surface mediated phase transfer of the pre-generated spherical nanoparticles to new crystal structures during the growth and nucleation stages [32].

To prepare cubic-shaped magnetic nanoparticles of magnesium ferrite with 25% replacement by calcium ions, a solution containing 3.1 mM oleic acid, 2 mM iron(III) chloride hexahydrate, 0.75 mM magnesium acetate tetrahydrate and 0.25 mM calcium acetate hydrate in 15 mL of octadecene was heated to 120 °C, under continuous magnetic stirring. After 1 h at these conditions, a condenser was attached to the system and the temperature was increased at a rate of 1 °C per min until reaching 200 °C. After 1 h, the mixture was heated at a rate of 5 °C per min until 290 °C and then left at this temperature for 60 min. The precipitated MNPs were washed by several cycles of magnetic decantation with THF and a solution of water/ethanol (1:1). In order to remove surface surfactant residues, a chemical calcination step was performed using an adapted protocol previously described [33]. Briefly, the total of nanoparticles was treated with a 35% hydrogen peroxide (H_2O_2) solution, in a proportion of 1 g of MNPs per 40 mL of solvent, for 48 h at room temperature, under continuous magnetic stirring. Considering this exothermic reaction, an ice bath was used. Finally, the mixed ferrite nanocubes were washed with ethanol and water by magnetic decantation and then dried.

2.2. Synthesis of Gold Nanorods

Gold nanorods (AuNRs) were synthesized using a chemical reduction protocol adapted from [34]. Briefly, an aqueous solution of 9141 µL containing 364.45 mg of CTAB was prepared and heated up to a temperature of 30-35 °C in a water bath, with magnetic stirring. Subsequently, 400 µL of gold(III) chloride solution (0.01 M), 75 µL of silver nitrate (0.15 mM), 364 µL of hydroquinone (0.14 M) and 20 µL of sodium borohydride (0.005 M) were added. All additions were performed within 5 min. intervals. During each interval, the next solution to be added was prepared. After 30 s following the last addition, magnetic stirring was removed, and the solution was maintained at 35 °C for 4 h.

The obtained nanoparticles were washed by centrifugation (10,000 rpm for 30 min), using a 2 mM CTAB solution. In order to promote the deposition of the gold nanorods, 50 µL of a monopotassium phosphate solution (1 M) were added. Finally, the purified nanoparticles were dispersed in a 2 mM CTAB solution for future use.

2.3. Coupling between MNPs and AuNRs

Magneto-plasmonic nanoparticles were obtained by coupling Ca/Mg mixed ferrite nanocubes and gold nanorods. Cysteine was employed as a "bridging" molecule between the two types of nanoparticles (Figure 1). The –OH groups at the surface of ferrite nanocubes were first activated with carbonyl-diimidazole (CDI). Through the amine group (–NH_2) in cysteine, a carbamate linkage was formed. The pending thiol groups (–SH) of cysteine can strongly chemisorb to the gold surface, resulting in the coupling of the two types of nanoparticles. Hydrazine is additionally used to prevent aggregation of the cysteine functionalized ferrite nanocubes through formation of S–S bonds.

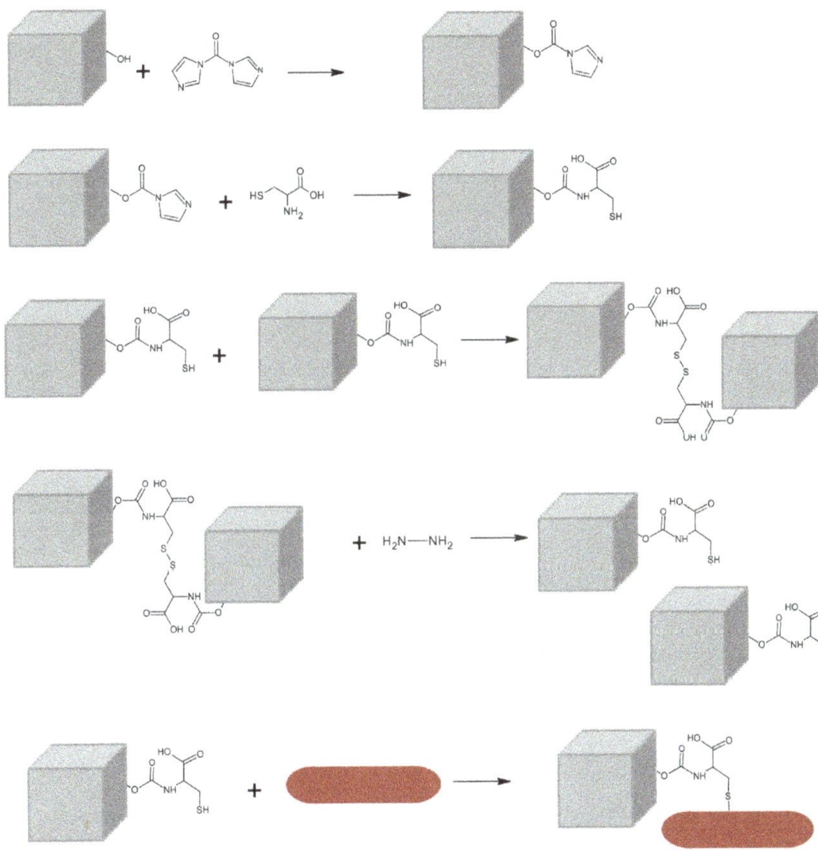

Figure 1. Schematic representation of the coupling between $Mg_{0.75}Ca_{0.25}Fe_2O_4$ nanocubes and gold nanorods, originating magneto-plasmonic nanoparticles as the final structure.

First, a 0.2 mg/mL solution of magnetic nanoparticles in dry dimethyl-sulfoxide (DMSO) was prepared. Subsequently, 3.14 mg of CDI was added, and the mixture was heated up to 60 °C, under magnetic stirring, and kept overnight at these conditions. The nanoparticles were subjected to three washing cycles with a mixture of ethanol/water (1:1) by magnetic decantation, to remove any remaining CDI molecules. Upon redispersion in dried DMSO, 2.5 mg of cysteine were added and 2.17 mg of imidazole-HCl was used as a reaction catalyst. The mixture was allowed to react for 1 h at the same temperature (60 °C) under magnetic stirring, and the washing procedure was repeated. The final step was the addition of 5 mL of a pre-prepared solution of gold nanorods along with 20 µL of hydrazine solution (8.24×10^{-4} M). The reaction was maintained for 1 h in the same conditions.

Finally, the magneto-plasmonic NPs were washed by magnetic decantation to remove all gold nanorods that did not attach to the magnetic nanoparticles. The supernatant and the sediment (coupled nanoparticles) were redispersed in 5 mL of 0.2 mM CTAB.

2.4. Preparation of Elastic Liposomes and Aqueous Elastic Magnetoliposomes

Several lipid formulations using DPPC as the main phospholipid and with different surfactants, namely Span-60 (SP60), Span-65 (SP65), Span-80 (SP80), Span-85 (SP85) and Tween-80 (TW80), were prepared. Different molar ratios of 70:30, 80:20 and 85:15 were investigated. The conventional ethanolic injection method was followed in the preparation of elastic liposomes, as described by Ribeiro et al. [35]. Briefly, an ethanolic lipid solution of 1 mM was injected, drop by drop and under continuous stirring, to a pre-heated ultrapure water (55 °C), inducing the vesicles' formation.

Aqueous elastic magneto-liposomes were prepared following a similar procedure, using the lipids DPPC and SP80 (85:15) in a final concentration of 1 mM. For this purpose, the lipid content was injected, under vortex, to 3 mL of an aqueous dispersion of magneto-plasmonic nanoparticles (1×10^{-4} M) above the main phospholipid phase transition temperature (55 °C). The non-encapsulated nanoparticles were removed by several cycles of magnetic decantation [36].

2.5. Structural and Magnetic Characterization

The composition and crystallographic structure of the mixed ferrite nanocubes were investigated by X-ray diffraction (XRD) using a PAN'alytical X'Pert PRO diffractometer (Malvern Panalytical Ltd., Malvern, UK), at the Electron Microscopy Unit of the University of Trás-os-Montes and Alto Douro (UTAD), Vila Real, Portugal. The sample was analysed operating with CuK_{α} radiation (λ = 0.154060 nm), in a Bragg–Brentano configuration. The XRD diffractogram was acquired and interpreted using Profex Software (version 4.3.6) (RMS Foundation, Bettlach, Switzerland) [37], with a BGMN Rietveld model [38].

Transmission electron microscopy (TEM) images of isolated and coupled magnetic and gold nanoparticles were recorded at the Centro de Apoio Científico-Tecnolóxico à Investigación (CACTI), University of Vigo, Spain, using a JEOL JEM-1010 high-contrast microscope, operating at 100 kV. The aqueous elastic magneto-plasmonic liposomes were also observed by TEM. Copper grids with carbon and Formvar were used for sample preparation and the NPs' dispersion was ultrasonicated before deposition. The images were processed using ImageJ software (version 1.53t, National Institutes of Health (NIH), Bethesda, MD, USA). Average sizes were estimated using the length and height of each particle and the corresponding histogram was fitted to a Gaussian distribution.

Dynamic light scattering (DLS) and electrophoretic light scattering (ELS) techniques were employed to determine the average hydrodynamic size and colloidal stability in the Litesizer^{TM} 500 DLS equipment from Anton-Paar (Anton-Paar GmbH, Graz, Austria) containing a laser diode of λ = 658 nm. Three independent measurements were performed for each sample, at room temperature, and the experimental data were processed using Kalliope software (Anton-Paar GmbH, Graz, Austria).

Magnetic response of $Mg_{0.75}Ca_{0.25}Fe_2O_4$ nanocubes and magneto-plasmonic nanoparticles was assessed at University of Aveiro (Portugal) using a MPMS3 Superconducting Quantum Interference Device (SQUID) magnetometer Quantum Design MPMS5XL (Quantum Design Inc., San Diego, CA, USA). The measurements were conducted at room temperature with applied magnetic fields in the convenient range for each sample. A specific magnetic field correction for the trapped flux in the superconducting coil was performed, resulting in an accuracy of residual less than 2 Oe. Corrections were applied to the results for geometric effects and additional measurements were made to rectify the remaining field of the superconducting coil.

2.6. Measurement of Hyperthermia Capability

The ability for photothermal therapy was evaluated using a home-made experimental irradiation setup at the Physics Centre of University of Minho (Braga, Portugal). This setup is equipped with a sample holder, a continuous laser light source with λ = 808 nm and 1 W/cm^2 of power density, and a T-type thermocouple connected to a digital multimeter (Agilent U1242A) for temperature measurement. The thermal energy dissipation of gold nanorods, magnetic NPs and magneto-plasmonic NPs dispersions (1 mg/mL) under NIR radiation was recorded for a period of time of 30 min. After this time, the laser was turned off and the cooling curves were recorded for another 30 min. The same procedure was employed for MPEL suspensions. For each sample, three independent measurements were performed.

The nanoparticles' ability to release heat when irradiated with NIR light was evaluated by the specific absorption rate (SAR). This is defined as the total energy absorbed per unit of mass (W/g), and is given by Equation (1),

$$\mathrm{SAR} = C\,\frac{\Delta T}{\Delta t} \times \frac{m_s}{m_m} \tag{1}$$

where C is the specific heat capacity of the solution (4.186 J g^{-1} K^{-1} for water), $\Delta T/\Delta t$ expresses the initial slope of the curve and m_s and m_m refer to the mass of the solution and of the magnetic/plasmonic content (g), respectively [39]. The SAR calculation was performed employing the initial linear slope method.

2.7. UV-Vis–NIR and Fluorescence Measurements

UV-Vis–NIR absorption spectra were obtained on a double-beam Shimadzu UV-3600 Plus UV-Vis–NIR spectrophotometer (Shimadzu Corporation, Kyoto, Japan). Fluorescence steady-state anisotropy studies on the ELs and MPELs were performed using a Fluorolog 3 spectrofluorometer (HORIBA Jobin Yvon IBH Ltd., Glasgow, UK), possessing double monochromators (excitation and emission) and Glan–Thompson polarizers. To acquire these spectra, the probe 1,6-diphenyl-1,3,5-hexatriene (DPH) was included in all the formulations, at a final concentration of 1×10^{-6} M. The prepared samples were excited at λ_{exc} = 350 nm, and the temperature varied between 25 °C and 55 °C, using a temperature-controlled cuvette holder. To assess the apparent microviscosity (η), the steady state fluorescence anisotropy (r) was calculated according to Equation (2):

$$r = \frac{I_{VV} - G I_{VH}}{I_{VV} + 2 G I_{VH}} \tag{2}$$

where I_{VV} and I_{VH} correspond to the fluorescence intensities obtained with vertical (V) and horizontal (H) polarization, for vertically polarized excitation light, respectively, and G is the instrument correction factor. The latter is given by Equation (3),

$$G = \frac{I_{HV}}{I_{HH}} \tag{3}$$

where I_{HV} and I_{HH} are the fluorescence intensities obtained with V and H polarization, for horizontally polarized excitation light, respectively. Considering a fluorescence lifetime for DPH of 11.4 ns in lipid membranes at room temperature [40,41], it is then possible to calculate η (P) from r values, following Equation (4) [40,41]. Three independent measurements were made, and the respective standard deviation (SD) was determined.

$$\eta = \frac{2.4\,r}{0.362 - r} \tag{4}$$

3. Results and Discussion

3.1. Magnetoplasmonic Nanoparticles' Characterization

Magneto-plasmonic nanoparticles, i.e., $Mg_{0.75}Ca_{0.25}Fe_2O_4$ nanocubes coupled with gold nanorods, were prepared. Magnesium/calcium ferrites have the advantage of enhanced biocompatibility and of forming cubic-shaped nanoparticles that possess better magnetic properties than the corresponding spherical nanoparticles [15], as already reported for shape-anisotropic nanoparticles of the same composition [17]. A comprehensive characterization of each particle was conducted, as well as the final nanostructure, in order to study individual properties and analyze the effectiveness of the coupling.

The XRD pattern of the magnesium ferrite MNPs with 25% replacement by calcium ions was processed by importing the CIF file with number 1011245 (space group 227), from the crystallographic open database (COD). Since this file is valid for $MgFe_2O_4$, adjustments to the unit cell composition were implemented to correspond with the actual structure of the NPs. More precisely, 25% of the crystallographic positions occupied by magnesium ions were considered to be the calcium ions' contribution. A variable inversion degree was assumed, considering that the distribution of Ca^{2+} and Mg^{2+} cations between tetrahedral and octahedral sites occurs so that the mixed ferrite stoichiometry is valid in both type of position. This led to a fit with χ^2 = 1.341, R_P = 12.31 with an inversion degree of 0.5. The obtained diffractogram (Figure 2a) provides evidence for the synthesis of $Mg_{0.75}Ca_{0.25}Fe_2O_4$ nanoparticles, revealing an average size of 41.3 nm, estimated by analyzing all the diffraction peaks, as implemented in BGMN model [38]. These peaks appeared at 2θ = 18.4° (1 1 1), 30.2° (2 2 0), 35.6° (3 1 1), 37.2° (2 2 2), 43.3° (4 0 0), 47.4° (3 3 1), 53.7° (4 2 2), 57.2° (5 1 1), 57.2° (3 3 3), 62.9° (4 4 0), 66.1° (5 3 1), 71.3° (6 2 0), 74.4° (5 3 3), 75.4° (6 2 2), 79.4° (4 4 4), 82.4° (7 1 1), 82.4° (5 5 1), 87.2° (6 4 2), 90.2° (7 3 1), 90.2° (5 5 3), 95.1° (8 0 0), 98.0° (7 3 3), corresponding to a lattice constant a = 8.392 Å.

Figure 2b,c present the absorption spectra of $Mg_{0.75}Ca_{0.25}Fe_2O_4$ nanoparticles and gold nanorods, respectively. The magnetic nanocube spectrum (Figure 2b) reveals a broad absorption, showing that the MNPs can absorb light at 808 nm (as highlighted by the red line). Absorption of $MgFe_2O_4$ in the NIR region was reported as arising from intra-atomic transitions among the 3d energy levels split by the crystal field [42]. This type of transitions is known to be forbidden for Fe^{3+} (d^5 high spin), but gains intensity through spin coupling interactions between Fe^{3+} atoms in close proximity [43]. At NIR wavelengths, the human tissues show minimal light absorption, which is a key feature for the main objective of this work. Several studies have documented a considerable photothermal potential exhibited by magnetic NPs under NIR laser irradiation [44–46], sparking significant interest and promise for the application of Ca/Mg ferrites in PTT. In fact, this aptitude has already been attested on different ferrite nanoparticles, being especially intensified in those based on alkaline earth metals [16,47,48].

Regarding the gold nanorods, the UV–Vis–NIR spectrum (Figure 2c) allows for conclusiio of their synthesis and the obtaining of information about their size and shape. Two absorption bands, the first around 500–550 nm and the second with a maximum at 880 nm, corresponding to the transverse plasmon and the longitudinal surface plasmon, respectively, are observed. Considering the position of the plasmonic bands, nanoparticles with a diameter of around 9 nm and length of 31 nm, are expected [34]. From these values (Figure 2c), gold nanorods with a diameter/length ratio of 3.5 are anticipated.

The coupling process of gold nanorods with $Mg_{0.75}Ca_{0.25}Fe_2O_4$ nanoparticles, allowing the obtaining of the magnetic and plasmonic components in a single nanosystem, was monitored by UV–Vis–NIR absorption spectroscopy. To these measurements, the gold solution was diluted to the same concentration that would be obtained in the supernatant solution, if no coupling has occurred. This solution was labeled as "diluted AuNR solution". The absorption spectra of this solution and of the supernatant containing the uncoupled AuNRs are presented in Figure 2d. From the obtained spectra, it was possible to conclude that the intensity of the bands of gold nanoparticles that did not couple with the magnetic nanoparticles is lower than that which would have been obtained if no coupling has happened. This is an indication that the coupling has effectively occurred. The spectrum of coupled nanoparticles is displayed in Figure 2e, evidencing the coupling by the very different relation between the transverse and longitudinal plasmon bands.

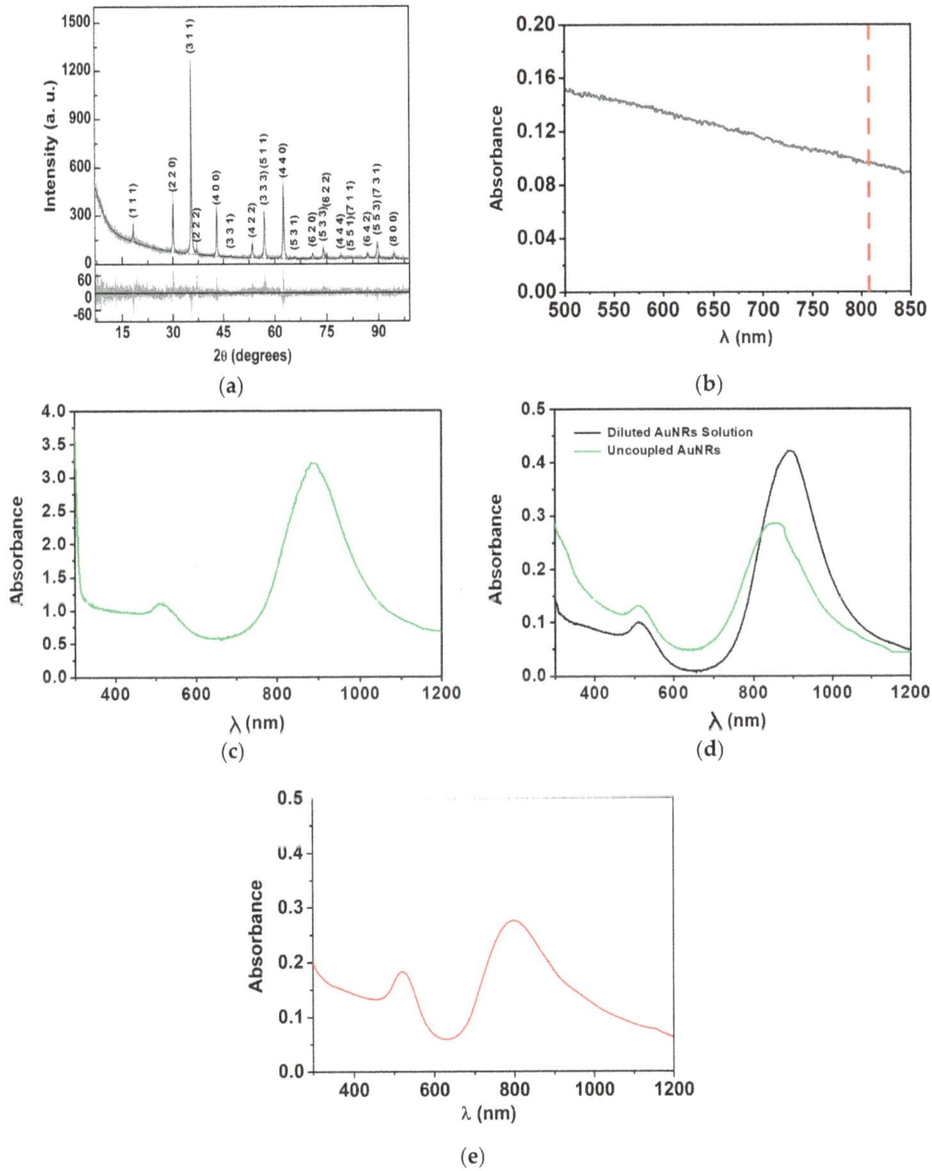

Figure 2. (**a**) X-ray diffractogram of $Mg_{0.75}Ca_{0.25}Fe_2O_4$ nanoparticles with the respective Rietveld analysis and Miller indices. The residuals of the fitting are displayed in the bottom panel. (**b**) UV-Vis–NIR absorption spectra of $Mg_{0.75}Ca_{0.25}Fe_2O_4$ nanoparticles dispersed in water (red line highlights absorption at 808 nm). (**c**) UV-Vis–NIR absorption spectra of gold nanorods dispersed in CTAB. (**d**) UV-Vis–NIR absorption spectra of diluted AuNRs solution and uncoupled AuNRs. (**e**) UV-Vis–NIR absorption spectrum of the coupled magneto-plasmonic nanoparticles.

The formation of magneto-plasmonic nanoparticles was assessed through the difference in maximum longitudinal plasmon absorbance of the initial gold nanorods used for coupling, designated as "diluted AuNR solution", and the maximum absorbance of the longitudinal plasmon of gold nanoparticles that did not bound to the magnetic nanoparticles, designated as "uncoupled AuNRs". Thus, the mass of bound gold was determined using Equation (5), corresponding to a coupling efficiency of 32%.

$$\frac{\text{Abs (diluted AuNRs solution)} - \text{Abs (uncoupled AuNRs)}}{\text{Abs (diluted AuNRs solution)}} \times \text{diluted AuNRs solution mass} \quad (5)$$

For accurate evaluation of the morphology and size distribution of magnetic and plasmonic NPs and of the coupling between them, a TEM characterization was performed. The obtained images are presented in Figure 3. The TEM images are in accordance with the expected, revealing the homogeneity of the prepared NPs and corroborating the successful coupling between the magnetic and plasmonic components, as predicted by the spectroscopic results. One the one hand, the mixed ferrite NPs generally exhibit a cubic shape (Figure 3a), with an average length of major and minor axes of 37 ± 3 nm and 31 ± 5 nm, respectively (Figure 3b), resulting in an aspect ratio ranging from 1.1 to 1.2. This value slightly above 1.0 reveals the presence of magnetic structures with a slight elongation, although some more spherical nanoparticles are also observed [17]. The results are in reasonable accordance with the average size estimated by XRD. One the other hand, the images obtained for plasmonic nanoparticles confirmed the formation of gold NPs in the shape of nanorods. Additionally, it is observed that these are uniform in shape and size. From Figure 3c, the distributions of longitudinal and transverse sizes of the nanorods were estimated as 46 ± 7 nm and 12 ± 1.6 nm, respectively. An aspect ratio of around 3.9 was calculated (Figure 3d), in general accordance with that obtained from UV–Vis–NIR absorption. This characterization was crucial to anticipate the geometry associated with the interaction of gold nanorods and magnetic nanoparticles. TEM images also allowed to confirm the formation of the magneto-plasmonic nanostructures. From Figure 3e, it is possible to observe the coupling of gold nanorods with magnetic nanoparticles.

The magnetic behavior of both MNPs and MPNPs was evaluated by SQUID, aiming to study the effect of the plasmonic component in the magnetic properties. Figure 4 shows the dependence of the magnetic moment (M) on the applied magnetic field (H), at room temperature. The coercive field (H_c), saturation magnetization (M_s), remnant magnetization (M_r) and the ratio between M_r/M_s (squareness value) were obtained and are summarized in Table 1.

(a)

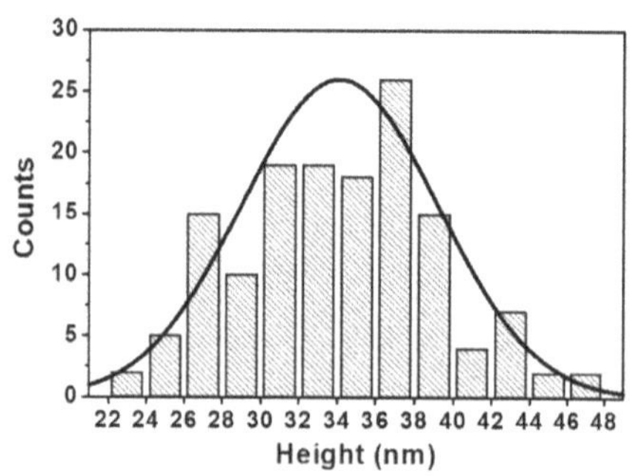

(b)

Figure 3. *Cont.*

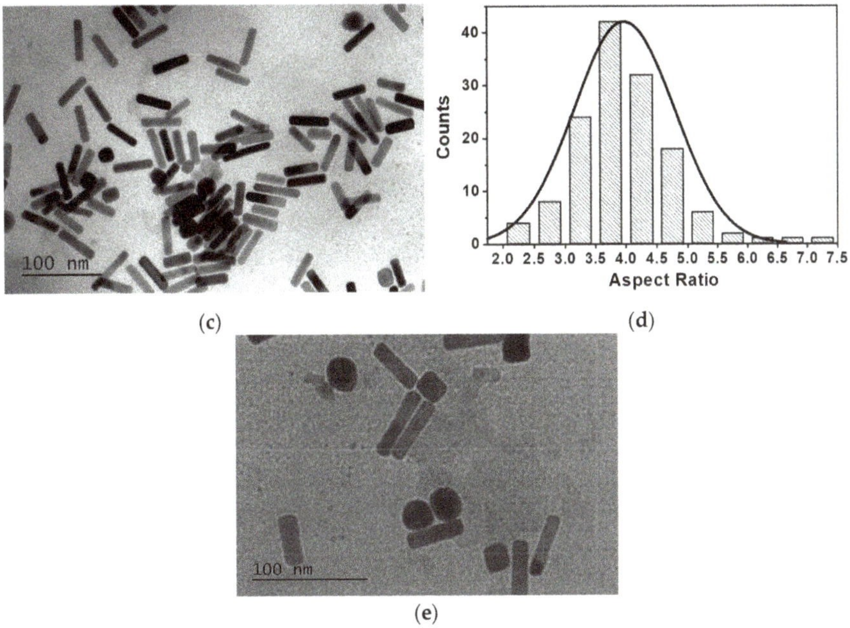

Figure 3. (**a**) TEM image of $Mg_{0.75}Ca_{0.25}Fe_2O_4$ nanocubes and (**b**) respective size histogram fitted to a Gaussian distribution. (**c**) TEM image of gold nanorods and (**d**) respective aspect ratio histogram fitted to a Gaussian distribution. (**e**) TEM image of magnetic nanoparticles coupled with plasmonic nanorods. Scale bar: 100 nm.

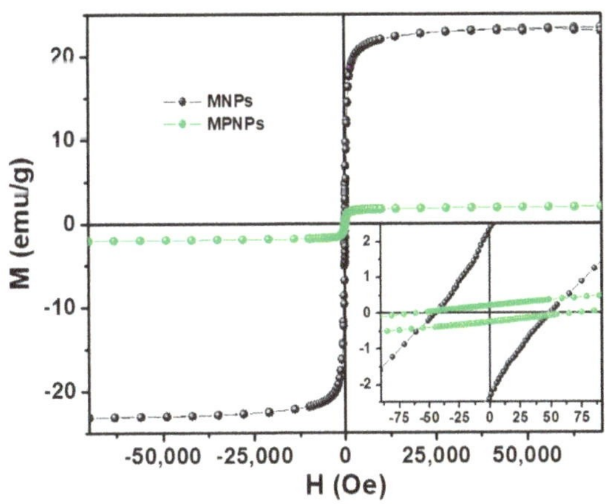

Figure 4. Magnetization hysteresis loops of $Mg_{0.75}Ca_{0.25}Fe_2O_4$ magnetic nanocubes (MNPs) before (grey) and after (green) coupling with gold nanorods (MPNPs), measured at T = 300 K. The inset provides a magnified view of the M-H loops in the low field region.

The obtained results clearly emphasize that the M–H curve of cubic-shaped magnetic NPs was significantly influenced after coupling with the gold nanorods due to their diamagnetic contribution. Although both hysteresis cycles display some attributes typical of particles with a superparamagnetic behavior, such as an almost closed loop and low remnant magnetization (2.4 and 0.23 emu/g for MNPs and MPNPs, respectively), the

M_s was highly compromised after the coupling with the plasmonic NPs. This parameter decreased from 23.5 to 2.2 emu/g, representing a reduction of more than tenfold. Even so, the squareness value of, approximately, 0.1 for both NPs points to a superparamagnetic nature at room temperature, indicating that 90% of the magnetization is lost upon the removal of the applied magnetic field [49].

Table 1. Coercive field, H_c (Oe), saturation magnetization, M_s (emu/g), remnant magnetization, M_r (emu/g), and M_r/M_s ratio for magnetic (MNPs) and magneto-plasmonic nanoparticles (MPNPs) at room temperature.

Samples	H_c (Oe)	M_s (emu/g)	M_r (emu/g)	M_r/M_s
MNPs	49.0	23.5	2.4	0.102
MPNPs	67.2	2.20	0.23	0.105

The photothermic potential of the prepared nanoparticles was assessed using NIR irradiation. In these measurements, each type of nanoparticle was evaluated separately. Thus, aqueous dispersions of neat MNPs, neat AuNRs and coupled MPNPs, at the same NP concentration (1 mg of NPs per mL of water), were prepared. For comparison, pure water was also tested to evaluate the heating of the medium in the absence of nanoparticles. The heating and cooling curves are shown in Figure 5.

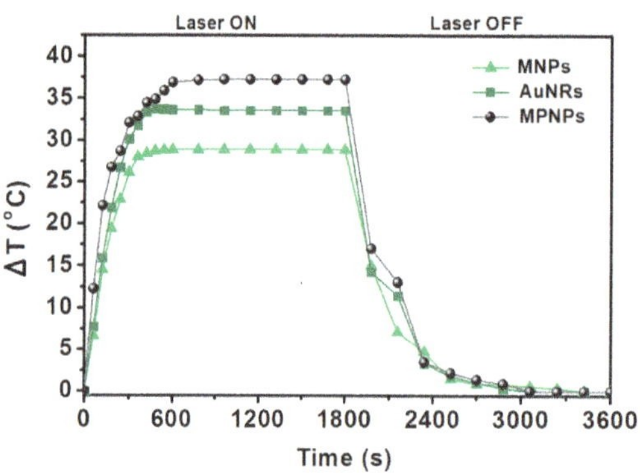

Figure 5. Heating profile of aqueous dispersions (1 mg/mL) of $Mg_{0.75}Ca_{0.25}Fe_2O_4$ magnetic nanocubes (MNPs), gold nanorods (AuNRs) and magnetic/plasmonic nanoparticles (MPNPs) under exposure to a laser light source with 808 nm wavelength and 1 W/cm^2 power density. ΔT is the increment from room temperature (20 °C).

SAR values were obtained using the initial slope method. The linear region of the curve covers the time during which the heating process occurs effectively, without any loss/exchange of heat with the environment, making it possible to extract the actual temperature increase caused by laser irradiation. The obtained results are summarized in Table 2.

Table 2. Specific absorption rate, SAR (W/g), for gold nanorods (AuNRs), $Mg_{0.75}Ca_{0.25}Fe_2O_4$ magnetic nanocubes (MNPs) and magneto-plasmonic nanoparticles (MPNPs), under a NIR laser light (808 nm wavelength and 1 W/cm^2 power density).

Sample	SAR (W/g)	ΔT (°C)
AuNRs	523.8 ± 6.7	33.6
MNPs	467.9 ± 21	28.9
MPNPs	680.7 ± 43	37.2

As expected, the gold nanorods showed better heating capacity compared to the magnetic nanoparticles, with corresponding SAR values of 523.8 W/g and 467.9 W/g, for AuNRs and MNPs, respectively. Regarding magnetic/plasmonic NPs, these exhibited a higher SAR when compared to individual components. This favorable behaviour shows the synergistic effect between the plasmonic and magnetic components in the nanosystem, potentiating the photothermal capability. The MPNP sample would still contain 68% of MNPs and some AuNRs arrested in the centrifugation pellet. However, the conclusion regarding the synergistic effect is the same and purification of MPNP samples would have resulted in an even higher SAR. This synergistic effect has been reported previously for gold nanorods and magnetite nanoparticles in a silica matrix [26,50], although a clear reasoning for such behavior was not presented and is still under investigation. One possibility is an increased intensity of the plasmon resonance in the NIR region (as can be observed in Figure 2e), if one takes into account the 32% coupling efficiency. This increase has been reported in theoretical calculations on related systems with a layer-by-layer geometry [51]. As $MgFe_2O_4$ is a known n-type semiconductor [52], another possibility is an efficient charge transfer from electrons in $MgFe_2O_4$ to gold, due to the close proximity of the two nanoparticles. Moreover, the maximum temperature variation is adequate for therapeutic action, since the temperature of photothermal hyperthermia (42–45 °C) is easily reached at physiological temperature.

3.2. *Magneto-Plasmonic Elastic Liposomes' Characterization*

3.2.1. Elastic Properties Investigation

The prepared nanoparticles were encapsulated in elastic liposomes to ensure the absorption of the nanoparticles into the skin. For topical applications, the thermosensitive phospholipid DPPC is preferred as the main component, in consequence of its low T_m (41 °C) [10]. To destabilize the vesicles and increase their deformability, SP60, SP65, SP80, SP85 and TW80 were used as EAs. Thus, the impact of the lipid composition on the degree of elasticity of the ELs was assessed using fluorescence anisotropy studies. For this, the membrane probe DPH was included into the lipid bilayer. This probe is widely used to monitor the membrane fluidity of biological systems, being an indirect reporter of their microviscosity [40]. Several formulations with different compositions were investigated regarding their micro-viscosity at 25 °C (room temperature) and 37 °C (body average temperature [53]). These temperature points were selected, as they fall below and above the average skin temperature range (between 33.5 °C and 36.9 °C) [54]. Considering that η is indirectly indicative of the degree of elasticity, it was assumed that the lower its value, the higher is the liposomes' fluidity and, consequently, their elasticity. Table 3 summarizes the results of the anisotropy experiments, using equation (4) for micro-viscosity estimation (considering that the variation of DPH fluorescence lifetime between 25 °C and 37 °C is very low [55] and could be neglected).

These data clearly indicate a correlation between the liposomes' microviscosity and the type of surfactant. Although to dissimilar degrees and excluding DPPC:TW80 (80:20) liposomes, it can be inferred that all the developed formulations exhibit elastic properties at 25 °C, as their fluidity increased in comparison to the system composed only by DPPC. Despite the variations observed in the three ratios, liposomes containing Tween-80 as edge activator are too viscous, potentially hindering their ability to cross the barriers of the skin. On the other hand, formulations composed by SP80 and SP85 have shown more potential as ELs, exhibiting the lowest η values at both temperatures. Considering the aim of this work, liposomes of DPPC:SP80 (85:15) were selected to encapsulate the MPNPs. This formulation stands out for its reduced microviscosity at 37 °C (1.3 ± 0.3 P), being that with the highest degree of elasticity. In this way, the loaded vesicles will be able to "squeeze" and pass through the skin pores by modulating its structure. In fact, Hussain et al. [9] proved that liposomes containing SP80 have an adequate vesicular size and high elasticity. Moreover, this formulation showed the highest drug permeation flux (89.74 ± 8.5 mg/cm^2/h) and proved to be able to deliver the drug 5-fluorouracil into the SC for enhanced skin delivery with minimal hemolysis. These results, together with those obtained here, point to the development of novel MPELs of DPPC:SP80 (85:15) loaded with magneto-plasmonic NPs.

Table 3. Fluorescence anisotropy, r, of DPH and micro-viscosity, η (P), of several lipid formulations at 25 °C and 37 °C.

		Temperature			
		25 °C		37 °C	
Formulation	**Ratio**	**r**	**η (P)**	**r**	**η (P)**
DPPC	---	0.277 ± 0.003	7.9 ± 0.3	0.236 ± 0.003	4.5 ± 0.2
DPPC:SP60	70:30	0.261 ± 0.015	6.4 ± 1.2	0.209 ± 0.020	3.3 ± 0.8
	80:20	0.249 ± 0.014	5.5 ± 1	0.242 ± 0.005	4.9 ± 0.3
	85:15	0.252 ± 0.005	5.5 ± 0.3	0.232 ± 0.014	4.4 ± 0.7
DPPC:SP65	70:30	0.222 ± 0.004	3.8 ± 0.2	0.197 ± 0.004	2.9 ± 0.1
	80:20	0.243 ± 0.003	4.9 ± 0.2	0.223 ± 0.019	3.9 ± 0.8
	85:15	0.230 ± 0.004	4.2 ± 0.2	0.226 ± 0.008	4.0 ± 0.5
DPPC:SP80	70:30	0.225 ± 0.015	4.0 ± 0.6	0.135 ± 0.002	1.4 ± 0.03
	80:20	0.248 ± 0.007	5.3 ± 0.5	0.144 ± 0.002	1.6 ± 0.04
	85:15 *	0.218 ± 0.003	3.63 ± 0.01	0.125 ± 0.018	1.3 ± 0.3
DPPC:SP85	70:30	0.207 ± 0.005	3.2 ± 0.2	0.112 ± 0.004	1.6 ± 0.05
	80:20	0.230 ± 0.010	4.2 ± 0.5	0.107 ± 0.002	1.3 ± 0.03
	85:15	0.231 ± 0.018	4.4 ± 0.9	0.135 ± 0.024	1.5 ± 0.4
DPPC:TW80	70:30	0.274 ± 0.006	7.5 ± 0.7	0.251 ± 0.002	5.4 ± 0.1
	80:20	0.282 ± 0.003	8.5 ± 0.4	0.236 ± 0.011	4.5 ± 0.5
	85:15	0.269 ± 0.015	7.2 ± 1	0.224 ± 0.008	3.9 ± 0.4

* Selected formulation.

3.2.2. Study of the Phase Transition Temperature of MPELs

As mentioned in Section 1, it is important to use edge activators as an ingredient to prepare ELs, considering that these play a crucial role in decreasing the phase transition temperature of typical liposomes and conferring them elastic properties. In this context, anisotropy studies were conducted in order to identify the transition temperature of the nanosystems and understand the influence of EAs on this parameter. For this assay, DPH fluorescence anisotropy was used to predict the properties of lipid bilayers at different temperatures (Figure 6).

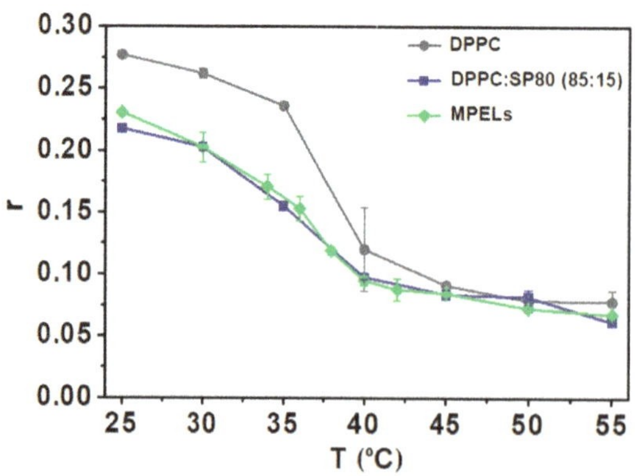

Figure 6. Temperature dependence of fluorescence anisotropy (r) of DPH incorporated into liposomes of DPPC:SP80 (85:15) and magneto-plasmonic elastic liposomes (MPELs). DPPC liposomes are shown for comparison.

Liposomes composed only of DPPC exhibit a T_m around 41 °C, as previously reported [40,55]. Similar results were obtained for fluorescence anisotropy of DPH in neat DPPC liposomes, as reported by Pereira et al. [56]. It is particularly highlighted that the inclusion of Span-80 into lipid vesicles is actually reflected in a notable decrease of the probe's fluorescence anisotropy, which indicates an increase in membrane fluidity. This observation validates the effectiveness of SP80 as edge activator in reducing the melting temperature of liposomes and, consequently, in promoting their elastic nature. Regarding MPELs, it is concluded that magneto-plasmonic NPs did not induce any significant changes in the system's phase transition temperature, of around 35 °C (inflexion point of the curve). As the liposomes' T_m should be localized in the range 25 °C to 35 °C, an improved permeation flux is assured when in direct contact with the skin. At this temperature, MPELs will adopt a more fluidic structure (liquid crystal state) capable of crossing the SC through the intercellular route, ensuring the "same time at same place" therapeutic strategy.

The recognition of the system's phase transition temperature is also of major interest to understand its role as mediator of cellular damage via phototherapy. For this type of therapeutic modality, temperatures in the range of 42 °C to 45 °C are generally effective in starting to accelerate biochemical reactions, leading to protein denaturation and production of reactive oxygen species (ROS) [57]. The T_m of the developed magneto-plasmonic liposomes suggests that these cytotoxic episodes would be triggered only upon penetration into the skin layers. It is expected that, when in contact with the skin at 35 °C, the fluidity of MPELs increases, allowing them to cross this barrier. By focusing a NIR laser light in the target site, a localized temperature rise would be promoted and, upon reaching that thermal interval, apoptotic events should be exploited [57]. Additionally, a thermosensitive nano-system with this T_m would be capable of retaining drugs at physiological conditions and selectively release their content at the target location only after penetration and reaching hyperthermic temperatures.

3.2.3. MPELs' Structural Characterization and Stability

Here, magneto-plasmonic elastic liposomes were prepared for the first time using the conventional ethanolic injection method. This technique was chosen for its easy synthesis, while producing homogeneous, monodisperse and stable solutions of small uni-lamellar vesicles without the need of additional steps as extrusion, that can lead to the liposomes' degradation and oxidation [58]. The TEM image of MPELs (Figure 7) shows structures with sizes of around 95 nm, with a dark core (the magneto-plasmonic core after the application of vacuum) surrounded by a thin layer (the lipid bilayer). The polydispersity index (PDI) below 0.3 (0.25 ± 0.015), measured by DLS, indicates generally monodisperse structures. For lipid-based nanocarriers, suspensions with standard PDI values below 0.3 are considered acceptable for biomedical applications [59]. MPELs size between 85 and 95 nm are in the defined range for biomedical applications (~50–200 nm) [60], making them promising candidates for the main objective of this work. Moreover, the liposomes' zeta potential (ζ), before and after the magneto-plasmonic NPs incorporation (−1.18 ± 0.157 eV for ELs and +19.56 ± 1.039 eV for MPELs), reveal that being a non-ionic surfactant SP80 did not cause changes in the near-neutral charge of the liposomes conferred by the zwitterionic lipid DPPC. However, the stability of the nanosystem was modified by adding the hyperthermia agent. The positive surface charge obtained in magneto-plasmonic elastic liposomes is a result of the capping agent used in the preparation of AuNRs. CTAB is a cationic surfactant that acts in the stabilization of nanorods during their synthesis forming a bilayer structure around the particles [61]. In addition, the obtained positive charge suggests that the magneto-plasmonic NPs must be located close to the surface of the liposomes.

A stability study was performed to determine whether these magneto-plasmonic elastic liposomes satisfy the requirements for commercial and clinical outcome. For that purpose, mean size and PDI variations were monitored over a 20-day storage period at 4 °C (Figure 8); considerable changes in these domains reveal signs of either aggregation or membrane rupture [62]. During this time, the MPELs size presented minimal shifts from the day 1 value (D_0 = 105.9 ± 4 nm; D_{20} = 106.7 ± 3 nm), suggesting that these should remain stable in long-term storage. In addition, no significant PDI deviations were found. Still, the MPELs' suspension shows an increasing pattern in that index, starting from the 8th day onward. With this background, the developed nanostructure is expected to, when in contact with the skin, easily adjust its shape and be rapidly absorbed, permeating the SC up to, at least, 20 days after synthesis.

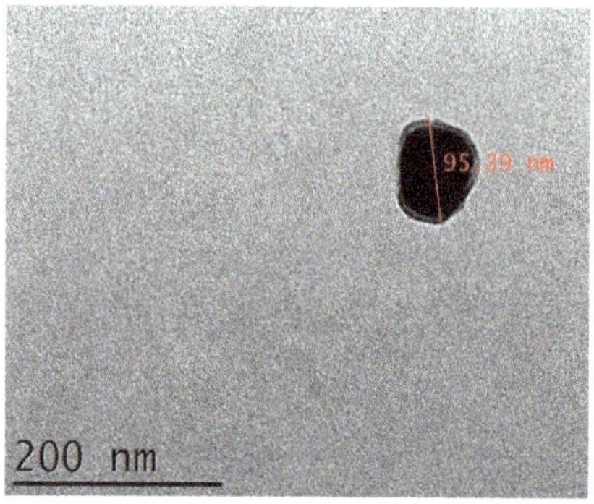

Figure 7. TEM image of DPPC:SP80 (85:15) magneto-plasmonic elastic liposomes (scale bar: 200 nm).

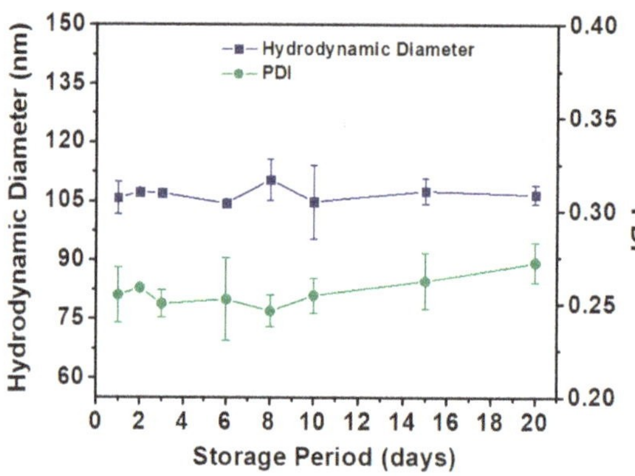

Figure 8. Stability of MPELs aqueous solution stored at 4 °C, represented as the variation of its hydrodynamic diameter and PDI over a 20-day period.

3.2.4. MPELs as Photothermal Hyperthermia Agents

The photothermal potential of the novel MPELs here developed was assessed under laser exposure at 808 nm. This specific wavelength is frequently employed in such experiments, as it falls within the first near-infrared (NIR-I) biological window (700–900 nm), i.e., spectral wavelengths capable of penetrating biological tissues with minimum absorbance and scattering [16]. Under these conditions, the ability for energy dissipation of magnetic/plasmonic NPs enclosed in elastic liposomes of DPPC:SP80 (85:15) was assessed (for the same concentration of MPNPs used before). The equivalent SAR and temperature variation (above room temperature), 616.9 ± 32 W/g and 33.5 °C, respectively (Figure 9), lead to a conclusion that the final nanostructure exhibits appealing features for performing efficient photothermal hyperthermia to treat topical pathologies.

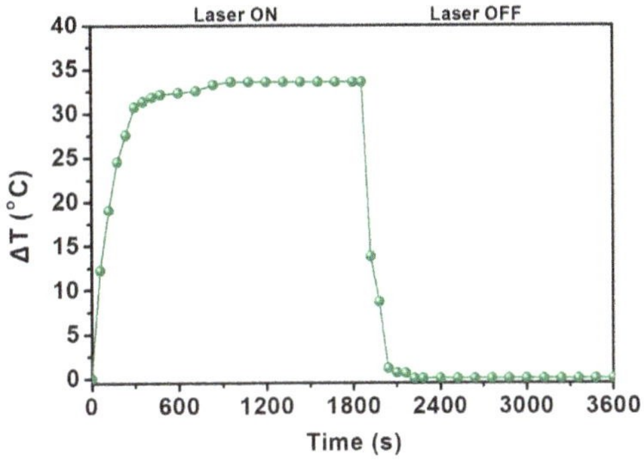

(a)

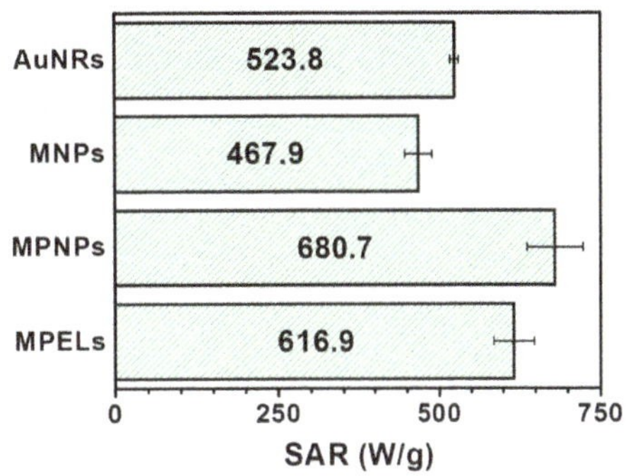

(b)

Figure 9. (**a**) Heating profile of an aqueous solution of magneto-plasmonic elastic liposomes under exposure to a laser light source with 808 nm wavelength and 1 W/cm^2 power density. ΔT is the increment from room temperature (20 °C). (**b**) Histogram of specific absorption rate, SAR (W/g), of gold nanorods (AuNRs), $Mg_{0.75}Ca_{0.25}Fe_2O_4$ magnetic nanocubes (MNPs), magnetic/plasmonic nanoparticles (MPNPs) and magneto-plasmonic elastic liposomes (MPELs).

4. Conclusions

This research aimed to design and prepare an innovative nanoplatform with ultra-deformable properties for applications in skin cancer diseases by photothermal hyperthermia. For that purpose, gold nanorods and cubic-shaped magnetic nanoparticles of magnesium ferrite with 25% replacement by calcium ions were synthetized and characterized through XRD, UV-Vis–NIR spectroscopy, TEM and SQUID. TEM images validated the synthesis, coupling and encapsulation of magneto-plasmonic nanoparticles into DPPC:SP80 (85:15) thermoelastic liposomes, with suitable properties for biomedical applications. Photothermal hyperthermia measurements under a NIR laser light demonstrated an encouraging prospect of MPELs in addressing topical disease conditions (SAR = 617 W/g) Therefore, this work provides valuable insights into the potential of the developed nanosystems for future potential application in skin cancer.

To the best of our knowledge, this is the first time that magneto-plasmonic elastic liposomes containing Ca/Mg ferrites coupled to gold nanorods were developed and characterized, and their photothermal capability evaluated.

Author Contributions: Conceptualization, A.R.O.R., P.J.G.C. and E.M.S.C.; methodology, A.R.F.P., A.M.B. and A.R.O.R.; validation, V.S.A., P.J.G.C., A.R.O.R. and E.M.S.C.; formal analysis, A.R.F.P., A.M.B., C.O.A., P.J.G.C. and A.R.O.R.; investigation, A.R.F.P., A.M.B. and C.O.A.; supervision, A.R.O.R., V.S.A., P.J.G.C. and E.M.S.C.; writing—original draft preparation, A.R.F.P. and A.M.B.; writing—review and editing, A.R.F.P., A.R.O.R., P.J.G.C. and E.M.S.C.; project administration, P.J.G.C. All authors have read and agreed to the published version of the manuscript.

Funding: This work was funded by the Portuguese Foundation for Science and Technology (FCT) in the framework of the Strategic Funding of CF-UM-UP: UIDB/04650/2020 (DOI: 10.54499/UIDB/04650/2020) and UIDP/04650/2020 (DOI: 10.54499/UIDP/04650/2020); of CICECO Aveiro Institute of Materials: UIDB/50011/2020 (DOI: 10.54499/UIDB/50011/2020), UIDP/50011/2020 (DOI: 10.54499/UIDP/50011/2020), and LA/P/0006/2020 (DOI:10.54499/ LA/P/0006/2020); and of Associate Laboratory Institute of Nanostructures, Nanomodelling and Nanofabrication—i3N: LISBOA-01–0247-FEDER-039985/POCI-01–0247-FEDER-039985, LA/P/0037/2020, UIDP/50025/2020 and UIDB/50025/2020.

Data Availability Statement: Data available upon request.

Conflicts of Interest: The authors declare no conflicts of interest.

References

1. Antimisiaris, S.G.; Marazioti, A.; Kannavou, M.; Natsaridis, E.; Gkartziou, F. Overcoming barriers by local drug delivery with liposomes. *Adv. Drug Deliv. Rev.* **2021**, *174*, 53–86. [CrossRef] [PubMed]
2. Souto, E.S.; Macedo, A.S.; Dias-Ferreira, J.; Cano, A.; Zielińska, A.; Matos, C.M. Elastic and Ultradeformable Liposomes for Transdermal Delivery of Active Pharmaceutical Ingredients (APIs). *Int. J. Mol. Sci.* **2021**, *22*, 9743. [CrossRef] [PubMed]
3. World Health Organization. Epidemiology and Management of Common Skin Diseases in Children in Developing Countries. Available online: https://apps.who.int/iris/bitstream/handle/10665/69229/WHO_FCH_CAH_05.12_eng.pdf (accessed on 7 August 2023).
4. Dar, J.M.; Khalid, S.; Varikuti, S.; Satoskar, A.R.; Khan, G.M. Nano-elastic liposomes as multidrug carrier of sodium stibogluconate and ketoconazole: A potential new approach for the topical treatment of cutaneous Leishmaniasis. *Eur. J. Pharm. Sci.* **2020**, *145*, 105256. [CrossRef] [PubMed]
5. Sadabad, R.K.; Xia, A.; Benkafadar, N.; Faniku, C.; Preciado, D.; Yang, S.; Valdez, T.A. Topical Delivery of Elastic Liposomal Vesicles for Treatment of Middle and Inner Ear Diseases. *ACS Appl. Bio Mater.* **2022**, *5*, 4849–4859. [CrossRef] [PubMed]
6. Hussain, A.; Haque, M.W.; Singh, S.K.; Ahmed, F.J. Optimized permeation enhancer for topical delivery of 5-fluorouracil-loaded elastic liposome using Design Expert: Part II. *Drug Deliv.* **2016**, *23*, 1242–1253. [CrossRef] [PubMed]
7. Carter, P.; Narasimhan, B.; Wang, Q. Biocompatible nanoparticles and vesicular systems in transdermal drug delivery for various skin diseases. *Int. J. Pharm.* **2019**, *555*, 49–62. [CrossRef] [PubMed]
8. Chen, J.; Lu, W.L.; Gu, W.; Lu, S.S.; Chen, Z.P.; Cai, B.C. Skin permeation behavior of elastic liposomes: Role of formulation ingredients. *Expert Opin. Drug Deliv.* **2013**, *10*, 845–856. [CrossRef] [PubMed]
9. Hussain, A.; Singh, S.; Sharma, D.; Webster, T.J.; Shafaat, K.; Faruk, A. Elastic liposomes as novel carriers: Recent advances in drug delivery. *Int. J. Nanomedicine* **2017**, *12*, 5087–5108. [CrossRef] [PubMed]
10. Demel, R.A.; De Kruyff, B. The function of sterols in membranes. *Biochim. Biophys. Acta BBA-Rev. Biomembr.* **1976**, *457*, 109–132. [CrossRef]
11. Bierman, W. The temperature of the skin surface. *J. Am. Med. Assoc.* **1936**, *106*, 1158. [CrossRef]
12. Materón, E.M.; Miyazaki, C.M.; Carr, O.; Joshi, N.; Picciani, P.H.S.; Dalmaschio, C.J.; Davis, F.; Shimizu, F.M. Magnetic nanoparticles in biomedical applications: A review. *Appl. Surf. Sci. Adv.* **2021**, *6*, 100163. [CrossRef]
13. Manohar, A.; Vijayakanth, V.; Vattikuti, S.V.P.; Manivasagan, P.; Jang, E.S.; Chintagumpala, K.; Kim, K.H. Ca-Doped $MgFe_2O_4$ Nanoparticles for Magnetic Hyperthermia and Their Cytotoxicity in Normal and Cancer Cell Lines. *ACS Appl. Nano Mater.* **2022**, *5*, 5847–5856. [CrossRef]
14. Hirazawa, H.; Kusamoto, S.; Aono, H.; Naohara, T.; Mori, K.; Hattori, Y.; Maehara, T.; Watanabe, Y. Preparation of fine $Mg_{1-X}Ca_XFe_2O_4$ powder using reverse coprecipitation method for thermal coagulation therapy in an ac magnetic field. *J. Alloys Compd.* **2008**, *461*, 467–473. [CrossRef]
15. Cardoso, B.D.; Rodrigues, A.R.O.; Almeida, B.G.; Amorim, C.O.; Amaral, V.S.; Castanheira, E.M.S.; Coutinho, P.J.G. Stealth Magnetoliposomes Based on Calcium-Substituted Magnesium Ferrite Nanoparticles for Curcumin Transport and Release. *Int. J. Mol. Sci.* **2020**, *21*, 3641. [CrossRef] [PubMed]
16. Cardoso, B.D.; Fernandes, D.E.M.; Amorim, C.O.; Amaral, V.S.; Coutinho, P.J.G.; Rodrigues, A.R.O.; Castanheira, E.M.S. Magnetoliposomes with Calcium-Doped Magnesium Ferrites Anchored in the Lipid Surface for Enhanced DOX Release. *Nanomaterials* **2023**, *13*, 2597. [CrossRef] [PubMed]
17. Cardoso, B.D.; Rodrigues, A.R.O.; Bañobre-López, M.; Almeida, B.G.; Amorim, C.O.; Amaral, V.S.; Coutinho, P.J.G.; Castanheira, E.M.S. Magnetoliposomes Based on Shape Anisotropic Calcium/Magnesium Ferrite Nanoparticles as Nanocarriers for Doxorubicin. *Pharmaceutics* **2021**, *13*, 1248. [CrossRef] [PubMed]
18. Nikitin, A.; Khramtsov, M.; Garanina, A.; Mogilnikov, P.; Sviridenkova, N.; Shchetinin, I.; Savchenko, A.; Abakumov, M.; Majouga, A. Synthesis of iron oxide nanorods for enhanced magnetic hyperthermia. *J. Magn. Magn. Mater.* **2019**, *469*, 443–449. [CrossRef]
19. Mona, L.P.; Songca, S.P.; Ajibade, P.A. Synthesis and encapsulation of iron oxide nanorods for application in magnetic hyperthermia and photothermal therapy. *Nanotechnol. Rev.* **2021**, *11*, 176–190. [CrossRef]
20. Lisjak, D.; Mertelj, A. Anisotropic magnetic nanoparticles: A review of their properties, syntheses and potential applications. *Prog. Mater. Sci.* **2018**, *95*, 286–328. [CrossRef]
21. Yougbaré, S.; Mutalik, C.; Chung, P.F.; Krisnawati, D.I.; Rinawati, F.; Irawan, H.; Kristanto, H.; Kuo, T.R. Gold Nanorod-Decorated Metallic MoS_2 Nanosheets for Synergistic Photothermal and Photodynamic Antibacterial Therapy. *Nanomaterials* **2021**, *11*, 3064. [CrossRef]
22. Pakravan, A.; Salehi, R.; Mahkam, M. Comparison study on the effect of gold nanoparticles shape in the forms of star, hallow, cage, rods, and Si-Au and Fe-Au core-shell on photothermal cancer treatment. *Photodiagnosis Photodyn. Ther.* **2021**, *33*, 102144. [CrossRef] [PubMed]
23. Joe, A.; Han, H.W.; Lim, Y.R.; Manivasagan, P.; Jang, E.S. Triphenylphosphonium-Functionalized Gold Nanorod/Zinc Oxide Core–Shell Nanocomposites for Mitochondrial-Targeted Phototherapy. *Pharmaceutics* **2024**, *16*, 284. [CrossRef] [PubMed]
24. Liao, S.; Yue, W.; Cai, S.; Tang, Q.; Lu, W.; Huang, L.; Qi, T.; Liao, J. Improvement of Gold Nanorods in Photothermal Therapy: Recent Progress and Perspective. *Front. Pharmacol.* **2021**, *12*, 664123. [CrossRef] [PubMed]
25. Zheng, J.; Cheng, X.; Zhang, H.; Bai, X.; Ai, R.; Shao, L.; Wang, J. Gold Nanorods: The Most Versatile Plasmonic Nanoparticles. *Chem. Rev.* **2021**, *121*, 13342–13453. [CrossRef] [PubMed]

26. Ovejero, J.G.; Casero, I.M.; de la Presa, P.; Mille, N.; Carrey, J.; Garcia, M.A.; Hernando, A.; Herrasti, P. Hybrid nanoparticles for magnetic and plasmonic hyperthermia. *Phys. Chem. Chem. Phys.* **2018**, *20*, 24065–24073. [CrossRef] [PubMed]
27. Rodrigues, A.R.O.; Santos, L.C.A.; Macedo, D.O.; Rio, I.S.R.; Pires, A.; Pereira, A.M.; Araújo, J.P.; Castanheira, E.M.S.; Coutinho, P.J.G. Plasmonic/magnetic liposomes based on nanoparticles with multicore-shell architecture for chemo/thermotherapy. *J. Phys. D Conf. Ser.* **2022**, *2407*, 012051. [CrossRef]
28. Sood, A.; Arora, V.; Shah, J.; Kotnala, R.K.; Jain, T.K. Multifunctional gold coated iron oxide core-shell nanoparticles stabilized using thiolated sodium alginate for biomedical applications. *Mater. Sci. Eng. C* **2017**, *80*, 274–281. [CrossRef] [PubMed]
29. Rodrigues, A.R.O.; Matos, J.O.G.; Nova Dias, A.M.; Almeida, B.G.; Pires, A.; Pereira, A.M.; Araújo, J.P.; Queiroz, M.J.R.P.; Castanheira, E.M.S.; Coutinho, P.J.G. Development of multifunctional liposomes containing magnetic/plasmonic $MnFe_2O_4$/Au core/shell nanoparticles. *Pharmaceutics* **2019**, *11*, 10. [CrossRef] [PubMed]
30. Wang, H.; Mu, Q.; Revia, R.; Wang, K.; Tian, B.; Lin, G.; Lee, W.; Hong, Y.-K.; Zhang, M. Iron oxide-carbon core-shell nanoparticles for dual-modal imaging-guided photothermal therapy. *J. Control. Release* **2018**, *289*, 70–78. [CrossRef]
31. Pacheco, A.R.F.; Cardoso, B.D.; Pires, A.; Pereira, A.M.; Araújo, J.P.; Carvalho, V.M.; Rodrigues, R.O.; Coutinho, P.J.G.; Castelo-Grande, T.; Augusto, P.A.; et al. Development of pH-Sensitive Magnetoliposomes Containing Shape Anisotropic Nanoparticles for Potential Application in Combined Cancer Therapy. *Nanomaterials* **2023**, *13*, 1051. [CrossRef]
32. Wang, Y.; Yang, H. Oleic acid as the capping agent in the synthesis of noble metal nanoparticles in imidazolium-based ionic liquids. *Chem. Commun.* **2006**, *24*, 2545–2547. [CrossRef] [PubMed]
33. López-Ramón, M.V.; Moreno-Castilla, C.; Fontecha-Cámara, M.A. Removal of the oleylamine capping agent from $MnFe_2O_4$ hollow spheres prepared by an Ostwald ripening mechanism. *Appl. Surf. Sci.* **2023**, *612*, 155796. [CrossRef]
34. Salavatov, N.A.; Dement'eva, O.V.; Mikhailichenko, A.I.; Rudoy, V.M. Some Aspects of Seedless Synthesis of Gold Nanorods. *Colloid J.* **2018**, *80*, 541–549. [CrossRef]
35. Ribeiro, B.C.; Alvarez, C.A.R.; Alves, B.C.; Rodrigues, J.M.; Queiroz, M.J.R.P.; Almeida, B.G.; Pires, A.; Pereira, A.M.; Araújo, J.P.; Coutinho, P.J.G.; et al. Development of Thermo- and pH-Sensitive Liposomal Magnetic Carriers for New Potential Antitumor Thienopyridine Derivatives. *Materials* **2022**, *15*, 1737. [CrossRef] [PubMed]
36. Zhang, S.; Niu, H.; Zhang, Y.; Liu, J.; Shi, Y.; Zhang, X.; Cai, Y. Biocompatible phosphatidylcholine bilayer coated on magnetic nanoparticles and their application in the extraction of several polycyclic aromatic hydrocarbons from environmental water and milk samples. *J. Chromatogr. A* **2012**, *1238*, 38–45. [CrossRef] [PubMed]
37. Doebelin, N.; Kleeberg, R. Profex: A graphical user interface for the Rietveld refinement program BGMN. *J. Appl. Crystallogr.* **2015**, *48*, 1573–1580. [CrossRef] [PubMed]
38. Bergmann, J.; Friedel, P.; Kleeberg, R. IUCr Commission on Powder Diffraction. *Newsletter* **1998**, *20*, 5–8.
39. Lanier, O.L.; Korotych, O.I.; Monsalve, A.G.; Wable, D.; Savliwala, S.; Grooms, N.W.F.; Nacea, C.; Tuitt, O.R.; Dobson, J. Evaluation of magnetic nanoparticles for magnetic fluid hyperthermia. *Int. J. Hyperth.* **2019**, *36*, 686–700. [CrossRef] [PubMed]
40. Poojari, C.; Wilkosz, N.; Lira, R.B.; Dimova, R.; Jurkiewicz, P.; Petka, R.; Kepczynski, M.; Róg, T. Behavior of the DPH fluorescence probe in membranes perturbed by drugs. *Chem. Phys. Lipids* **2019**, *223*, 104784. [CrossRef]
41. Pandey, B.N.; Mishra, K.P. Radiation induced oxidative damage modification by cholesterol in liposomal membrane. *Radiat. Phys. Chem.* **1999**, *54*, 481–489. [CrossRef]
42. Pailhé, N.; Wattiaux, A.; Gaudon, M.; Demourgues, A. Correlation between structural features and vis–NIR spectra of α-Fe_2O_3 hematite and AFe_2O_4 spinel oxides (A = Mg, Zn). *J. Solid State Chem.* **2008**, *181*, 1040. [CrossRef]
43. Granone, L.I.; Ulpe, A.C.; Robben, L.; Klimke, S.; Jahns, M.; Renz, F.; Gesing, T.M.; Bredow, T.; Dillert, R.; Bahnemann, D.W. Effect of the degree of inversion on optical properties of spinel $ZnFe_2O_4$. *Phys. Chem. Chem. Phys.* **2018**, *20*, 28267. [CrossRef] [PubMed]
44. Peng, H.; Tang, J.; Zheng, R.; Guo, G.; Donga, A.; Wang, Y.; Yang, W. Nuclear-Targeted Multifunctional Magnetic Nanoparticles for Photothermal Therapy. *Adv. Healthc. Mater.* **2017**, *6*, 1601289. [CrossRef] [PubMed]
45. Zhang, F.; Lu, G.; Wen, X.; Li, F.; Ji, X.; Li, Q.; Wu, M.; Cheng, Q.; Yu, Y.; Tang, J.; et al. Magnetic nanoparticles coated with polyphenols for spatio-temporally controlled cancer photothermal/immunotherapy. *J. Controlled Release* **2020**, *326*, 131–139. [CrossRef] [PubMed]
46. Xing, M.; Mohapatra, J.; Beatty, J.; Elkins, J.; Pandey, N.K.; Chalise, A.; Chen, W.; Jin, M.; Liu, J.P. Iron-based magnetic nanoparticles for multimodal hyperthermia heating. *J. Alloys Compd.* **2021**, *871*, 159475. [CrossRef]
47. Qiu, E.; Chen, X.; Yang, D.P.; Regulacio, M.D.; Ramos, R.M.C.R.; Luo, Z.; Wu, Y.L.; Lin, M.; Li, Z.; Loh, X.J.; et al. Fabricating Dual-Functional Plasmonic–Magnetic Au@$MgFe_2O_4$ Nanohybrids for Photothermal Therapy and Magnetic Resonance Imaging. *ACS Omega* **2022**, *7*, 2031–2040. [CrossRef] [PubMed]
48. Manivasagan, P.; Ashokkumar, S.; Manohar, A.; Joe, A.; Han, H.W.; Seo, S.H.; Thambi, T.; Duong, H.S.; Kaushik, N.K.; Kim, K.H.; et al. Biocompatible Calcium Ion-Doped Magnesium Ferrite Nanoparticles as a New Family of Photothermal Therapeutic Materials for Cancer Treatment. *Pharmaceutics* **2023**, *15*, 1555. [CrossRef] [PubMed]
49. Rodrigues, A.R.O.; Ramos, J.M.F.; Gomes, I.T.; Almeida, B.G.; Araújo, J.P.; Queiroz, M.J.R.P.; Coutinho, P.J.G.; Castanheira, E.M.S. Magnetoliposomes based on manganese ferrite nanoparticles as nanocarriers for antitumor drugs. *RSC Adv.* **2016**, *6*, 17302–17313. [CrossRef]
50. Encarnación, C.; Jungwirth, F.; Vila-Liarte, D.; Renero-Lecuna, C.; Kavak, S.; Orue, I.; Wilhelm, C.; Bals, S.; Henriksen-Lacey, M.; Aberasturi, D.J.; et al. Hybrid Core–Shell Nanoparticles for Cell-Specific Magnetic Separation and Photothermal Heating. *J. Mater. Chem. B* **2023**, *11*, 5574–5585. [CrossRef]

51. Brullot, W.; Verbiest, T. Optical Properties of Magnetic-Plasmonic Nanoparticle Multilayers. *Proc. SPIE* **2014**, *9163*, 362–374. [CrossRef]
52. Guo, J.; Shi, L.; Wu, L.; Pan, S.; Yuan, X.; Zhao, J. Spin-Polarized Electron Transport in Highly Reduced $MgFe_2O_4$-δ. *Mater. Res. Express* **2018**, *5*, 126301. [CrossRef]
53. Cabanac, M.; Massonnet, B. Thermoregulatory responses as a function of core temperature in humans. *J. Physiol.* **1977**, *265*, 587–596. [CrossRef] [PubMed]
54. Constantinou, A.; Oikonomou, S.; Konstantinou, C.; Makris, K.C. A randomized cross-over trial investigating differences in 24-h personal air and skin temperatures using wearable sensors between two climatologically contrasting settings. *Sci. Rep.* **2021**, *11*, 22020. [CrossRef] [PubMed]
55. Lentz, B.R. Membrane "fluidity" as detected by diphenylhexatriene probes. *Chem. Phys. Lipids* **1989**, *50*, 171–190. [CrossRef]
56. Pereira, M.; Rodrigues, A.R.O.; Amaral, L.; Côrte-Real, M.; Santos-Pereira, C.; Castanheira, E.M.S. Bovine Lactoferrin-Loaded Plasmonic Magnetoliposomes for Antifungal Therapeutic Applications. *Pharmaceutics* **2023**, *15*, 2162. [CrossRef]
57. Xie, Z.; Fan, T.; An, J.; Choi, W.; Duo, Y.; Ge, Y.; Zhang, B.; Nie, G.; Xie, N.; Zheng, T.; et al. Emerging combination strategies with phototherapy in cancer nanomedicine. *Chem. Soc. Rev.* **2020**, *49*, 8065–8087. [CrossRef]
58. Gouda, A.; Sakr, O.S.; Nasr, M.; Sammour, O. Ethanol injection technique for liposomes formulation: An insight into development, influencing factors, challenges and applications. *J. Drug Deliv. Sci. Technol.* **2021**, *61*, 102174. [CrossRef]
59. Danaei, M.; Dehghankhold, M.; Ataei, S.; Davarani, F.H.; Javanmard, R.; Dokhani, A.; Khorasani, S.; Mozafari, M.R. Impact of Particle Size and Polydispersity Index on the Clinical Applications of Lipidic Nanocarrier Systems. *Pharmaceutics* **2018**, *10*, 57. [CrossRef]
60. Woodle, M.C.; Lasic, D.D. Sterically stabilized liposomes. *Biochim. Biophys. Acta BBA-Rev. Biomembr.* **1992**, *1113*, 171–199. [CrossRef]
61. Shi, X.; Perry, H.L.; Wilton-Ely, J.D.E.T. Strategies for the functionalization of gold nanorods to reduce toxicity and aid clinical translation. *Nanotheranostics* **2021**, *5*, 155–165. [CrossRef]
62. Teixeira, P.V.; Adega, F.; Martins-Lopes, P.; Machado, R.; Lopes, C.C.; Lúcio, M. pH-Responsive Hybrid Nanoassemblies for Cancer Treatment: Formulation Development, Optimization, and In Vitro Therapeutic Performance. *Pharmaceutics* **2023**, *15*, 326. [CrossRef] [PubMed]

nanomaterials

MDPI

Article

Utilizing Constant Energy Difference between sp-Peak and C 1s Core Level in Photoelectron Spectra for Unambiguous Identification and Quantification of Diamond Phase in Nanodiamonds

Oleksandr Romanyuk [1,*], Štěpán Stehlík [1,2], Josef Zemek [1], Kateřina Aubrechtová Dragounová [1,3] and Alexander Kromka [1]

1 Institute of Physics of the Czech Academy of Sciences, Cukrovarnická 10, 162 00 Prague, Czech Republic; stehlik@fzu.cz (Š.S.); zemek@fzu.cz (J.Z.); dragounova@fzu.cz (K.A.D.); kromka@fzu.cz (A.K.)

2 New Technologies—Research Centre, University of West Bohemia, Univerzitní 8, 306 14 Pilsen, Czech Republic

3 Faculty of Nuclear Sciences and Physical Engineering, Czech Technical University in Prague, Břehová 7, 115 19 Prague, Czech Republic

* Correspondence: romanyuk@fzu.cz

Abstract: The modification of nanodiamond (ND) surfaces has significant applications in sensing devices, drug delivery, bioimaging, and tissue engineering. Precise control of the diamond phase composition and bond configurations during ND processing and surface finalization is crucial. In this study, we conducted a comparative analysis of the graphitization process in various types of hydrogenated NDs, considering differences in ND size and quality. We prepared three types of hydrogenated NDs: high-pressure high-temperature NDs (HPHT ND-H; 0–30 nm), conventional detonation nanodiamonds (DND-H; ~5 nm), and size- and nitrogen-reduced hydrogenated nanodiamonds (snr-DND-H; 2–3 nm). The samples underwent annealing in an ultra-high vacuum and sputtering by Ar cluster ion beam (ArCIB). Samples were investigated by in situ X-ray photoelectron spectroscopy (XPS), in situ ultraviolet photoelectron spectroscopy (UPS), and Raman spectroscopy (RS). Our investigation revealed that the graphitization temperature of NDs ranges from 600 °C to 700 °C and depends on the size and crystallinity of the NDs. Smaller DND particles with a high density of defects exhibit a lower graphitization temperature. We revealed a constant energy difference of 271.3 eV between the sp-peak in the valence band spectra (at around 13.7 eV) and the sp^3 component in the C 1s core level spectra (at 285.0 eV). The identification of this energy difference helps in calibrating charge shifts and serves the unambiguous identification of the sp^3 bond contribution in the C 1s spectra obtained from ND samples. Results were validated through reference measurements on hydrogenated single crystal C(111)-H and highly-ordered pyrolytic graphite (HOPG).

Keywords: nanodiamonds; hydrogenation; XPS; UPS; Raman spectroscopy; surface graphitization; Ar cluster ion beam sputtering

1. Introduction

Nanodiamonds (NDs) have potential applications in a wide range of technologies that exploit their unique electronic and surface properties. Their crystalline diamond core imparts structural stability and biocompatibility, while the ability to tailor surface chemistry and functional groups allows controllable interaction with various systems. As a catalytically active material, NDs show promise for photo- and electrocatalytic reactions like nitrogen reduction [1] or hydrogen production [2]. They may also find use in energy storage as electrode materials in batteries and supercapacitors [3]. Their tunable electronic bandgaps and band edge positions relative to vacuum through modification of size, defects, non-diamond carbon phases, and surface chemistry [4] could enable applications in optoelectronics such as light-emitting diodes, photovoltaics, and electron emitters [5]. Biomedical applications [6] are envisioned as well, taking advantage of biocompatibility and the broad options to functionalize nanodiamond surfaces for drug delivery or imaging.

Several types of nanodiamonds are commonly studied and utilized, mainly differentiated primarily by their production method and the resulting structural characteristics. Detonation nanodiamonds (DND) are synthesized via controlled explosions of carbon-containing explosives and are the most abundantly available type. Due to their explosive formation, detonation nanodiamonds tend to be sub-10 nm in size and highly polycrystalline and roundish in shape. They also often contain a significant portion of non-diamond carbon impurities incorporated during their preparation process [7]. In contrast, high-pressure high-temperature (HPHT) nanodiamonds [8] are produced by milling of synthetic diamond microcrystals grown by the HPHT process. HPHT NDs are generally larger than DND NDs, with particles exhibiting uniform monocrystalline and highly irregular shapes. HPHT NDs also demonstrate reduced and surface-limited non-diamond sp^2 carbon phase content relative to DNDs [9].

Realization of the full potential of NDs requires functionalization and understanding of the properties of ND surfaces. Basic surface chemistry modifications (oxidation, hydrogenation) can be achieved by convenient thermal treatments which can be followed by more complex chemical functionalization required for specific uses [4,10]. Like for diamond single crystals, surface hydrogenation of HPHT NDs causes the true negative electron affinity (NEA) [4] making NDs an efficient photoemitter [11], and facilitates disintegration of DNDs [9,12].

X-ray Photoemission Spectroscopy (XPS) is a suitable method for the investigation of chemical and electronic properties of ND surfaces, particularly to reveal the surface composition and chemical bond types, to assess the sp^2/sp^3 carbon ratio, and to specify the surface terminations or changes in the surface electronic properties of NDs [4,13–16]. However, there are several partially solved or unsolved problems in the analysis of ND surfaces. Diamond is a wide-bandgap material (5.5 eV) and therefore highly electrically resistive. Its surface becomes positively charged during measurements, causing recorded spectra to shift towards higher binding energy (BE). In the case of homogeneous and time-stable surface charging, the spectra are shifted by a constant value. To compensate for the surface charging, current spectrometers are usually equipped with low-energy flood guns. Precise compensation and spectra calibration, however, is sometimes not straightforward.

One can quickly detect surface charging by changing the input power of the X-ray source [16] or, more precisely, by applying a negative bias voltage to the sample holder [14]. The use of flood guns can easily lead to under- or over-compensation of the analyzed surface [17]. However, in diamond samples, selecting an appropriate internal standard poses a challenge. For this purpose, the C 1s peak maximum is usually employed [18], along with the C sp^2 contribution located at the position of the C 1s line recorded from graphite [19], or the O 1s peak position [20]. Regardless of the uncertainty associated with this correction, the described referencing is affected simultaneously by the presence of surface upward or downward band bending due to the surface graphitization and hydrogenation of NDs [14]. Moreover, the analysis of heterogeneous samples often leads to heterogeneous surface charging [21,22] and such analysis can be tricky or even impossible.

The sp^3-hybridized bonds contribution in C 1s peaks are identified as the highest intensity component at around 285.0 eV (after spectra calibration) for diamonds [12,23]. Density functional theory, as utilized by Fujimoto et al. [21] supports the assignment of the sp^3 contribution at a lower binding energy of the C 1s line. The authors argue that the conflicting assignment is a result of surface charging of diamond. Haerle et al. [24] and subsequently Titantah et al. [25] employed first-principles calculations to determine a core-level position and shift between the sp^3 and sp^2 contributions for amorphous carbon or carbon-based materials, yielding a value of ~1 eV, independent of mass density of carbon phase. Utilizing this value, the C sp^2/C sp^3 phase ratios align with measurements of the sp^2/sp^3 component areas in the C 1s line [24] and with the findings obtained by high-energy electron energy-loss spectroscopy (HREELS) [25]. This assignment gains further support from experimental investigations on crystalline diamonds annealed at high temperatures under ultra-high vacuum conditions [12,26], as well as from studies about interaction of energetic ions with diamond surfaces [19,27]. A progressively increasing sp^2 bond-related component at the low binding energy side of the C 1s peaks as temperature and/or ion dose increase can serve as indicator of diamond graphitization process.

The line shape of C 1s spectra recorded from oxygen-free diamond surfaces sometimes exhibits asymmetry. In other words, there is an additional component shifted positively by 0.5–0.9 eV with respect to the sp^3 component in C 1s spectrum. Hence, this spectral intensity cannot be attributed to carbon bonded to oxygen. In the existing literature, it is attributed to photoemission from carbon atoms situated within various defects [28–30], from hydrogen bonded to carbon atoms, denoted as C-H_x [15], where $x \geq 2$ [19,31,32]. High intensity of this component was observed in the C 1s spectra of DND samples [29,33]. Considering this complexity, calibrating the peak maxima of the C 1s line by assigning either the positions to the sp^2, sp^3 components or to the sp^3, C-H_x components is not straightforward due to the similarity of the chemical shifts in both cases. For the correct assignment of the individual C 1s peak components, the initiation of the phase transition between the C sp^3 phase of diamond and the C sp^2 graphite phase, could help to resolve the problem.

The graphitization of monocrystalline diamond surfaces [34], nanocrystalline diamond (NCD) films [35], and DNDs [29] during annealing under ultra-high vacuum (UHV) conditions has been previously studied. The temperature required to start the surface graphitization reaction on a monocrystalline diamond (111) surface under UHV conditions was reported to be approximately 850–900 °C [34]. A similar temperature range for graphitization was observed for DND particles: below 900 °C, the ND surface undergoes reconstruction into isolated graphitic domains, evident from the emergence of the sp^2 carbon peak as observed by XPS, while the diamond core remains unaltered. Bulk graphitization of ND cores occurs above 900 °C. It should be noted that significantly lower graphitization temperatures (<500 °C) were observed in the presence of a Ni atoms (catalyst layer) on a NCD surface [35], i.e., graphitization temperature also depends on the diamond surface chemistry.

The graphitization of NCD films can also be reached through monoatomic Ar ion sputtering and Ar cluster ion beam (ArCIB) sputtering [19,36], leading to an increase in concentration of the C sp^2 phase on the NCD surface. In contrast to monoatomic Ar ion sputtering, no Ar atoms were incorporated into the NCD film using the ArCIB sputtering. Thus, no interaction between Ar ions and C atoms [37], nor formation of Ar-C bonds [36] are expected by using ArCIB sputtering.

The recent investigation of graphitization on DND-H samples through laser irradiation was reported [33]. The study indicated an increase in the concentration of the C sp^2 phase with longer irradiation times, while the concentration of defects or C-N bonds remained nearly constant. Although an exact temperature was not provided, the results suggest that laser irradiation induced a localized temperature increase of the DND samples, which was adequate to initiate graphitization of the surface layers of DND. This temperature increase is likely below 900 °C, as suggested by previous thermal studies [29].

Photoemission from the valence bands (VB) of diamonds [38–40] and NDs [14,15,41] reveals spectral features originating from the joint sp-electron and p-electron partial density of states [38]. The sharp sp-peak between 12–14 eV has been primarily attributed to carbon sp^3 bonds [39,42]. Graphitization of the microcrystalline diamond film, however, results in a reduction of the sp-peak intensity, but the peak intensity is not completely eliminated for the graphitized carbon [41] or graphene [43]. Therefore, the use of C sp^3 notation for the sp-peak could be misleading. On the other side, the position of the sp-peak could be used for the calibration of the VB spectra measured from diamond films with a mixture of C sp^2/C sp^3 phases [44].

In the present study, we examined the graphitization of three different types of hydrogenated NDs using XPS, UPS, and Raman spectroscopy. For the calibration purposes, we employed reference samples of hydrogenated single crystal diamond C(111)-H and freshly cleaved HOPG. The ND samples were subjected to annealing in UHV and sputtering using an ArCIB. We also establish a correlation between the sp-peak position in the VB and the position of the sp^3 component in the C 1s line and employ it for the calibration of photoelectron spectra from ND samples.

2. Experimental Details

As HPHT NDs, we used monocrystalline synthetic NDs from Pureon (Lengwil, Switzerland), featuring a size range of 0–30 nm. To reduce the non-diamond carbon content, the as received HPHT NDs were subjected to annealing in ambient air atmosphere at 450 °C for 5 h, resulting in purified and oxidized HPHT NDs [9]. Next, the HPHT NDs were annealed in hydrogen gas at a flow rate of 5 L/min at atmospheric pressure for 6 h, at an annealing temperature of 800 °C [45]. This sample is further labeled as *HPHT ND-H*.

As DNDs we used two different materials. First we used conventional DNDs from New Metals and Chemicals Corp. (Tokyo, Japan) with a nominal size of about 5 nm and 2 at.% of nitrogen [46]. The as-received DNDs were first annealed in ambient air atmosphere at 450 °C for 30 min and then hydrogenated in hydrogen gas at a flow rate of 5 L/min at atmospheric pressure for 6 h at an annealing temperature of 700 °C. This sample is further labeled as *DND-H*.

The second DND material was provided by OZM Research Ltd. (Hrochův Týnec, Czech Republic), featuring a nominal size of approximately 2–3 nm [estimated by transmission electron microscopy (TEM) and small-angle X-ray scattering (SAXS) measurements, not shown here] and a reduced nitrogen content as shown below. The particles were first annealed in air at 350 °C for 30 min and then hydrogenated at 700 °C for 3 h at a flow rate of 5 L/min at atmospheric pressure of hydrogen. This sample is further labeled as *snr-DND-H* where *s*, *n*, *r* stand for size-nitrogen-reduced NDs.

ND powders were first dispersed in water (1.0 mg of ND powder in 1.0 mL of deionized water) by sonication using a rod-type sonicator (Hielscher UP200s, Teltow, Germany) at 200 W for 1 h. The ND colloids were drop casted on gold bulk substrates and dried before loading in UHV chamber for XPS and UPS measurements. We ensured sufficient thickness of ND layers on the conducting substrate. The as-prepared ND samples (for measurements at room temperature, RT samples) were transparent and became non-transparent after annealing, qualitatively signalizing increase of their C sp^2 content [see Figure S1 in Supplementary Materials].

Monocrystalline diamond C(111) and HOPG samples were used as reference of C sp^3 and C sp^2 phase materials. In order to clean C(111) surface and to obtain a hydrogenated diamond surface, a monocrystalline C(111) sample was exposed to a 10 min hydrogen plasma treatment in a multimode clamshell cavity reactor (SDS6K, Seki Diamond Systems, Seki City, Japan). The process conditions were as follows: H_2 gas flow rate of 300 sccm, pressure of 4 kPa (30 Torr), microwave power of 2 kW, and the average process temperature of 465 ± 5 °C. The contamination-free surface of the HOPG was prepared through exfoliation. The process involved attaching the 10 × 10 mm^2 bulk HOPG sample to the sample holder. Scotch tape was then applied multiple times to remove the topmost contaminated layers. Once the largest area of the HOPG remained on the scotch tape, the bulk sample with freshly-prepared surface was promptly inserted into the spectrometer chamber.

Raman spectra were measured using a Renishaw InVia Reflex Raman spectrometer (Renishaw, Wotton-under-Edge, UK) with an excitation wavelength of 442 nm (Dual Wavelength He–Cd laser, model IK5651R-G, Kimmon Koha, Tokyo, Japan) and air-cooled CCD camera. The samples were exposed by a continuous wave laser (the intensity power of 0.33 mW) focused to a spot of 1 µm exposure time was set for 100 s, and spectra were sampled 42 times. Spectra were measured in a backscattering setup with a 100× objective (Leica, Wetzlar, Germany, NA = 0.9) in the confocal mode with a grating of 2400 lines/mm.

XPS spectra were acquired with an AXIS-Supra photoelectron spectrometer (Kratos Analytical Ltd., Manchester, UK) using monochromatized Al Kα radiation (1486.6 eV, 150 W, analyzed area of 0.7 × 0.3 mm^2). The X-ray incidence angle and photoelectron emission angle were 36° and 90°, respectively. The high-resolution core level spectra were recorded with a pass energy of 10 eV and with an overall energy resolution of 0.45 eV. The valence band XPS spectra (VB-XPS) should be measured with the same pass energy to avoid possible charge-related shifts for non-conductive samples. Low energy electron flood gun was used for neutralization of charge on surfaces. The higher-resolution spectra were measured with an energy step of 0.1 eV.

Sample annealing up to 800 °C was carried out in situ in an XPS chamber with temperature increments of 100 °C. After each annealing step, in situ XPS spectra were acquired at a given temperature. Throughout the annealing process, the pressure in the XPS chamber remained below 2.0×10^{-7} Torr. To maintain UHV conditions, the sample temperature increased gradually. It took approximately 20–40 min to raise the temperature by 100 °C, and a similar period was needed to acquire the photoelectron spectra.

Ar cluster ion beam (ArCIB) sputtering was conducted on annealed samples at room temperature, following the cooling of the samples from 800 °C to room temperature (more than 4 h was needed for cooling to room temperature). Sputtering was performed with the Kratos Gas Cluster Ion Source (Minibeam 6) with a cluster size of 1000 atoms and the Ar ion beam energy of 5 keV. The exposure area of the ArCIB was set to 2×2 mm^2, and the incident angle of the ion beam was set to 45° from the surface normal. The sample surfaces underwent sputtering for a duration of 60 min. The UPS spectra were acquired utilizing He II light excitation (40.8 eV) with a pass energy of 40 eV and an energy step size of 0.1 eV. Note, the He I was less suitable for resolving sp-peak in the valence spectra because it overlapped with the secondary electron emission intensity. Therefore, He II with a broader valence spectra range was used.

Quantification of atomic composition was carried out by analyzing the integrated high-resolution peak areas, following Shirley's electron inelastic background subtraction. We employed the ESCApe software (Kratos Analytical Ltd., ver. 1.4.0.1149), utilizing the incorporated atomic sensitivity factors, and employed the KolXPD software (ver. 1.7.0.8) for the core-level peak treatment [47]. Fitting of the XPS and UPS spectra was carried out by Voigt and Gaussian functions, respectively. The width of the Lorenzian function in the Voigt function was consistently maintained in all spectra.

3. Results and Discussion

Prior to analyzing the ND samples, photoelectron spectroscopy measurements were carried out on the reference C(111)-H and HOPG samples, with defined majority of the C sp^3 and C sp^2 phases, respectively. Figure 1 displays the measured high-resolution C 1s spectra, valence band XPS spectra (VB-XPS), and UPS He II spectra from the C(111)-H and HOPG samples. The reference sample surfaces were relatively clean: an O concentration of 0.8 at.% was measured on the C(111)-H surface and no O was present on HOPG surface. The position of the C 1s peak of C(111)-H sample was calibrated to 285.0 eV. The VB-XPS spectrum in Figure 1c underwent the same calibration binding energy shift as for C 1s line. For HOPG spectra in Figure 1b,d, no calibration was needed (i.e., no surface band bending was observed, the as-measured XPS and UPS spectra are presented in figure). The binding energy of the sp^2 component in HOPG was at 284.4 eV. The C 1s spectrum of the C(111)-H sample was deconvoluted into components with the chemical shifts of −0.8 eV (for sp^2 component), 0.5 eV (for C_x component), and 1.1–1.5 eV (for C-O/C-OH) relative to the position of the sp^3 component. The C 1s spectrum of HOPG sample was fitted by one sp^2 (asymmetrical, Doniach-Sunjic) component, confirming clean and well-ordered HOPG surface.

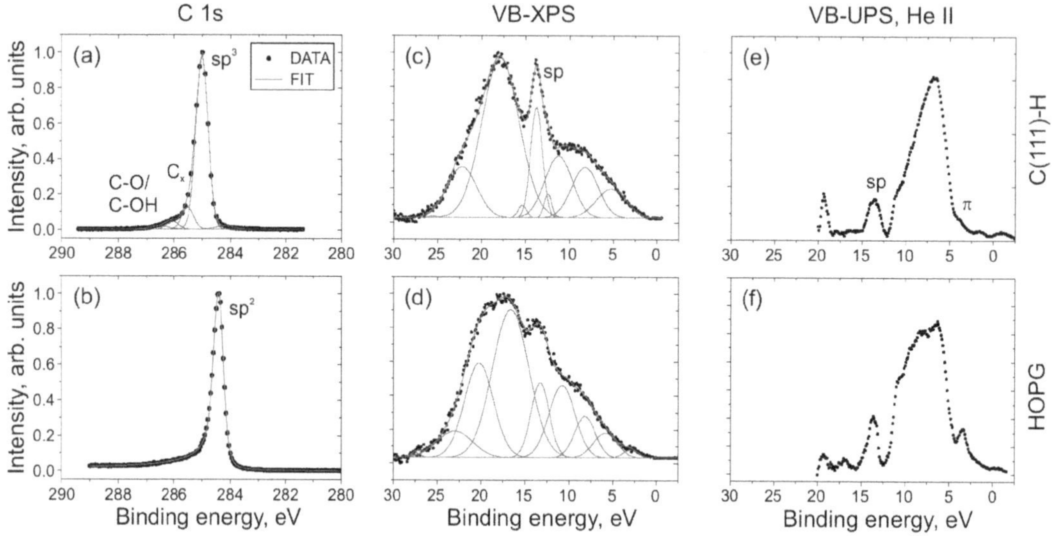

Figure 1. Photoelectron emission spectra from the hydrogenated monocrystalline C(111)-H diamond (upper row) and HOPG (bottom row) surfaces. (**a**,**b**) High-resolution C 1s XPS spectra, (**c**,**d**) valence band XPS spectra, (**e**,**f**) valence band UPS He II spectra. Peak components are indicated by thin grey lines. sp^3 and sp^2 components are shifted with respect to each other, whereas sp-band positions are similar for all samples.

In previous studies, eight components were utilized for fitting VB-XPS spectra of diamond and HOPG samples [44,48]. In this study, we used a similar component distribution in the VB spectra but with additional components near the sp-peak to capture the shape of the peak more accurately. The sp-peak maximum position was at 13.7 ± 0.1 eV for both samples. We should emphasize, however, that the measured position of peaks could be affected by surface band bending [11], but the relative energy difference between C 1s and sp-peak should not. The measured energy differences between the C 1s peak components and the sp-peak positions were 271.3 ± 0.1 eV (for sp^3) and 270.7 ± 0.1 eV (for sp^2). These relative binding energy differences were used for identification of the sp^3 and sp^2 components in the C 1s spectra from ND samples.

The UPS He II spectra (after the background subtraction) are depicted in Figure 1e for the C(111)-H sample and Figure 1f for the HOPG sample. The sp-peaks were present on both diamond and HOPG UPS spectra [44] despite of the absence of C sp^3 phase in HOPG [see Figure 1b]. The UPS spectra were then calibrated [44] to the position of the sp-peak at 13.7 eV (note, charge-induced shifts could be different for XPS and UPS spectra because of difference in excitation energies). Unlike the VB-XPS spectra, the UPS spectra exhibit an amplified contribution from the p-electrons within the range of 0–11 eV [38,39]. The π-band peak at 3.6 eV is clearly discernible in Figure 1f for the HOPG sample. So, an intensified π-band peak indicates the presence of an C sp^2 phase on the diamond on a surface.

In Figure 2a–c the C 1s peaks from HPHT ND-H, DND-H, and snr-DND-H samples are displayed. The corresponding survey spectra and N 1s spectra from DND samples are included in Supplementary Materials (Figures S2 and S4). The spectra were measured from as-prepared samples at room temperature (RT, red), from annealed samples at 800 °C (green), and from sputtered samples (ArCIB at RT, blue). All spectra were calibrated with respect to the position of the sp^3 component at 285.0 eV. Fitted components are shown for RT samples exclusively. The core level shifts and component assignments to bond types are similar to those of the reference sample in Figure 1 and in agreement with our prior studies [4,9,19,33]. The C_x component of DND-H sample spectra also involves contributions from C-N bonds. The concentration of C-N bonds is small, however, and therefore these components were not resolved by the fit. The values of the chemical core level shifts and the measured atomic compositions of NDs are included in Supplementary Materials (Table S1 and Figure S3).

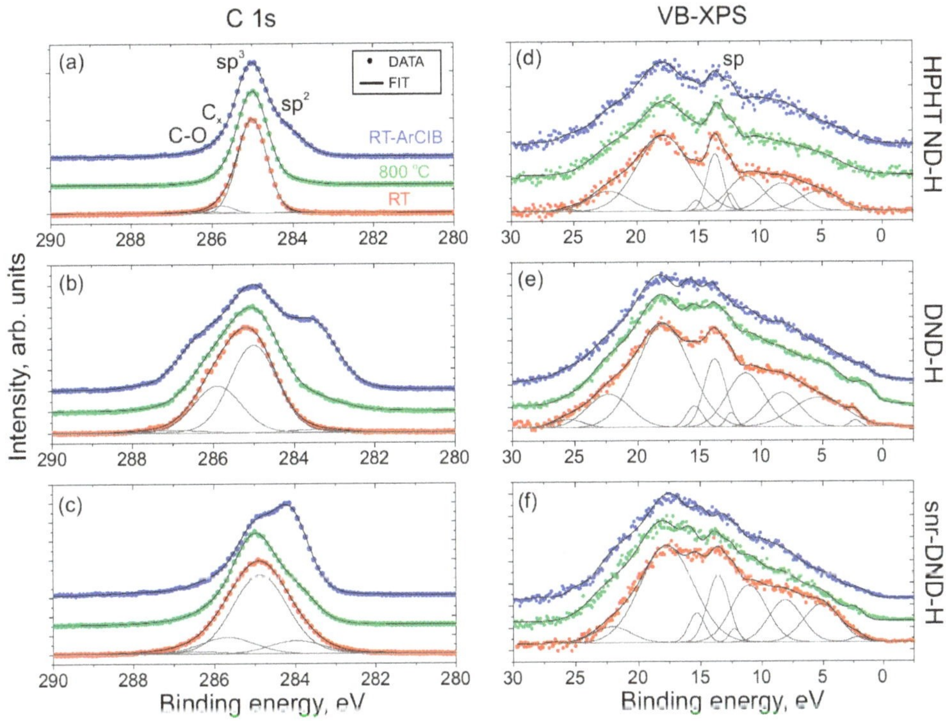

Figure 2. (**a–c**) XPS high-resolution C 1s and (**d–f**) valence band spectra of HPHT ND-H, DND-H, and snr-DND-H samples. Samples were measured at room temperature (as-prepared, RT), annealed at 800 °C in UHV, and sputtered by ArCIB at RT. Peak components are shown by grey lines. An increase in the sp^2 phase of carbon has been observed after sample annealing and sputtering. The sp-peak in the valence band is smeared out after graphitization.

A negative chemical shift of −1.5 eV was obtained for the sp^2 component of sputtered DND-H samples (with 2.0–2.5 at% of N atoms in DND-H). Smaller shifts of −0.8 eV and −0.9 eV were derived for snr-DND-H samples. Similar shifts of about −1.4 eV for the sp^2 components were previously measured on graphitized DND samples with N contaminants [29] and on defective N-doped graphene layers [49]. Shifts were identified as due to hybridization of C sp^2 bonds close to vacancies [49]. This could correlate with sputtering of DND-H samples (introducing vacancies in C sp^2 phase) in a proximity of N atoms.

Examination of the C 1s peaks disclosed partial graphitization of the sample surfaces: following sample annealing, the areas of sp^2 components exhibited slight augmentation, and a more notable increase in the sp^2 component area was observed after ArCIB sputtering, where more pronounced amorphization of NDs and increase of sp^2 bond concentration is expected. After sputtering, sample surface morphology has also changed, specifically for HPHT ND-H (see Figure S5 in Supplementary Materials). The increase of the sp^2 component area varied among different types of ND particles, affirming the link between graphitization efficiency and NP types (or size): the concentration of the graphitic phase was higher after annealing and sputtering for the smaller and more defective DND samples.

Figure 2d–f show VB-XPS spectra of RT (as-received), annealed, and sputtered samples. The VB spectra of the as-received HPHT ND-H sample [in (d), red dots] are very similar to the VB spectra of the diamond monocrystal sample [Figure 1c] because the emission from the diamond core dominates and the C sp^2 phase concentration is small. The sp-peaks are broader in RT DND-H and snr-DND-H VB spectra [Figure 2e,f, red dots], where concentration of C sp^2 phase is higher. A similar broadening was observed between the C(111)-H and HOPG samples in Figure 1c,d. The shape of the sp-peak therefore depends on the concentration of C sp^2 and C sp^3 phases. It is important to note that similarly to the reference bulk samples, the binding energy difference between the sp^3 components and the sp-peaks was preserved (271.3 ± 0.1 eV for all as-prepared ND samples).

Annealing and sputtering induced further broadening of sp-peaks [green and blue curves in Figure 2d–f], but the sp-peak was still preserved and no contribution from the π-band peak was resolved on the graphitized HPHT ND-H samples [Figure 2d]. More prominent graphitization has occurred for smaller DND-H and snr-DND-H particles. In particular, the sp-peaks have smeared out and the π-peak at around 3.6 eV appeared on the VB-XPS spectra [Figure 2e,f].

Figure 3 illustrates the dependence of the sp^2/sp^3 component area ratios on annealing temperature as derived from the C 1s spectra analysis. Initial concentration of C sp^2 phase on as-prepared HPHT ND-H sample was smaller than for DND samples. Annealing up to 700 °C did not alter the C sp^2/C sp^3 ratio significantly. At >700 °C, graphitization starts, leading to an increase in sp^2/sp^3 phase concentration ratio.

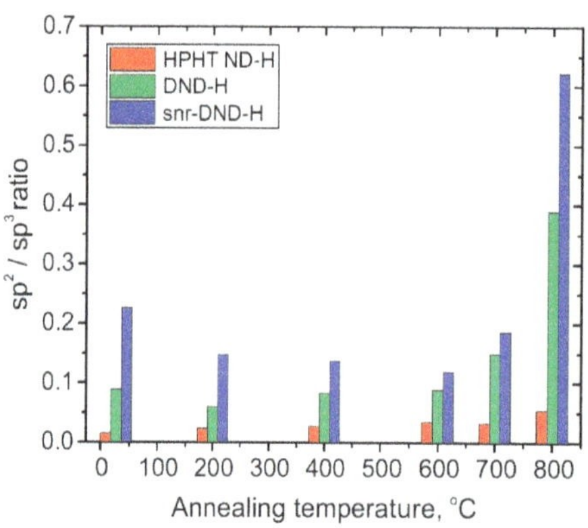

Figure 3. Dependence of the sp^2/sp^3 component ratio in the C 1s peak on the annealing temperature. An increase in sp^2 phase concentration was observed at temperatures above 700 °C and 600 °C for HPHT ND-H and DND-H, SND-H samples, respectively.

The initial concentration of the C sp^2 phase was higher for DND and snr-DND-H samples, which is consistent with the higher concentration of contaminations observed for DND samples (see Figure S3 in Supplementary Materials). Up to 600 °C, the sp^2/sp^3 ratio remained below ~0.1 for DND-H samples but decreased for snr-DND-H samples. In the latter cases, oxygen desorption could be associated with a reduction in the C sp^2 phase on the surface. After 600 °C, the sp^2/sp^3 ratio increased for both types of DND samples. Note, the sp^2/sp^3 concentration was higher for the smallest snr-DND-H particles than for DND-H particles.

In terms of DND, our results are comparable to previous study, where graphitization temperatures of >700 °C for DND surfaces and >900 °C for DND cores have been reported [29]. Here, we show that the surface rearrangement of DND-H starts already at the temperature > 600 °C. We assign this minor difference to a slight difference in ND particle size and crystallinity of NDs. In contrast to DND, the HPHT ND-H surface rearrangement starts at the higher temperature > 700 °C. The surface rearrangement eventually leading to the ND graphitization is triggered by the thermal desorption of the surface functional groups, which is accompanied by the formation of surface dangling bonds. In vacuum this leads to the formation of the sp^2 bonds between the adjacent carbon atoms and formation of C-sp^2 patches. However, in atmosphere, the surface dangling bonds can be passivated by surrounding atoms such as hydrogen [50]. Indeed, the observed graphitization onset temperature match well with the temperature of thermal hydrogenation, i.e., 600–700 °C for DND [51] and 700–800 °C for HPHT ND [15,52]. Our findings confirm that the initial graphitization temperature of ND depends on the type and size of studied NDs, i.e., the temperature is lower for smaller and more defective DNDs.

Figure 4 displays UPS He II spectra measured on (a) HPHT ND-H, (b) DND-H, and (c) snr-DND-H samples at room temperature (RT, red line) and after annealing in UHV (800 °C, green line). The UPS spectra were calibrated to the sp-peak positions (and π-band positions). Due to the higher surface sensitivity of UPS compared to VB-XPS, the enhancement of the π-band was more clearly distinguishable, especially for HPHT ND-H samples. The sp-peak deteriorated with the decrease in NP size (resulting in a lower concentration of C sp^3 core phase) and with graphitization (leading to an increase in C sp^2 phase on NP surfaces).

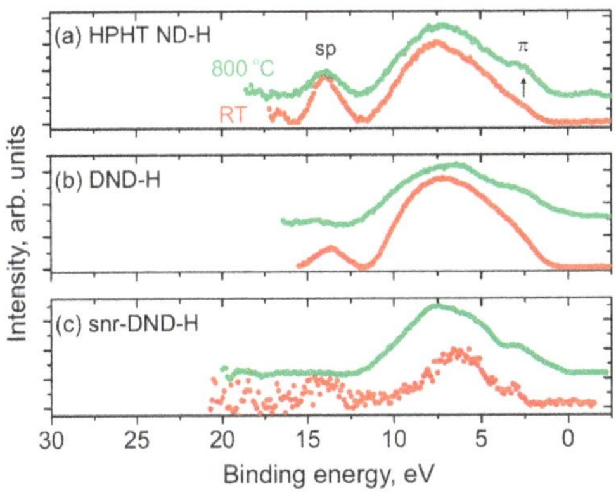

Figure 4. UPS He II spectra of (**a**) HPHT ND-H, (**b**) DND-H, and (**c**) snr-DND-H samples measured at RT (red curves) and after annealing at 800 °C (green curves). The intensity of the sp peak decreased, and the intensity of the π peak increased after annealing (graphitization).

In Figure 4c, the π-band peak was well-resolved and sp-peak was very weak on as-prepared snr-DND-H sample at RT (with the highest C sp^2 phase concentration). After annealing, the intensity of the π-band peak increased, and the sp-peak completely disappeared. We can assume that the surface of the snr-DND-H sample was almost completely covered by the C sp^2 phase layer.

The appearance of the π-band peak and variation in the sp-peak intensities with sp^2/sp^3 ratio change in NDs agree with the previous results obtained for the diamond films [41,44,48]. Therefore, we would like to emphasize that a combination of XPS, VB-XPS, and UPS measurements has a great potential for the spectra calibration and resolving different phases of carbon films and diamond NPs.

Figure 5 shows Raman spectra of as-prepared and sputtered (a) HPHT ND-H, (b) DND-H, and (c) snr-DND-H samples measured by 442 nm excitation laser. The diamond peak in HPHT ND-H spectra at 1332 cm^{-1} is very sharp (blue rectangular), confirming the dominated C sp^3 core structure of as-prepared HPHT ND-H sample (measured at RT, red line). In as-prepared DND-H, the diamond peak is broader and shifted to lower wavenumbers (1327–1328 cm^{-1}) compared to bulk diamonds. A weaker and broader diamond peak at about 1340 cm^{-1} was also resolved on snr-DND-H RT sample, but the spectrum is rather noisy due to very small ND particle size (the diamond sp^3 peak of the snr-DND-H sample could be better resolved on UV-Raman spectrum, see Figure S6 in Supplementary Materials). In accordance with the XPS results the C sp^2 carbon phase content increases in the line HPHT ND-H, DND-H, snr-DND-H.

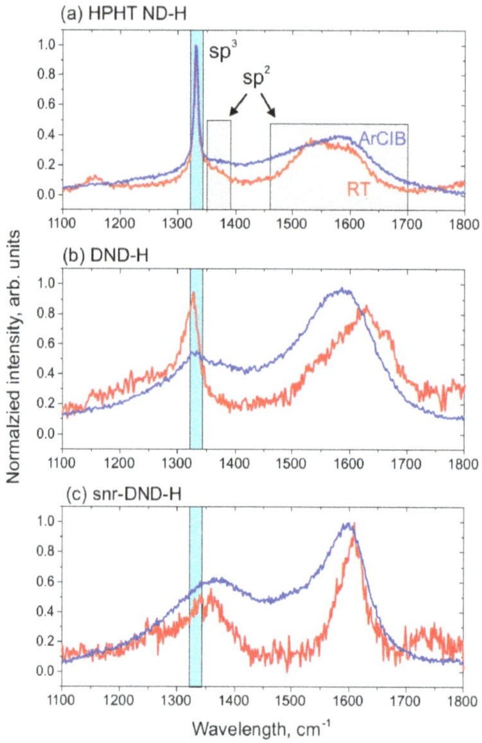

Figure 5. Raman spectra of as-prepared and ArCIB sputtered (**a**) HPHT ND-H, (**b**) DND-H, and (**c**) snr-DND-H samples. All samples were measured at RT. Sputtering induced disorder on the surface and broadening of the diamond peaks as well as the sp^2-related peaks (D-bands, G-bands).

In addition to the diamond peak, all the spectra contain a D-band peak at 1350–1360 cm^{-1} and G-band peak between 1500–1700 cm^{-1} corresponding to the C sp^2 carbon phase (defects and disordered C sp^2 phase, grey rectangular). Sputtering induced a reduction of C sp^3 peaks and an increase of D-band and G-band peaks in all the samples. In DND samples a redshift of the G-band was observed which indicates a change in the C sp^2 carbon surface arrangement, probably from more or less isolated defects to more continuous domains or patches [53].

Raman spectroscopy and XPS measurements have confirmed the dependence of the graphitization process on the type and size of NDs. Previous studies have attributed the temperature-induced graphitization of NDs to a reduction in their surface energy [54], achieved by the formation of fullerene-like or bucky diamond C sp^2 structures on the H-free C sp^3 ND surfaces [55]. This C sp^2-type surface reconstruction eliminates the unfavorable surface dangling bonds on clean ND surfaces. Therefore, a sequence involving ND surface cleaning or purification, H desorption, and the formation of surface dangling bonds is expected to precede the C sp^3 to C sp^2 phase transition, i.e., ND graphitization.

Our comparative study of NDs with different sizes, ratios of defects, contamination, and initial C sp^2 concentrations indicates that initial defect density in NDs plays a significant role in the graphitization process. The well-ordered HPHT ND-H particles demonstrated greater resistance to annealing, surface functional group desorption, followed by graphitization. Conversely, defects and potential strain in smaller DNDs could facilitate desorption of hydrogen or related moieties at lower temperatures. Consequently, smaller, and defective NDs may exhibit higher graphitization rates, potentially leading to the formation of multiple graphitic (onion-like) C sp^2 shells on C sp^3 DND cores, rather than C sp^2-type surface reconstructions.

4. Conclusions

We investigated the graphitization of three different types of hydrogenated NDs (top-down or bottom-up HPHT, and DNDs) using XPS, UPS, and Raman spectroscopy. To accurately assign the C sp3 and C sp2 phase features in the ND samples, we first collected referential photoelectron spectra of hydrogenated monocrystalline diamond C(111)-H and HOPG. Utilizing the temperature-dependent in situ XPS measurements, we observed that the graphitization onset temperatures for employed NDs varied. In particular, the graphitization onset temperatures decreased with decreasing ND size; >700 °C for ~18 nm HPHT ND-H, >600 °C for 5 nm DND-H and 2–3 nm snr-DND-H. Additionally to the sp^2-related component areas in the C 1s peaks, we observed signs of graphitization such as the broadening of the sp-peak in the valence band, an increase in the π-band peak intensity, and a decrease in the 1332 cm^{-1} Raman diamond peak intensity. The C sp^2 related contributions in the spectra and the correct assignment were independently verified through the ND surface amorphization after ArCIB sputtering. We found that the energy separation between the C 1s sp^3 peak and valence band sp-peak remained constant at 271.3 ± 0.1 eV, enabling clear assignment of the sp^2 and sp^3 components in the C 1s spectra as well as calibration of UPS spectra. Overall our findings suggest that the combination of XPS, VB-XPS and UPS can help to resolve uncertainties in identifying of sp^2 or sp^3 components in C 1 s peak, and provide a solid basis for qualitative and quantitative investigation of ND modifications, here demonstrated on ND graphitization.

Supplementary Materials: The following supporting information can be downloaded at: https://www.mdpi.com/article/10.3390/nano14070590/s1, Figure S1: Picture of as-prepared and graphitized DND-H sample, Figure S2: Survey XPS spectra, Figure S3: The measured atomic concentrations of oxygen and nitrogen, Figure S4: N 1s core level spectra and concentration, Figure S5: SEM images, Figure S6: UV-Raman spectra, Table S1: Core level shifts. Refs [56–58] are cited in Supplementary Materials.

Author Contributions: Conceptualization, O.R.; Methodology, J.Z.; Investigation, O.R., Š.S., J.Z. and K.A.D.; Writing—original draft, O.R.; Writing—review & editing, Š.S., J.Z., K.A.D. and A.K.; Supervision, O.R.; Project administration, A.K.; Funding acquisition, Š.S. All authors have read and agreed to the published version of the manuscript.

Funding: This work was supported by the Ministry of Education, Youth and Sports of the Czech Republic, grant No. LUASK22147, the Czech Academy of Sciences—Strategy AV21, and by the Czech Science Foundation (GACR) grant number 23-04876S. We acknowledge CzechNanoLab Research Infrastructure supported by MEYS CR (LM2023051) and the OP JAC financed by ESIF and the MEYS SENDISO—CZ.02.01.01/00/22_008/0004596. All authors thank R. Jackiv, O. Babcenko, P. Jiricek, and A. Vlk for their support in experimental work and fruitful discussion of the results.

Data Availability Statement: Data are contained within the article and Supplementary Materials.

Conflicts of Interest: The authors declare no conflicts of interest.

References

1. Zhu, D.; Zhang, L.; Ruther, R.E.; Hamers, R.J. Photo-illuminated diamond as a solid-state source of solvated electrons in water for nitrogen reduction. *Nat. Mater.* **2013**, *12*, 836–841. [CrossRef]
2. Marchal, C.; Saoudi, L.; Girard, H.A.; Keller, V.; Arnault, J. Oxidized Detonation Nanodiamonds Act as an Efficient Metal-Free Photocatalyst to Produce Hydrogen Under Solar Irradiation. *Adv. Energy Sustain. Res.* **2023**, *5*, 2300260. [CrossRef]
3. Zhou, Y.; Zhao, K.; Zhang, J.; Zhu, Y.; Ma, Y.; Zhang, H.; Song, D.; Shi, X.; Zhang, L.; Ding, Y. Synergistic effects of nanodiamond modified separators toward highly stable and safe lithium metal batteries. *J. Mater. Chem. A* **2021**, *9*, 16046–16055. [CrossRef]
4. Miliaieva, D.; Djoumessi, A.S.; Čermák, J.; Kolářová, K.; Schaal, M.; Otto, F.; Shagieva, E.; Romanyuk, O.; Pangrác, J.; Kuliček, J.; et al. Absolute energy levels in nanodiamonds of different origins and surface chemistries. *Nanoscale Adv.* **2023**, *5*, 4402–4414. [CrossRef]

5. Afandi, A.; Howkins, A.; Boyd, I.W.; Jackman, R.B. Nanodiamonds for device applications: An investigation of the properties of boron-doped detonation nanodiamonds. *Sci. Rep.* **2018**, *8*, 3270. [CrossRef]
6. Qin, J.-X.; Yang, X.-G.; Lv, C.-F.; Li, Y.-Z.; Liu, K.-K.; Zang, J.-H.; Yang, X.; Dong, L.; Shan, C.-X. Nanodiamonds: Synthesis, properties, and applications in nanomedicine. *Mater. Des.* **2021**, *210*, 110091. [CrossRef]
7. Turner, S.; Shenderova, O.; Da Pieve, F.; Lu, Y.; Yücelen, E.; Verbeeck, J.; Lamoen, D.; Van Tendeloo, G. Aberration-corrected microscopy and spectroscopy analysis of pristine, nitrogen containing detonation nanodiamond: Microscopy and spectroscopy analysis of pristine, nitrogen containing DND. *Phys. Status Solidi A* **2013**, *210*, 1976–1984. [CrossRef]
8. Shenderova, O.; Nunn, N. Production and purification of nanodiamonds. In *Nanodiamonds;* Elsevier: Amsterdam, The Netherlands, 2017; pp. 25–56. [CrossRef]
9. Stehlik, S.; Mermoux, M.; Schummer, B.; Vanek, O.; Kolarova, K.; Stenclova, P.; Vlk, A.; Ledinsky, M.; Pfeifer, R.; Romanyuk, O.; et al. Size Effects on Surface Chemistry and Raman Spectra of Sub-5 nm Oxidized High-Pressure High-Temperature and Detonation Nanodiamonds. *J. Phys. Chem. C* **2021**, *125*, 5647–5669. [CrossRef]
10. Arnault, J.C.; Girard, H.A. Hydrogenated nanodiamonds: Synthesis and surface properties. *Curr. Opin. Solid State Mater. Sci.* **2017**, *21*, 10–16. [CrossRef]
11. Maier, F.; Ristein, J.; Ley, L. Electron. affinity of plasma-hydrogenated and chemically oxidized diamond (100) surfaces. *Phys. Rev. B* **2001**, *64*, 165411. [CrossRef]
12. Dideikin, A.T.; Aleksenskii, A.E.; Baidakova, M.V.; Brunkov, P.N.; Brzhezinskaya, M.; Davydov, V.Y.; Levitskii, V.S.; Kidalov, S.V.; Kukushkina, Y.A.; Kirilenko, D.A.; et al. Rehybridization of carbon on facets of detonation diamond nanocrystals and forming hydrosols of individual particles. *Carbon* **2017**, *122*, 737–745. [CrossRef]
13. Arnault, J.C. X-ray Photoemission Spectroscopy applied to nanodiamonds: From surface chemistry to in situ reactivity. *Diam. Relat. Mater.* **2018**, *84*, 157–168. [CrossRef]
14. Yeganeh, M.; Coxon, P.R.; Brieva, A.C.; Dhanak, V.R.; Šiller, L.; Butenko, Y.V. Atomic hydrogen treatment of nanodiamond powder studied with photoemission spectroscopy. *Phys. Rev. B* **2007**, *75*, 155404. [CrossRef]
15. Saoudi, L.; Girard, H.A.; Larquet, E.; Mermoux, M.; Leroy, J.; Arnault, J.-C. Colloidal stability over months of highly crystalline high-pressure high-temperature hydrogenated nanodiamonds in water. *Carbon* **2023**, *202*, 438–449. [CrossRef]
16. Kono, S.; Kageura, T.; Hayashi, Y.; Ri, S.-G.; Teraji, T.; Takeuchi, D.; Ogura, M.; Kodama, H.; Sawabe, A.; Inaba, M.; et al. Carbon 1s X-ray photoelectron spectra of realistic samples of hydrogen-terminated and oxygen-terminated CVD diamond (111) and (001). *Diam. Relat. Mater.* **2019**, *93*, 105–130. [CrossRef]
17. Baer, D.R.; Artyushkova, K.; Cohen, H.; Easton, C.D.; Engelhard, M.; Gengenbach, T.R.; Greczynski, G.; Mack, P.; Morgan, D.J.; Roberts, A. XPS guide: Charge neutralization and binding energy referencing for insulating samples. *J. Vac. Sci. Technol. A Vac. Surf. Film.* **2020**, *38*, 031204. [CrossRef]
18. Kozak, H.; Artemenko, A.; Čermák, J.; Švrček, V.; Kromka, A.; Rezek, B. Oxidation and reduction of nanodiamond particles in colloidal solutions by laser irradiation or radio-frequency plasma treatment. *Vib. Spectrosc.* **2016**, *83*, 108–114. [CrossRef]
19. Romanyuk, O.; Zemek, J.; Houdková, J.; Babčenko, O.; Shagieva, E.; Beranová, K.; Kromka, A.; Jiříček, P. Effects of monoatomic and cluster bombardment with Ar ion beam on the surface of hydrogenated nanocrystalline diamond. *Diam. Relat. Mater.* **2023**, *133*, 109748. [CrossRef]
20. Cuenca, J.A.; Thomas, E.L.H.; Mandal, S.; Morgan, D.J.; Lloret, F.; Araujo, D.; Williams, O.A.; Porch, A. Microwave Permittivity of Trace sp^2 Carbon Impurities in Sub-Micron Diamond Powders. *ACS Omega* **2018**, *3*, 2183–2192. [CrossRef]
21. Fujimoto, A.; Yamada, Y.; Koinuma, M.; Sato, S. Origins of sp^3C peaks in C $_{1s}$ X-ray Photoelectron Spectra of Carbon Materials. *Anal. Chem.* **2016**, *88*, 6110–6114. [CrossRef] [PubMed]
22. Wang, Z.-Q.; Yang, Y.-G.; Tai, L.; Lau, L.W.-M.; Zhou, D. Mitigating surface charging in XPS using an in-situ sub-nanometer gold coating technique. *Mater. Charact.* **2023**, *196*, 112663. [CrossRef]
23. Filippi, M.; Calliari, L.; Pucella, G.; Verona-Rinati, G. Temperature evolution of the surface region of CVD diamond: An electron spectroscopy study. *Surf. Sci.* **2004**, *573*, 225–236. [CrossRef]
24. Haerle, R.; Riedo, E.; Pasquarello, A.; Baldereschi, A. sp^2/sp^3 hybridization ratio in amorphous carbon from C 1 s core-level shifts: X-ray photoelectron spectroscopy and first-principles calculation. *Phys. Rev. B* **2001**, *65*, 045101. [CrossRef]
25. Titantah, J.T.; Lamoen, D. sp^3/sp^2 characterization of carbon materials from first-principles calculations: X-ray photoelectron versus high energy electron energy-loss spectroscopy techniques. *Carbon* **2005**, *43*, 1311–1316. [CrossRef]
26. Kumaragurubaran, S.; Yamada, T.; Shikata, S. Core Level Photoelectron. Spectroscopic Study on Oxidized Phosphorus-Doped (100) Diamond Surfaces after Vacuum Annealing. *Jpn. J. Appl. Phys.* **2009**, *48*, 011602. [CrossRef]
27. Lau, W.M.; Huang, L.J.; Bello, I.; Yiu, Y.M.; Lee, S.-T. Modification of surface band bending of diamond by low energy argon and carbon ion bombardment. *J. Appl. Phys.* **1994**, *75*, 3385–3391. [CrossRef]
28. Ley, L.; Graupner, R.; Cui, J.B.; Ristein, J. Electronic properties of single crystalline diamond surfaces. *Carbon* **1999**, *37*, 793–799. [CrossRef]
29. Petit, T.; Arnault, J.-C.; Girard, H.A.; Sennour, M.; Bergonzo, P. Early stages of surface graphitization on nanodiamond probed by x-ray photoelectron spectroscopy. *Phys. Rev. B* **2011**, *84*, 233407. [CrossRef]
30. Petit, T.; Arnault, J.-C.; Girard, H.A.; Sennour, M.; Kang, T.-Y.; Cheng, C.-L.; Bergonzo, P. Oxygen hole doping of nanodiamond. *Nanoscale* **2012**, *4*, 6792–6799. [CrossRef]

31. Graupner, R.; Maier, F.; Ristein, J.; Ley, L.; Jung, C. High-resolution surface-sensitive C 1 s core-level spectra of clean and hydrogen-terminated diamond (100) and (111) surfaces. *Phys. Rev. B* **1998**, *57*, 12397–12409. [CrossRef]
32. Liu, J.W.; Liao, M.Y.; Imura, M.; Koide, Y. Band offsets of Al_2O_3 and HfO_2 oxides deposited by atomic layer deposition technique on hydrogenated diamond. *Appl. Phys. Lett.* **2012**, *101*, 252108. [CrossRef]
33. Bydzovska, I.; Shagieva, E.; Gordeev, I.; Romanyuk, O.; Nemeckova, Z.; Henych, J.; Ondic, L.; Kromka, A.; Stehlik, S. Laser-Induced Modification of Hydrogenated Detonation Nanodiamonds in Ethanol. *Nanomaterials* **2021**, *11*, 2251. [CrossRef] [PubMed]
34. Ogawa, S.; Yamada, T.; Ishizduka, S.; Yoshigoe, A.; Hasegawa, M.; Teraoka, Y.; Takakuwa, Y. Vacuum Annealing Formation of Graphene on Diamond C(111) Surfaces Studied by Real-Time Photoelectron Spectroscopy. *Jpn. J. Appl. Phys.* **2012**, *51*, 11PF02. [CrossRef]
35. Romanyuk, O.; Varga, M.; Tulic, S.; Izak, T.; Jiricek, P.; Kromka, A.; Skakalova, V.; Rezek, B. Study of Ni-Catalyzed Graphitization Process of Diamond by in Situ X-ray Photoelectron Spectroscopy. *J. Phys. Chem. C* **2018**, *122*, 6629–6636. [CrossRef] [PubMed]
36. Veyan, J.-F.; de Obaldia, E.; Alcantar-Peña, J.J.; Montes-Gutierrez, J.; Arellano-Jimenez, M.J.; Yacaman, M.J.; Auciello, O. Argon atoms insertion in diamond: New insights in the identification of carbon C 1s peak in X-ray photoelectron spectroscopy analysis. *Carbon* **2018**, *134*, 29–36. [CrossRef]
37. Greczynski, G.; Hultman, L. Critical method evaluation refutes the Ar 2p signal of implanted Ar for referencing X-ray photoelectron spectra. *Appl. Surf. Sci.* **2023**, *635*, 157598. [CrossRef]
38. Gaowei, M.; Muller, E.M.; Rumaiz, A.K.; Weiland, C.; Cockayne, E.; Jordan-Sweet, J.; Smedley, J.; Woicik, J.C. Annealing dependence of diamond-metal Schottky barrier heights probed by hard X-ray photoelectron spectroscopy. *Appl. Phys. Lett.* **2012**, *100*, 201606. [CrossRef]
39. Kern, G.; Hafner, J.; Kresse, G. Atomic and electronic structure of diamond (111) surfaces I. Reconstruction and hydrogen-induced de-reconstruction of the one dangling-bond surface. *Surf. Sci.* **1996**, *366*, 445–463. [CrossRef]
40. Wilson, J.I.B.; Walton, J.S.; Beamson, G. Analysis of chemical vapour deposited diamond films by X-ray photoelectron spectroscopy. *J. Electron Spectrosc. Relat. Phenom.* **2001**, *121*, 183–201. [CrossRef]
41. Kozakov, A.T.; Kochur, A.G.; Kumar, N.; Panda, K.; Nikolskii, A.V.; Sidashov, A.V. Determination of sp^2 and sp^3 phase fractions on the surface of diamond films from C1s, valence band X-ray photoelectron spectra and CKVV X-ray-excited Auger spectra. *Appl. Surf. Sci.* **2021**, *536*, 147807. [CrossRef]
42. Seshan, V.; Murthy, D.H.K.; Castellanos-Gomez, A.; Sachdeva, S.; Ahmad, H.A.; Janssens, S.D.; Janssen, W.; Haenen, K.; van der Zant, H.S.J.; Sudhölter, E.J.R.; et al. Contactless Photoconductance Study on Undoped and Doped Nanocrystalline Diamond Films. *ACS Appl. Mater. Interfaces* **2014**, *6*, 11368–11375. [CrossRef]
43. Speranza, G. Characterization of Carbon Nanostructures by Photoelectron Spectroscopies. *Materials* **2022**, *15*, 4434. [CrossRef]
44. Speranza, G.; Calliari, L.; Laidani, N.; Anderle, M. Semi-quantitative description of C hybridization via s- and p-partial density of states probing: An electron spectroscopy study. *Diam. Relat. Mater.* **2000**, *9*, 1856–1861. [CrossRef]
45. Mikesova, J.; Miliaieva, D.; Stenclova, P.; Kindermann, M.; Vuckova, T.; Madlikova, M.; Fabry, M.; Veverka, V.; Schimer, J.; Krejci, P.; et al. Nanodiamonds as traps for fibroblast growth factors: Parameters influencing the interaction. *Carbon* **2022**, *195*, 372–386. [CrossRef]
46. Stehlik, S.; Henych, J.; Stenclova, P.; Kral, R.; Zemenova, P.; Pangrac, J.; Vanek, O.; Kromka, A.; Rezek, B. Size and nitrogen inhomogeneity in detonation and laser synthesized primary nanodiamond particles revealed via salt-assisted deaggregation. *Carbon* **2021**, *171*, 230–239. [CrossRef]
47. KolXPD. Software for Spectroscopy Data Measurement and Processing. Available online: https://www.kolibrik.net/en/solutions-products/kolxpd (accessed on 23 March 2024).
48. Speranza, G.; Laidani, N. Measurement of the relative abundance of sp^2 and sp^3 hybridised atoms in carbon based materials by XPS: A critical approach. Part I. *Diam. Relat. Mater.* **2004**, *13*, 445–450. [CrossRef]
49. Rabchinskii, M.K.; Saveliev, S.D.; Stolyarova, D.Y.; Brzhezinskaya, M.; Kirilenko, D.A.; Baidakova, M.V.; Ryzhkov, S.A.; Shnitov, V.V.; Sysoev, V.V.; Brunkov, P.N. Modulating nitrogen species via N-doping and post annealing of graphene derivatives: XPS and XAS examination. *Carbon* **2021**, *182*, 593–604. [CrossRef]
50. Ahmed, A.-I.; Mandal, S.; Gines, L.; Williams, O.A.; Cheng, C.-L. Low temperature catalytic reactivity of nanodiamond in molecular hydrogen. *Carbon* **2016**, *110*, 438–442. [CrossRef]
51. Kondo, T.; Neitzel, I.; Mochalin, V.N.; Urai, J.; Yuasa, M.; Gogotsi, Y. Electrical conductivity of thermally hydrogenated nanodiamond powders. *J. Appl. Phys.* **2013**, *113*, 214307. [CrossRef]
52. Kolarova, K.; Bydzovska, I.; Romanyuk, O.; Shagieva, E.; Ukraintsev, E.; Kromka, A.; Rezek, B.; Stehlik, S. Hydrogenation of HPHT nanodiamonds and their nanoscale interaction with chitosan. *Diam. Relat. Mater.* **2023**, *134*, 109754. [CrossRef]
53. Mermoux, M.; Chang, S.; Girard, H.A.; Arnault, J.-C. Raman spectroscopy study of detonation nanodiamond. *Diam. Relat. Mater.* **2018**, *87*, 248–260. [CrossRef]
54. Sorkin, A.; Tay, B.; Su, H. Three-Stage Transformation Pathway from Nanodiamonds to Fullerenes. *J. Phys. Chem. A* **2011**, *115*, 8327–8334. [CrossRef] [PubMed]
55. Raty, J.-Y.; Galli, G.; Bostedt, C.; Van Buuren, T.; Terminello, L. Quantum Confinement and Fullerenelike Surface Reconstructions in Nanodiamonds. *Phys. Rev. Lett.* **2003**, *90*, 037401. [CrossRef] [PubMed]
56. Dhanak, V.R.; Butenko, Y.V.; Brieva, A.C.; Coxon, P.R.; Alves, L.; Šiller, L. Chemical Functionalization of Nanodiamond by Amino Groups: An X-Ray Photoelectron Spectroscopy Study. *J. Nanosci. Nanotech.* **2012**, *12*, 3084–3090. [CrossRef]

57. Moulder, J.F.; Chastain, J. (Eds.) *Handbook of X-ray Photoelectron Spectroscopy: A Reference Book of Standard Spectra for Identification and Interpretation of XPS Data, Update*; Perkin-Elmer Corporation: Eden Prairie, MN, USA, 1992.
58. Ducrozet, F.; Girard, H.A.; Jianu, T.; Peulon, S.; Brun, E.; Sicard-Roselli, C.; Arnault, J.-C. Unintentional formation of nitrate and nitrite ions during nanodiamonds sonication: A source of radical and electron scavengers. *Colloids Surf. A Physicochem. Eng. Asp.* **2023**, *663*, 131087. [CrossRef]

MDPI AG
Grosspeteranlage 5
4052 Basel
Switzerland
Tel.: +41 61 683 77 34

Nanomaterials Editorial Office
E-mail: nanomaterials@mdpi.com
www.mdpi.com/journal/nanomaterials

www.ingramcontent.com/pod-product-compliance
Lightning Source LLC
Chambersburg PA
CBHW040742250926
51850CB00007BB/1378
9783725874149